Chemistry Was Their Life

Pioneer British Women Chemists, 1880–1949

Chemistry Was Their Life

Pioneer British Women Chemists, 1880–1949

by

Marelene Rayner-Canham
Geoff Rayner-Canham
Memorial University of Newfoundland, Canada

ICP
Imperial College Press

Published by

Imperial College Press
57 Shelton Street
Covent Garden
London WC2H 9HE

Distributed by

World Scientific Publishing Co. Pte. Ltd.
5 Toh Tuck Link, Singapore 596224
USA office: 27 Warren Street, Suite 401-402, Hackensack, NJ 07601
UK office: 57 Shelton Street, Covent Garden, London WC2H 9HE

British Library Cataloguing-in-Publication Data
A catalogue record for this book is available from the British Library.

CHEMISTRY WAS THEIR LIFE
Pioneer British Women Chemists, 1880–1949

ISBN-13 978-1-86094-986-9
ISBN-10 1-86094-986-X

Typeset by Stallion Press
Email: enquiries@stallionpress.com

Printed in Singapore by World Scientific Printers

Acknowledgements

Our research could never have been successful without the help of the archivists at each of the many institutions we have visited to undertake the research. They responded promptly to our e-mail enquiries, and during our visits we found the rarely accessed material needed for our investigations. In addition, we must express our gratitude to the long-suffering Ms. Elizabeth Behrens, Associate University Librarian, Sir Wilfred Grenfell College (SWGC), Memorial University (MUN), Corner Brook, Newfoundland, Canada, who, over decades, has tracked down copies of articles from long-defunct journals and found interlibrary loan sources for equally obscure late 19th- and early 20th-century books. For each major quotation, we have done our best to obtain copyright permission.

Through the many years that our research was in progress, the Administration at SWGC and the Research Office at MUN have been very supportive. In particular, we thank Memorial University for a MUN Vice-President's/Social Sciences and Humanities Research Council grant for U.K. travel in 1998, and Sir Wilfred Grenfell College for a SWGC Principal's research grant to support U.K. travel in 2001. Memorial University has a campus at Old Harlow, Essex, and we have been fortunate to sojourn there each time we have had the opportunity to visit the U.K. to hunt down primary material. In this context, we wish to thank the staff at the Harlow Campus for their hospitality and helpfulness. Our travel needs, primarily the indispensable BritRail passes, were organised by the unflappable Ms. Frances Drover, LeGrow's Travel, Corner Brook.

It is vital to have other pairs of eyes survey the manuscript, and we have been particularly fortunate in finding three individuals who generously gave up their time to provide a different perspective on shortcomings of our first drafts. They were Dr. David Waddington, Professor (retired), Chemical Education, University of York, York; Dr. Hannah Gay, Professor (retired), History of Science, Simon Fraser University, Burnaby, British Columbia, Canada and Imperial College, London; and Dr. David Edgerton, Hans Rausing Professor, Centre for the History of Science, Technology and Medicine, Imperial College, London, all of whom gave helpful comments on the manuscript. Though we like to think we have reasonable competence in the English language, we wisely asked a professional English scholar, Ms. Heather (Doody) Wellman, to read the final manuscript and make constructive grammatical and punctuation corrections.

Finally, we are grateful to Imperial College Press and World Scientific Publishing for publishing (and hosting the accompanying website) what we contend is an important missing piece of the historical record in British chemistry. In particular, Ms. Lizzie Bennett, Imperial College Press, is thanked for her enthusiasm and encouragement; and Ms. Wanda Tan, World Scientific Publishing Company, is thanked for her excellent editing and proofreading.

Contents

Introduction

In 1947, the Chemical Society published a book titled *British Chemists*.[1] This book contained the compiled biographies of prominent British chemists. None were women, and the accounts read as if women were essentially absent from the chemical enterprise in the latter decades of the 19th century and the first half of the 20th century.

Nothing could be further from the truth: there were significant numbers of women chemistry students and women chemists. For example, we have identified 896 who were members of the Royal Institute of Chemistry and/or the Chemical Society during our time frame of 1880–1949. In this book, we provide biographical accounts of 141 women chemists, together with brief notes on an additional 21.

Documentation

For the most part, these women chemists did not gain any recognition. In those times, most considered it unthinkable for a woman to occupy anything more than a supporting role. The majority of the women remained unmarried; thus, when they died, no record of them survived. Any relative would probably have been unaware of the scientific contributions of their deceased aunt and hence destroyed her papers and correspondence. For this reason, many of our accounts raise more questions than provide answers. Some of the women simply "disappear" from the record, perhaps through marriage, perhaps death, or perhaps a change in career direction brought about by a lack of opportunity to practise their chosen profession.

Deaths in the early part of the 20th century were usually given some coverage, even a short article in the *Journal of the Institute of Chemistry*. Some of the Oxbridge women graduates, when deceased, rated a brief biography in the appropriate college newsletter. A few were fortunate to have another women chemist write an obituary of them. Paradoxically, the longevity of most of these women meant they did not die until the late decades of the 20th century, when a one-line name and date of death in *Chemistry in Britain* was all they received.

To a certain extent, it is the sparseness of records that has limited us in considering which women chemists to include and which to exclude. However, we decided some women played a role in the narrative even when very little information could be found about them. For this reason, we ask the reader to tolerate some of the short "bare essentials" biographical accounts.

The Pioneer Women Chemists

For the women chemists for whom we have biographical information, what comes through most strongly is their enthusiasm and dedication. It is for this reason we have titled the book *Chemistry Was Their Life*. The women chemists of the 1880s to 1920s saw themselves as the pioneers: they had to succeed for the sake of the young women who followed them. At the same time, they truly found chemistry enthralling — it was indeed the centre of their lives and, for those in academia, they were determined to convey this belief to their students.

Where we have documentation, the women chemists readily acknowledge their appreciation of their supervisors, such as biochemist F. Gowland Hopkins (see Chap. 8), for allowing them to follow their chosen path in an environment where other prominent male chemists refused to countenance a woman student.

In Marcia Bonta's study of the lives of pioneer American botanists, she found this same attitude:

> Because most women naturalists believed that the work was all that mattered, they seemed to feel little or no rivalry toward the more powerful males in their fields and were pleased and grateful for whatever help these men gave them.[2]

The quantity of research accomplished by these women was enormous. In this book, we only have space to list the books authored by the women chemists we have chosen to include. In the accompanying website, the publications of each of the 896 pioneer women chemists are listed (see: www.worldscibooks.com/histsci/p538.html). So why has their work been forgotten? Much was published under their names alone, and was never retrieved when historians were looking under the names of the "great men"; and when the women's research was co-authored with one of the "great men," then they were overshadowed by their more famous supervisor. An example of the latter is Mollie Barr (see Chap. 13). Of the 36 publications of the biochemist Alexander Glenny F. R. S. between 1932 and 1955, 25 were co-authored by Mollie Barr,[3] yet Barr's lifetime of contributions to the field has not been recognised and little can be found about her life history.

The underrecognition of women's contributions in science has been well documented by Margaret Rossiter.[4] Rossiter was building on the contribution of the sociologist, Robert K. Merton. Merton identified the "Matthew effect," the minimisation of the role of secondary contributors in science, naming the effect after the Biblical quotation of Matthew 13:12, which stated that "whosoever hath not, from him shall be taken away even that he hath." Rossiter contended that the effect was far more endemic and serious for women scientists, for whom she coined the term the "Matilda effect" after the American suffragist, Matilda J. Gage, who both experienced it and articulated the phenomenon.

For consistency, all women chemists are discussed under their birth name (their married name is provided in parentheses).

The change in name upon marriage has always been problematic for academic women. Sharon Bertsch McGrayne remarked the change in name even caused confusion in the case of Dorothy Crowfoot (Mrs. Hodgkin): "Dorothy published her penicillin studies under her maiden name 'Crowfoot' and announced vitamin B_{12} as 'Hodgkin.' Years later some scientists still did not know that the Crowfoot of penicillin fame was the Hodgkin of B_{12} fame."[5]

The reader will note that it was several males, especially William Tilden and William Ramsay, who vociferously promoted the right of women to be admitted to the Chemical Society (see Chap. 2). Sylvia Strauss has described the importance of sympathetic men to the suffrage campaign in Britain in her book, *Traitors to the Masculine Cause: The Men's Campaigns for Women's Rights*.[6] In the more general context of women's access to university, we highlight several male crusaders including William Shean (see Chap. 3), and Henry Sidgwick and Arthur Sidgwick (see Chap. 6). The admission of women to the Pharmaceutical Society would never have succeeded at the time without the efforts of Robert Hampton (see Chap. 10). Augustus Vernon Harcourt was another champion of women, both in the context of the Chemical Society and for the admission of women to chemistry lectures at Oxford (see Chap. 6).

Setting the Scene

As we show in Chap. 1, access of women to university could not have been possible until secondary schools existed that provided academic studies for girls. Certain schools played a particularly prominent role: North London Collegiate School for Girls (NLCS) and the related schools of the Girls' Public Day School Company (GPDSC); Cheltenham Ladies' College (CLC); King Edward VI High School for Girls, Birmingham (KEVI); and Manchester High School for Girls (MHSG). Not only did they provide an academic education, but they also saw the essentiality of teaching girls about science, and specifically chemistry. In subsequent chapters, we have reported the school attended by each woman chemist, when known, and the reader will note the frequency with which these schools are mentioned.

We also introduce an issue which bedevilled the teaching of chemistry to women: whether it should be academic chemistry to enable matriculating girls to take their place alongside men in university laboratories, or domestic chemistry that would be relevant to women's lives. We revisit the issue in Chap. 3 in the context of King's College of Household and Social Science and of Battersea Polytechnic.

In Chap. 2, we summarise the access of women to professional societies. Each society handled the admission of women in a different way, with the long battle for admission to the Chemical Society taking up a substantial portion of the chapter. In addition to describing the long struggle, we highlight the women who signed the 1904 petition for admission. Most of those women, together with many of the signatories of a letter in 1909 to *Chemical News*, seem to have been the "movers and shakers" of their time.

Chapter 2 is the first chapter in which individual biographies of women chemists appear. We have endeavoured to place each biography in the most relevant narrative. In some cases, this is in the chapter of the institution from which the woman graduated; in others, where she undertook research or became employed; in others, in the context of her speciality or time frame. Where possible, we have cross-referenced the biographical accounts to linking locations in other chapters.

The Colleges and Universities

The next five chapters focus on the British universities and colleges. In each case, we have attempted to provide the reader a brief historical context of each institution without overwhelming the true focus of the book: the women chemists and their experiences. The biographical accounts in these chapters focus mainly upon women chemists who joined the staff at those institutions, together with a selection of profiles of women chemistry students who later made notable contributions.

In some cases, we have found rich background material in student magazines of the 1880s to 1920s that give an insight into the lives of women students in those early days. We have employed quotes from these magazines to document the women's experiences both in general and specifically within the culture of the chemistry departments.

At the women's colleges (see Chaps. 4 and 6), the commentaries we have included highlight the regulations and formal dinners in the early years that were more like those at authoritarian boarding schools. At the same time, the women students had a sense of freedom and empowerment they lacked in the outside world. With the demise of the British women's colleges,[7] it is crucial for the reader to appreciate the unique "insulating" environment of the time. Of the women's colleges, three institutions played an especially important role in educating women chemists: Bedford College, London; Royal Holloway College, London; and Newnham College, Cambridge. Throughout the book the reader will note the frequency with which these names appear.

Though the (co-educational) universities (see Chaps. 3, 5, and 7) all claimed to welcome women students, their daily experience was often otherwise. We have included a selection of comments from student magazines by male correspondents that indicate significant hostility to the presence of women. Carol Dyhouse has summarised the situation:

> In many cases they [women students] were excluded from membership of existing societies and student unions, and found it necessary (or expedient) to form their own. The presence of a minority group of women frequently served to under*line* rather than under*mine* the norms of the dominant male culture, and male students often went in for exaggerated displays of masculinity, particularly in informal settings, where

the presence of women might supply both a target for aggression and an audience. Women students generally responded by keeping a very low profile.[8]

Favoured Fields for Women Chemists

One of our previous studies indicated that women scientists tended to "cluster" in certain areas.[9] There were two areas in which women chemists flourished: biochemistry (see Chap. 8) and crystallography (see Chap. 9). In those chapters, we focus not only on the women and their work, but also the environment in which they worked and the mentors who contributed to their advancement. These were the only two fields in which pioneering women chemists received the accolade of election to Fellowship of the Royal Society (see Table 0.1).

In Chap. 10, we provide an account of the pioneering women pharmacists, some of whom had been active chemists. Even more than crystallography and biochemistry, pharmacy was a direction in which women chemists could find employment. In addition, the fight for the admission of women to

Table 0.1. Women in the chemical sciences elected Fellows of the Royal Society, 1945–1980.

Name	Subject	Date	Chapter
Kathleen Yardley (Mrs. Lonsdale)	Crystallography	1945	9
Marjory Stephenson	Biochemistry	1945	8
Dorothy Crowfoot (Mrs. Hodgkin)	Crystallography	1947	9
Dorothy Needham	Biochemistry	1948	8
Rosalind Henley (Mrs. Pitt-Rivers)	Biochemistry	1954	4
Helen Archbold (Mrs. Porter)	Biochemistry	1956	3
Patricia Clarke	Biochemistry	1976	13
Elsie Widdowson	Biochemistry	1976	13

the Pharmaceutical Society has interesting parallels to, and differences from, the admission of women to the Chemical Society.

Options for Women Chemists

Up to this point, we have used institutional and subject narratives in which to embed the biographical accounts. These encompass a relatively small number of women chemists of the period. In the last three chapters, we look at the roles of women chemists from other perspectives.

Chapter 11 focuses on the effects of marriage on women chemists. Though a significant proportion stayed single — particularly those who stayed in academia — many did marry, of whom most relinquished their career. A significant proportion of the women who did marry married a chemist. Of the working chemistry couples, some wives continued with their original field while the majority joined their husband's research group. The degree of recognition of the women's work depended very much on their marriage partner. Some wives were cited as co-authors, while it would seem others received no credit for their collaboration.

The First World War proved a turning point in the acceptance of women chemists, as it did for educated women in the wider society. Therefore, we devote the whole of Chap. 12 to the different roles women chemists played in this war, especially in the production of fine chemicals in academic laboratories and in the many H. M. Factories around the country. With the destruction of many records after the First World War, it is unlikely we will ever determine the full role of women chemists during this period.

It may come as a surprise to many readers, but the proportion, and even the total number, of women students at university reached a maximum in the late 1920s; the trend applied equally to women chemists.[10] In Chap. 13, we examine this phenomenon and provide observations on the subsequent decline.

We give an overview of the different employment avenues for women chemists in the interwar period and choose exemplars for each. The chapter concludes with examples of women chemists who used the opportunities of the Second World War to develop career directions, though the effect of that war on women chemists seems to have been less momentous than that of the 1914–1918 conflict. The late 1940s seem to be an appropriate place to end the book, for as Evelyn Fox Keller has observed, the mid-20th century represented the "nadir of the history of women in science."[11]

Commentary

Each chapter ends with a commentary addressing some specific aspect of that particular chapter. In this commentary, we would like to consider that this book will finally bring awareness of the forgotten role of British women chemists in the late 19th and the first half of the 20th centuries. At last, in addition to the HIStory of British chemists, we now have HERstory.

Notes

1. Findlay, A. and Mills, W. H. (eds.) (1947). *British Chemists.* Chemical Society, London.
2. Bonta, M. M. (1991). *Women in the Field: America's Pioneering Women Naturalists.* Texas A&M Press, College Station, Texas, pp. xii–xiii.
3. Oakley, C. L. (1966). Alexander Thomas Glenny, 1882–1965. *Biographical Memoirs of Fellows of the Royal Society* **12**: 163–180.
4 Rossiter, M. W. (1993). The ~~Matthew~~ Matilda effect in science. *Social Studies of Science* **23**: 325–341.
5. McGrayne, S. B. (1993). *Nobel Prize Women in Science.* Birch Lane Press, New York, p. 248.
6. Strauss, S. (1982). *Traitors to the Masculine Cause: The Men's Campaigns for Women's Rights.* Greenwood Press, Westport, Connecticut.

7. Rosen, A. (2003). *The Transformation of British Life, 1950–2000: A Social History*. Manchester University Press, Manchester, p. 84; Howarth, J. (1994). Review article: Women's colleges — The latest lost cause? *Oxford Review of Education* **20**: 143–147.
8. Dyhouse, C. (1995). The British Federation of University Women and the status of women in universities, 1907–1939. *Women's History Review* **4**: 465–484.
9. Rayner-Canham, M. F. and Rayner-Canham, G. W. (1996). Women's fields of chemistry: 1900–1920. *Journal of Chemical Education* **73**: 136–138.
10. Rayner-Canham, M. F. and Rayner-Canham, G. W. (1996). Women in chemistry: Participation during the early twentieth century. *Journal of Chemical Education* **73**: 203–205.
11. Fox Keller, E. (1987). The gender/science system: Or, is sex to gender as nature is to science? *Hypatia* **2**: 40.

Chapter 1

Getting an Education

Access to education was crucial to the advancement of women in science. These days one takes for granted that girls should have the same secondary and tertiary education as boys, but up to the last two decades of the 19th century such ideas were heretical. It is therefore crucial to our saga on the pioneering women chemists that we spend a whole chapter on the many barriers to a girl's education and the ways in which these challenges were overcome. As we shall see, certain individuals — "the agents of change" — played crucial roles.[1]

Education for Girls

Even before there were any academic schools for girls, some young women yearned for an education. As Gillian Avery has noted:

> A passionate desire for learning seems to have been characteristic of many of the early nineteenth-century girls ... With dedicated self-discipline many girls worked at home, teaching themselves Latin, Greek, and German, reading Dante, translating Schiller, in between the demands that their families made upon them.[2]

The first controversy was that of the most appropriate type of secondary education for girls: whether the educational content should involve academic subjects, or simply focus on those activities that would make girls into better wives and mothers.[3] In 1868, the antifeminist writer, Sarah Sewell, contended:

> ... profoundly educated women rarely make good wives or mothers. The pride of knowledge does not amalgamate well with the every-day matter of fact rearing of children, and women who have stored their minds with Latin and Greek seldom have much knowledge of pies and puddings, nor do they enjoy the hard and uninteresting work of attending to the wants of little children.[4]

Lydia Becker made the contrary case in 1869, arguing that a training in science would be particularly beneficial for girls:

> Prevalent opinions and customs impose on women so much more monotonous and colourless lives, and deprive them of so much of the natural and healthy excitement enjoyed by the other sex in its free intercourse with the world ... many women might be saved from the evil of the life of intellectual vacuity, to which their present position renders them so peculiarly liable, if they had a thorough training in some branch of science, and the opportunity of carrying it on as a serious pursuit.[5]

There was no dispute about education for working-class girls. The aim was simple: to prepare them for marriage and children in the context of religious devotion. The possibility of any further aspirations was firmly dismissed in this comment of 1861:

> ... it is to be hoped that no desire to make girls little Newtons, little Captain Cooks, little Livingstones, little Mozarts and Handels, and little Sir Joshua Reynoldses, will ever take us too low for keeping in sight the object of teaching them to make and mend shirts, to make and mend pinafores, and darn stockings and socks.[6]

Pioneering Schools

Quaker schools initially led the way. The Quaker religious principles allowed women to play an active role in both the public and private spheres of the Religious Society of Friends.[7] In the history of The Mount School, a Quaker school in York, Winifred Sturge and Theodora Clark comment in general on the reason why Quaker schools for girls benefited from the austere aspects of the Quaker movement:

> Looking back on the [eighteen] sixties, we may feel satisfied that among girls' schools of the country, Quaker schools were indubitably in the front rank. The very narrowness of the curriculum — no music, dancing, or singing, no fine needlework — left space and time available for better grounding in history, arithmetic, and geography, and for relatively wider reading in English literature.[8]

Edward Grubb taught chemistry at the School (in addition to Latin, mathematics, and psychology), and Sturge and Clark commented: "To his lectures on chemistry his audience came in a mood of prophetic sympathy, awaiting the experiment: 'Will it? Won't it?' It generally wouldn't! Why should it? For before the laboratory was built in 1884 there was no scientific equipment worth the name."[8]

There were two Quaker women in particular, Priscilla Wakefield and Maria Hack, who popularised science and promoted scientific literacy in the early 19th century.[9] Science education became accepted for both Quaker boys and girls. In Bristol, the Quaker Dr. Thomas Pole gave lectures in general science for school children and young adults, the series starting with a course on chemistry.[7] At Susanna Corder's Quaker School for Girls in Stoke Newington, she acquired the assistance of the scientist and political radical, William Allen.[10] As one of the students, Louisa Stewart, later recalled: "William Allen gave the girls lessons in his own house in chemistry ..."[11]; while

Jane Heath, at Sarah and Harriet Hoare's Quaker school in Frenchay, wrote in 1820 to her mother:

> We rise a little before seven and study Geography till eight with dissected maps. After breakfast we make our beds and into school again by nine ... I have then to write a page of English and Natural History, lectures on Chemistry, Botany, etc., besides parsing and a slateful of exercises so that I cannot always finish before dinner.[12]

During the latter half of the 19th century, a range of academic schools for middle-class girls sprang into existence. We will describe a few of the most important of these in the following sections. Contemporary accounts of life at these schools from the 1860s to the 1920s usually paint a rosy picture; however, as Sara Delamont has uncovered, life was far from idyllic: the schools demanded total silence by the girls at all times during school hours, often including playtimes and breaks.[13] The headmistresses and teachers at these schools met fierce opposition from several directions, as Delamont notes:

> The feminist pioneers who opened academic secondary schools for young girls in the second half of the nineteenth century did so against a body of medical opinion, religious orthodoxy, and a widespread belief among their potential clientele (that is, middle- and upper-middle-class parents) that such institutions were dangerous. A pupil, or a member of staff, at an academic secondary school was held to be in physical danger (her health would suffer, she might become subfertile or die of brain fever); in moral danger (away from the control of her mother she might meet anybody) and liable to forfeit her marriage prospects for men who would not want a wife who knew algebra.[14]

The dedication and fearlessness of these pioneering headmistresses cannot be overemphasised. As a result of their efforts, and the girls' schools they founded, late 19th-century girls were able to obtain an academic education and hence have the requirements to enter university.[15] And it was very young shoulders who bore this burden incredibly well, as Nonita Glenday and Mary Price have noted: "Most of the headmistresses were in their early twenties, very young for the responsibilities they were carrying, and had led sheltered lives."[16]

In our studies, four particular schools provided models for the science education of girls: North London Collegiate and its offspring, the Girl's Public Day Schools; Cheltenham Ladies' College; King Edward VI School for Girls; and Manchester High School.

North London Collegiate School

It was Frances Mary Buss who founded the North London Collegiate School for Ladies, as it was first named, later called the North London Collegiate School for Girls (NLCS).[17] Buss, born in 1827, was the daughter of Robert William Buss, an engraver and illustrator, and Frances Fleetwood. When only 14 years old, Buss started her own teaching career at the Mrs. Wyand's School on Mornington Place, the school she herself had attended. Four years later, she and her mother opened a preparatory school for young children in Kentish Town. From 1849 to 1850, she attended the night classes at Queen's College for Women, Harley Street (see Chap. 4), from which she obtained teaching certificates in French, German, and geography.

Having acquired formal qualifications, Buss opened the NLCS in 1850 as a day school for the education of daughters of the local middle-class community, often of limited means, such as the daughters of clerks or tradesmen.[18] One of the students, Molly Hughes, had been delighted to move from a fashionable "snobby" school to NLCS: "Now at the North London I sensed at once a different atmosphere. No one asked where you lived, how much pocket-money you had, or what your father was — he might be a bishop or a ratcatcher."[19]

Buss wanted her school to give girls a very different education from that prevalent at the other girls' schools: she wanted her pupils to have equal opportunities to boys, and that included real science:

> It [NLCS] also set out to offer their daughters an education quite different from anything available anywhere else. There was, for example, the inclusion of science within the curriculum. Science was not really taken seriously by most girls' schools until well into the twentieth century. Robert Buss [Frances Buss's father] made a memorable science teacher as Annie Martinelli, an early pupil later remembered: "His talents were simply wonderful. His Chemistry series was marvellous, especially for smells and explosions." He was the first of a series of outstanding science teachers at the School where learning by rote was replaced from the beginning by encouraging pupils to learn by thinking for themselves.[20]

After Buss died in 1894, the strong science program at NLCS was continued by the second Headmistress, Sophia (Sophie) Willcock (Mrs. Bryant). Born in 1850, Willcock, daughter of Rev. W. A. Willcock (Fellow of Trinity College, Dublin), entered Bedford College in 1866. She married Dr. W. Hicks Bryant in 1869, and upon his death in 1870, Willcock obtained a teaching position at a school for ladies in Highgate. In 1875, she joined the staff at NLCS and was appointed Headmistress in 1895. During her "reign," the presence of laboratory facilities at NLCS was specifically mentioned in the girls' magazine, *Girl's Realm*: "Beyond is a chemical laboratory well fitted up and large enough for twenty-four girls to work together at one time."[21] She was Head until 1918, dying in 1922 after becoming lost on a mountaineering holiday in the Alps.

Buss' father, Robert, had taught the science classes in the early years, with the first qualified science teacher, Grace Heath,[22] being hired in 1888. Heath had been a chemistry student with Henry Armstrong (see Chap. 2) at the Central Technical College. Unfortunately, Heath died in 1895, her position passing to Rose Stern (see Chap. 2). Stern ensured that chemistry was a prominent subject at NLCS. One of her students reported how well the NLCS lab was, compared with that of Holloway College during a school visit there in 1897: "A hasty peep into the laboratory showed that it was very much like our own school laboratory, but, as someone triumphantly remarked, much more untidy, …."[23] There was a hand-written and illustrated student science magazine, *The Searchlight*, which mentioned Chemistry Club social events such as: "On Thursday, July 11th [1912], Miss Stern, Miss Drummond and the Science Sixth gave a party in the Old Laboratory. We drank tea out of beakers, and stirred it with long glass wands…."[24]

The Girls' Public Day Schools

NLCS became the model for the girls' schools financed by the Girls' Public Day School Company (GPDSC), later the Girls' Public Day School Trust.[25] The formation of the GPDSC was the greatest success of the National Union for Improving the Education of Women of all Classes, widely known as the Women's Educational Union. The funds raised by the Company were used to found independent, affordable, academically selective girls' schools (see Table 1.1). These schools had an influence on the educational opportunities for girls far beyond their numbers, particularly in providing many of the entrants to universities.

In the early years, the Heads and Assistants of GPDSC schools were required to visit NLCS before taking command of their own school in order to study and reproduce the methods and organisation of the NLCS.[26] In fact, the renown of NLCS as a model was such that each term there were visitors not only from the British Isles, but also from the European countries, Canada, Australia, the United States, China, and Japan.

Table 1.1. The Girls' Public Day School Company schools opened 1873–1901.

Birkenhead High School	Notting Hill & Ealing High School
Blackheath High School	Oxford High School
Brighton & Hove High School	Paddington & Maida Vale High School[a]
Bromley High School	Portsmouth High School
Carlisle High School[a]	Putney High School
Central Newcastle High School	Royal High School, Bath
Clapham High School (merged)	Sheffield High School
Hackney (later Clapton) High School[a]	Shrewsbury High School
Croydon High School	South Hampstead High School
Dover High School	Streatham Hill High School (merged)
Dulwich High School[a]	Sutton High School
East Liverpool High School (merged)	Swansea High School[a]
Gateshead High School (merged)	Sydenham High School
Highbury and Islington High School[a]	Tunbridge Wells High School[a]
Ipswich High School	Weymouth High School[a]
Liverpool High School	Wimbledon High School
Newton Abbot High School[a]	
Nottingham High School	

[a] Those schools later closed or exited the GPDSC.

The importance of Buss and the NLCS in providing the template for the GPDSC cannot be overemphasised. In 1900, there was a formal three-day Jubilee celebration of the founding of NLCS, including a service at St. Paul's Cathedral. An account of the several events in *The Magazine of the Manchester High School* ended with the following passage:

> Thus was bought to a close one of the most interesting events of our time — the Jubilee of the first "High School for Girls," the success of which has meant so much for all of us; for to Miss Buss's efforts we owe the inception of the whole movement, and were it not for her valiant struggles against much opposition, the establishment of Girls' High Schools, and the opening of the Universities to women, might have been greatly retarded.[27]

Cheltenham Ladies' College

The Young Ladies' College, Cheltenham (CLC), was opened in 1854.[28] Though it, too, was to offer an excellent training in science, it was very different to NLCS; the CLC was designed as a boarding establishment catering to a socially-elite clientele. Initially, the school followed the traditional view of the education of young women as was enunciated by the first report issued by the Governors:

> The school intends to provide an education based upon religious principles which, preserving the modesty and gentleness of the female character, should so far cultivate a girl's intellectual powers as to fit her for the discharge of those responsible duties which devolve upon her as a wife, mother and friend, the natural companion and helpmate for man.[29]

Over the first 4 years, the enrolment declined. Then the Governors appointed Dorothea Beale[30] as the second Principal and everything changed. Beale, the daughter of Miles Beale, a surgeon, and Dorothea Complin, was initially educated by a governess, but then she was sent away to school — an experience which convinced her of the need of a radical overhaul of girls' education:

> It was a school considered much above average for sound instruction; our mistresses had taken pains to arrange various schemes of knowledge: yet what miserable teaching we had in so many subjects; history was learned by committing to memory little manuals; rules of arithmetic were taught, but the principles were never explained.[31]

In 1848, Beale, like Buss, attended Queen's College for Women where she excelled, being appointed their first mathematics tutor after her graduation. However, she became dissatisfied with the College and left to become Head Teacher at Casterton

School, Cumbria. Her attempts to reform that school failed and she resigned within 1 year. During the next year, Beale wrote *A Textbook of General History*, and the success of this book contributed to her appointment as Principal of CLC. Here, at last, she was able to put her strong principles into action.

The construction of science laboratories was one of Beale's priorities. This was accomplished as noted in a description of the school in a 1900 issue of *Girl's Realm*: "in the Science Department there is a laboratory for physics and two for chemistry ..."[32]; while a new Science wing was added in 1904 that was designed by Millicent Taylor, the woman chemist who linked CLC and University College, Bristol (see Chap. 5).[33]

Beale of CLC and Buss of NLCS were two of the most important figures in women's education in Britain. They had met at Queen's College for Women and, from then on, their lives were intertwined. For example, in 1874, they were both instrumental in the formation of the Association of Head Mistresses, Buss being the first President while Beale was the first Chairman. Their names were even coupled in an anonymous rhyme:

> Miss Buss and Miss Beale
> Cupid's darts do not feel.
> How different from us,
> Miss Beale and Miss Buss.[34]

The CLC differed from the other girls' high schools in that, during the early period, students could actually complete an external B.Sc (London) degree. In 1904, Sophie Bryant visited CLC and described the two alternative futures for the College as was reported by Millicent Taylor: "Mrs. Bryant expressed the view that the Cheltenham Ladies College could (1) develop as a public school sending more and more students to university or (2) become a 'great University College of the West of England; perhaps the centre of a Women's University'."[33] The issue of a "Women's University" periodically arose, usually to be rejected by the large majority of women on the grounds that men would immediately label it as "inferior" to the male-based universities and it would serve to ghetto-ise women. Thus, CLC adopted the first course, ceasing to cater to degree students while University College, Bristol, rose in prominence as the centre for higher education in the southwest of England (see Chap. 5).

Beale remained as Principal of CLC until her death in 1906 — a "reign" of 50 years. The philosophy changed under the next Headmistress, Lilian M. Faithfull, who had been Vice-Principal at King's College, Ladies' Department (see Chap. 3). Academic success was no longer the prime focus. Faithfull gave her views in 1911:

> In recent years there has been a widespread movement to bring the education of our girls into relation with their work as home-makers. The old "blue-stocking type, who prided herself on not knowing how to sew or mend, and who thought cooking menial and beneath her, no longer appeals to anyone ... we want our girls to grow up into sensible, methodical, practical women, able to direct intelligently and practically the manifold duties of home ..."[35]

King Edward VI High School for Girls

In 1547, the Act of Suppression required the confiscation of all assets of religious guilds except for an amount of land with an annual income of £21 if the guild supported a school. The Guild of the Holy Cross in Birmingham had no school, but it persuaded the Earl of Northumberland (also the Lord of the Manor of Birmingham) to release land for the creation of a school. As a result, a Foundation was created in 1552 under King Edward VI for a Free Grammar School for boys. In 1883, the Charity Commissioners agreed to allow the opening of more schools under the same name, including the King Edward VI High School for Girls (KEVI).[36]

The first Head was Edith Elizabeth Maria Creak. By the time of her arrival at KEVI at age 27, she was already a veteran, having been appointed the first headmistress of Brighton and Hove Public Day School (a GPDSC school) before her 20th birthday. The School biographer, Winifred Vardy, noted: "To Miss Creak belongs the honour of being a pioneer in the teaching of science to girls. Though her own training [with Clough at Newnham College, Cambridge] had been mathematical and classical, she seems to have foreseen the value of scientific knowledge for women."[37]

Creak hired three dedicated and enthusiastic science staff, all Newnham graduates: Lizzie Davison, Alice Celia Slater, and Josephine Bingham. Another School biographer, Rachel Waterhouse, commented:

> Miss Davison and Miss Slater took charge of science, each stayed for thirty-one years at the School, and to them belongs almost all the credit for the great scientific successes achieved by Edwardians during the whole of that period. King Edward's Girls High School became the outstanding girls school for science in the country.[38]

Vardy also quotes a former student:

> Miss Davison also used to take the XIIth Class, little girls of 8 to 10, in the principal gases. "She did all the demonstrations, which according to modern ideas was bad," writes a pupil, "but she made it so interesting that I used to be impelled to tell my small brother all about it each week, and what she taught us *sticks*."[39]

Some KEVI students obtained permission to attend the men's classes of practical physiology at Mason College (later the University of Birmingham, see Chap. 5). Another former student commented:

> This meant our entering the sacred medical department of the College, and one of the demonstrators, who has since become

> known for his research work, announced that he would leave if those little girls from the High School came to his laboratory. He did not keep his word. The "little girls" referred to were Professor Winifred Cullis, Dr. Ida Smedley Maclean, and Hadda Hough. A year later Mary Phillp and Hadda Hough attended also the Chemistry Department of the College, both for lectures and for practical work in Organic Chemistry.[40]

A new building for KEVI enabled students to complete all of their science studies within the school and, as Vardy commented: "... the Science Classes had the best laboratories of any Girl's School in England."[37] The new building was opened by Eleanor Sidgwick (see Chap. 6), Principal of Newnham College, on 18 November 1896. In her address, Sidgwick not only commented on the flow of graduates from Newnham to teach at Birmingham, but also the flow of new students in the opposite direction:

> This school [KEVI] has sent up students to Newnham College every year, or almost every year, since there have been here any girls old enough to seek a university education.... I am certain that it has been an advantage of Newnham. If we were to lose it we should lose an important link with the education of the country, and we should also lose some of our best students.[41]

Manchester High School for Girls

In Manchester, Lydia Becker was a fervent believer that girls should be educated in science.[38] Becker became better known for her leadership of the Women's Suffrage Movement, but in fact before her shift into politics she was planning a scientific career. When the Education Act (1870) created School Boards, Becker was elected to the Board for Manchester, where she led the fight for science in girls' schools and the offering of scholarships for girls. To illustrate the effect of her initiative, in 1881, two (unnamed) girls gained first class honours in Practical

Chemistry in the science examinations in the Central (Manchester) Higher Grade Schools.[42]

Manchester High School for Girls (MHSG) was another school to graduate several of the pioneering women chemists mentioned in this book. The first Principal, Elizabeth Day, who had also attended Queen's College with Buss and Beale, was a believer in academic education, though she "... had grave doubts as to the suitability of science teaching, especially chemistry as then taught ...."[43]

However, her famous successor, Sara Burstall, swung the emphasis back towards science. In fact, the appointment of Burstall as the successor to Day was, in part, due to the desire among some of the governors that more science be taught at the school. Burstall commented that by the 1920s: "We had ... four specialist teachers on the staff, all first class honours graduates in chemistry, physics, botany and zoology, and many Old Girls were students in universities or science graduates."[44]

Just as the girls at NLCS considered their chemistry laboratory to be at least equal to that at Holloway College (see above), so the girls at MHSG, visiting Cambridge in 1901, were of the opinion that theirs was superior to Newnham's laboratory: "The Newnham Chemistry Laboratory was inspected on Monday morning, but was agreed to compare very unfavourably with the Chemical Laboratory of the Manchester High School, whatever the standard of work may be."[45]

The life of Burstall also illustrates the linkages amongst these powerful Headmistresses.[46] Burstall came from a poor family, but was awarded a scholarship to NLCS. From there, she proceeded to Girton College, Cambridge, then after a period as a teacher with Miss Buss at NLCS, she was offered the Headship at MHSG. Thus, MHSG, too, became modelled upon NLCS lines.

Chemistry in the Girls' Schools

Until the last two decades of the 19th century, Jane Marcet's *Conversations on Chemistry*[47] was the key chemistry resource for both governesses and science teachers in girls' schools. However, in that era, the girls were expected to memorise the book from cover to cover and regurgitate upon demand.[48] By the 1880s, the new progressive era had arrived with an emphasis on understanding rather than rote learning. In a definitive study of education for girls (one co-author being Beale) published in 1898, there was a chapter on *The Teaching of Chemistry* by Clare de Brereton Evans (see Chap. 4). In this chapter, de Brereton Evans argued that junior as well as senior girls needed exposure to practical chemistry:

> For success in examinations it is now necessary to have a certain amount of practical knowledge of chemistry and examination classes are therefore given some practical training, but this reform still remains to be extended universally to the junior classes, which need even more than the senior ones that the teaching should be objective: a child may learn and repeat correctly a dozen times that water is composed of oxygen and hydrogen, and the thirteenth time she will assure you that its constituents are oxygen and nitrogen; but let her make the gases herself, test them and get to know them as individuals, and mistakes of this kind will become impossible.[49]

As we described above, it was NLCS and KEVI that pioneered the introduction of chemistry laboratories. A few other schools followed their lead, including Redland High School, Bristol,[50] and St. Swithun's School, Winchester.[51] St. Swithun's acquired a chemistry laboratory in 1895, after an open flask releasing chlorine gas was left deliberately in a classroom prior to a tour by the Administrative Committee (see Fig. 1.1). One of the students taking the chemistry practical examination in 1897 reminisced:

> In those days a "don" was in charge in cap and gown. An enterprising examiner had given red phosphorus as the unknown substance. About ten minutes after we had commenced a nervous

Acid
The practical approach and independent research are introduced early in our science syllabuses . . .

Fig. 1.1. Pen-and-ink sketch from Bain, P. (1984). *St. Swithun's: A Centenary History*. Phillimore, Chichester, p. xvii.

> candidate dropped a glowing match on the "unknown" — result, a wild flare and we all "knew". Hardly had the invigilator extinguished this when it was discovered that a pile of dusters was on fire; this in turn was extinguished. Then suddenly the bottom came out of a medicine bottle improvised to contain sodium hydrate, devastating a varnished table and all the candidates' papers. Wearily our friend came for a third time to the rescue, remarking, "My life is insured — I only hope yours are!"[52]

In the 1895 Bryce Commission Report on Secondary Education, it was stated that in a mixed class of 14 year olds in Oldham: "There were 20 girls and 48 boys at work at the time of my visit, all doing practical chemistry together in a splendid laboratory, fitted up with all the new improvements; uncommonly bright and happy they looked."[53] In Liverpool, it was noted in the Report that all scientific subjects requiring apparatus were taught in the Central Higher Grade School, requiring the girls to walk there and back for their science lessons. Teachers at the girls' schools planned to discontinue the arrangement. However, girls at six of the local schools petitioned to be allowed to continue with the sharing arrangement to ensure that the standard of their science education matched that of the boys.

But the fervour for science education for girls seems to have abated in the early decades of the 20th century. In 1912, the Headmistress of Sacred Heart School, Hammersmith, described how the educational reforms of the later decades of the 19th century had emphasised the teaching of natural science. She added:

> So laboratories were fitted up at great expense, and teachers with university degrees were sought after. The height of the tide seemed to be reached in 1904 and 1905 ... Then disillusion seems to have set in and the tide began to ebb. It appeared that the results were small and poor in proportion to expectation and to the outlay on laboratories ... The conscientious accuracy that was to come of measuring a millimetre and weighing a milligramme was disappointing, and also the fluent readiness to give an account of observations made, the desired accuracy of expression, the caution in drawing inferences. The links between this teaching and after life did not seem to be satisfactorily established ... It begins to be whispered that even in some boys' schools the laboratory is only used under compulsion or by exceptional students, and the wave seems likely to go down as rapidly as it rose. Probably for girls the strongest argument against experimental science taught in laboratories is that it has so little connexion with after life.[54]

Which Schools Produced Women Chemistry Students?

Few of the first generation of women students at university studied science, and the majority of those came from a small number of schools.[55] Janet Howarth reported on the secondary school origins of residence women students attending either Oxford or Cambridge Colleges during the periods 1891–1893 and 1911–1913.[56] Her results showed that in the earlier period, the English secondary schools from which at least 10 women had come were (in order of decreasing numbers) North London Collegiate, Notting Hill High School (a GPDSC school), Croydon High School (a GPDSC school), Kensington High School (a GPDSC school), Cheltenham Ladies College, and Manchester High School. During the later period, the order was Cheltenham Ladies College, St. Paul's High School, Clapham High School (a GPDSC school), St. Felix School, Wycombe Abbey, Oxford High School (a GPDSC school), King Edward VI High School, Bedford High School, and North London Collegiate.

We have collected biographical information on the 898 women chemists who became Associates or Fellows of the Royal Institute of Chemistry and/or Fellows of the Chemical Society between 1880–1949, and we were able to ascertain the secondary schools attended for 284 of them. Our analysis differs from that of Howarth's in that we are looking at students who attended either Redbrick or Oxbridge universities and who had specifically chosen a chemistry career (Table 1.2). Nevertheless, there are strong similarities. In light of our earlier comments

Table 1.2. Most common English Schools attended by British women chemists who were Associates or Fellows of the Royal Institute of Chemistry and/or Fellows of the Chemical Society, 1880–1949.

School	Number of students
King Edward VI High School for Girls	20
North London Collegiate	11
Cheltenham Ladies College	10
Manchester High School for Girls	10
Clapham High School for Girls (GPDSC)	7
Croydon High School for Girls (GPDSC)	5
South Hampstead High School for Girls (GPDSC)	4

about the science emphasis, it is no surprise that KEVI, NLCS, CLC, and MHSG top the list. What is also interesting is that the remaining schools were members of the GPDSC, based on the North London Collegiate model.

However, one cannot generalise and say that all GPDSC schools were well equipped for chemistry. Some indeed were, such as Gateshead High School, which had a chemistry laboratory as early as 1886. On the other hand, Notting Hill High School was lacking in science facilities during the early years. Harriette Chick (see Chap. 2) recalled that: "we had no science laboratories at all and girls like myself who wanted to do Science had nothing. We learned Science almost theoretically but it was exceptionally well done."[57] The success of some Notting Hill chemistry students (three students in our study) was more a result of the inspired teaching of Mary M. Adamson, the first B.Sc. graduate of the University of London.

It was not only the GPDSC schools that excelled in science. Malvern Girls' College (also three students in our study) was another school proud of its science programme, as a former student, Grace Phillips, described in her biography of the Headmistress of the time, Iris Brooks:

> Miss Brook's knowledge of Science almost equalled the vacuity of her information about Physiology and Mathematics. Nevertheless, she admired wholeheartedly the work of the Science Staff. She also rejoiced when Malvern Girls' College acquired such a reputation for achievement in scientific circles that it became the first Girls' school to be awarded £10,000 from the Industrial Fund for the Advancement of Scientific Education in Schools.[58]

What Sort of Science?

Throughout the later 19th and early 20th centuries, there was an ambivalence about the purpose of an academic education for girls.[59] Should the school topics be aimed at the majority who were going to be wives and mothers, or the minority who were going to pursue careers? The same discourse was trying to define the more appropriate chemistry for girls: should girls study domestic science for their role as a homemaker, or academic science — including chemistry — to provide them with career options?[60]

A leading exponent of domestic science for girls was Arthur Smithells, Professor of Chemistry at Leeds University (see Chap. 5). Smithells was part of the "Science for All" movement, which was concerned with the low level of scientific awareness among the general population.[61] Members of the Movement contended that humanisation of science was the answer, in which scientific principles were related to people's daily lives. Smithells saw domestic science as a means of bringing an applied aspect that would, in particular, be appropriate in the education of girls.

What was domestic science? This, in itself, was a contentious issue, and in its teaching there seemed to be two extremes, as a report of 1911 indicated:

> The course may consist of little more than an ordinary course of Experimental Science ... with certain items in the Chemistry section of it, which fit in with housecraft requirements, tacked on the end; or it may consist of a course in which the housecraft bias is dominant throughout, a course consisting in fact of a kind of "Applied Science" from the outset.[62]

Having fought hard for getting girls an academic education equal to that of boys, many women scientists opposed domestic science as a turning back of the clock, limiting girls' aspirations and opportunities to that of domesticity. Ida Freund (see Chap. 6), Lecturer in Chemistry at Newnham College, critiqued the whole idea of domestic science in the feminist publication, *The Englishwoman*:

> It was erroneous to think that through the study of the scientific processes underlying housecraft and especially cookery,

> you can teach science, that is, give a valuable mental training which should enable the pupils in after life to judge whether an alleged connection between effect and cause has been established or not.[63]

Debate continued throughout 1911 and 1912. Lucy Hall and Ida Grünbaum, Science Lecturers at Avery Hill (Teachers') Training College, Eltham, supported Smithells[64]; while Hilda J. Hartle (see Chap. 13) of Homerton College, Cambridge, pointed out: "The science of cookery and of laundry work is yet in its infancy. No literature of the subject exists. Not even the most brilliant organic chemist can be said to 'know' the chemistry of foods, still less can such a subject be within the grasp of students in training."[65]

Six months later, Hartle wrote a scathing attack[66] on a book of domestic science written by R. H. Jones, Head of the Chemistry Department, Harris Institute, Preston, in which he claimed that: "science can be directly and adequately taught in the kitchen and that a previous training in elementary science is not indispensable." Hartle not only attacked the book in principle, but also pointed out the many errors in science that it contained, particularly "loose phraseology" such as: "distilled water contains nothing and is quite soft; fruit contains as a rule 80% of water and this is in pure condition."[66]

Though most women science educators favoured the pure classical sciences, there were a few who argued for a hybrid solution, one being Margaret Seward (Mrs. McKillop — see Chap. 3) of King's College for Women. Seward contended that pure and applied chemistry could be interwoven: "It was perfectly possible to make the outlines of the changes occurring in for example saponification clear to a class that had been trained in elementary chemistry and had further made acquaintance with some typical organic compounds."[67]

Rose Stern (see Chap. 2), the science teacher at NLCS, argued that using household chemicals would enable girls to better appreciate academic chemistry: "... every good teacher in science in a girls' school should look for examples for experiments from substances which are known to the pupils, for example, there is no reason why washing soda should not be used instead of another carbonate, and Epsom salts as a type of sulphate."[68]

However, the tide was against the "domestic science as science" movement in schools. William Tilden, a strong supporter of women chemists (see Chap. 2) and long-time President of the Chemical Society, gave an address in 1911 to the Science branch of the London Section of the Assistant Mistresses in which he stated: "Applications are not easy, and there is no satisfactory book on the chemistry of common life. Hence it is doubtful whether domestic science can form a proper University faculty."[69]

Faithfull concurred:

> The foundations of a knowledge of chemistry and physics should be built up on a well-ordered system which must not be subordinated from the outset to the requirements of home science. The teaching of science during the school years should be such as to prove equally useful to the pupil who elects to take at a later stage a university course in science and to the pupil who enters upon the home science course.[70]

There was a final bout of correspondence in 1914 — one which would resonate even today. It was initiated by Isabel C. Fortey, sister of chemist Emily C. Fortey (see Chap. 5):

> The review in your August number of Mrs. White's "First Book of Experimental Science for Girls" begins with the words, "To those who have studied the problem, it has long been obvious that girls require a course in science quite different from that which it has been customary to provide for boys." Such a statement will not be allowed to pass unchallenged because Miss Freund is dead.[71]

Fortey then argued that a girl had to do the same science as her brother in order to compete with him at science studies in university, or if she wanted to become a doctor. In reply, Jesse White contended that she was not in favour of teaching girls domestic science, but that science should be taught as relevant to girls' lives rather than assuming "... the only girls who really matter are the few who will go on to university."[72]

In the end, with the antipathy of most of the women science teachers, domestic science ceased to be an acceptable alternative science subject for girls. As illustration, Sutton High School closed its domestic science department in 1916.[59] By 1918, domestic science had been relegated to a low-status nonacademic subject.[60] The high-school debate was over.

University for Women?

Looking back at the late 19th century, it is astonishing to see how quickly the battle moved from the question of the admission of girls to secondary school education to that of university education. This is due to a couple of factors. Firstly, a number of organisations sprang up across the country to proselytise where possible and badger where necessary.[73] There were numerous local schoolmistresses and ladies' educational associations, including the Manchester Board of Schoolmistresses, the London Association of Schoolmistresses, the Leeds Ladies' Educational Association, and the Hampshire Association for Promoting Female Education. Then there were regional organisations, such as the North of England Council, together with the national organisation, the Women's Educational Union (the organisation instrumental in the founding of the GPDSC; see above). Each of these organisations was lobbying between 1865 and 1885 for the higher education of (especially middle-class) women.

The New Girl

A second factor was the production of a more assertive generation of middle-class girls who saw wider horizons for themselves.

Sally Mitchell has contended that the introduction of the school uniform promoted among young girls a liberating sense of "girlhood":

> I wonder, however, whether late-twentieth-century schoolgirls' joy in getting rid of school uniforms when they leave adolescence comes anywhere close to the turn-of-the-century girl's dramatic liberation when she first dressed in a costume distinctly her own, which marked her as neither child nor woman, had pockets, made it possible to run and climb, and let her add a boy-style shirt and tie.[74]

Victorian and Edwardian middle-class girls discovered role models through the new magazines for girls that appeared: the *Atalanta* (1887–1898), the *Girl's Realm* (1898–1915), and the *Girl's Own Paper* (1880–1927).[75] Though the viewpoint depended upon the editor of the time, these magazines were very forward-thinking and quite adult in their messages. The issues contained a significant proportion of intellectual material, including articles on science topics and reviews of colleges and universities for young women.

Some of the content was quite subversive. For example, an issue of the *Girl's Realm* of 1914 carried an article, "The Woman of the Future," which expounded at length that: "One of the greatest social crimes that has ever been committed was man's enslavement of woman." The author anticipated women's future role:

> Woman is taking to herself a new significance. She is discovering that she, as well as man, has another message for humanity besides that presented by wife and mother, and that henceforth she is to fulfil an entirely new mission in civilization. ... The new woman protests against having her life absorbed in ministering to man, in being exploited wholly for his benefit.[76]

With the girls' magazines urging them forth, many of the young girls of the 1880s to 1920s saw it almost a duty of their generation to see a university education as their next goal.

University Women: For and Against

Many of the arguments which had been used for or against an academic secondary education for girls became re-used and expanded for the new issue of women's admission to university. There were three arguments by proponents of a university education for women, particularly a scientific education. These were that women would become better wives and mothers if they had a scientific background; that simple justice and equal opportunity should permit women to enter scientific careers; and that women could perform certain scientific work better than men as a result of their superior patience and manual dexterity.[77] Yet many women themselves questioned the role of higher education for their gender, in particular, whether greater knowledge conflicted with the ultimate goal of domesticity.[78]

There was "scientific" evidence that could be used against women's advancement, especially the theory of evolution. For example, the sociologist Herbert Spencer had concluded that the difference between the sexes could best be understood in terms of "a somewhat earlier-arrest of individual evolution in women than men."[79] In fact, many scientists of the time had discovered proof in their research of women's "intellectual inadequacies"[80] and, of particular importance, Charles Darwin himself had found "scientific" evidence of female inferiority: "It is generally admitted that with women the powers of intuition, of rapid perception, and perhaps of imitation, are more strongly marked than in man; but some, at least, of these faculties are characteristic of the lower races, and therefore of a past and lower state of civilization."[81]

By the 1870s, the medical field had added its voice to the undesirability of advanced education for women. Edward H. Clarke, a former professor at Harvard Medical School, had described in his book, *Sex in Education*, how the health of many girls had been severely damaged by education.[82] In the 1880s, Henry Maudsley of University College, London, and John Thorburn, Professor of Obstetrics at Owens College, Manchester, contended that for women, the stress of education could have dire consequences.[83] When a woman student at Owens College (later, University of Manchester) died of tuberculosis, Thorburn publicly announced that the death was due to "overeducation." It was not until the 1890s that data conclusively showed that there was no indication of ill health among women students as a result of university studies.

There was also a concern that, if women's intellects were developed, it would result in their masculinisation, with women abandoning marriage and motherhood for academia and professions. Such a certain eventuality was put in verse in the pages of the magazine *Punch*:

> O pedants of these later days, who go on undiscerning,
> To overload a woman's brain and cram our girls with learning,
> You'll make a woman half a man, the souls of parents vexing,
> To find that all the gentle sex this process is unsexing.
> Leave one or two nice girls before the sex your system smothers.
> Or what on earth will poor men do for sweethearts, wives and mothers?[84]

There was the particular fear that educated women would not produce offspring — in fact, of the first generation of Girton College graduates, only 16 of the 35 married, and of these, only 7 had children.[85] It was popularly believed that such an outcome would cause the decline of civilisation. This common view was concisely expounded by Robert Lawson Tait, sometime President of the British Gynaecological Society: "To leave only the inferior women to perpetuate the species will do more to deteriorate the human race than all the individual victories at Girton will do to benefit it."[86]

At the same time, voices were raised to support women's entry. As early as 1870, the journal *Nature* expounded in an editorial on the virtue of the scientific education of women — albeit in the context of improving their homemaking talents: "Few have yet realized the enormous gain that will accrue to society from the scientific education of our women. ... What insight would a knowledge of chemistry afford into the wholesomeness or unwholesomeness of different articles of food!"[87]

Eleanor Sidgwick, in a lecture in 1896, saw university education as opening career opportunities:

> Women ought to have an independent career, because nothing can be more depressing or demoralising than waiting for the marriage which may never come; it is bad for them physically, intellectually, and morally; and moreover, nothing can be more apt to lead to unhappy marriages than the temptation to marry merely for the sake of a career.[88]

Yet, despite the naysayers, access to university turned from a dream to a reality. Even the supporters, such as Faithfull, marvelled at the speed of change:

> Perhaps it is safe to say that there is no movement which has made a more wonderful advance in the last twenty years than that which concerns the higher education of women. The very phrase twenty years ago made people shake their heads ominously, and prophesy untold evils to women's character, women's health, women's influence, to women's work in their homes and to society at large. ... Few people probably realize the courage and independence of those pioneers, who were certain to be dubbed "blue," and looked askance at as alarmingly intellectual and "advanced." Now-a-days all that is changed, it is hardly too much to say that throughout England college education is regarded as a desirable continuation of a girls school education.[89]

Who Went to University?

To decide to go to college was a brave act in itself. It required a strong self-image, particularly if the family held to the conventional view that a well-brought-up Victorian or Edwardian girl should stay quietly at home until a suitor appeared on the horizon. And, in most countries it was an avenue open only to daughters of the expanding middle class, a rapidly growing business and professional sector of society.[90] The daughters of the poor were simply financially unable to attend university; while the daughters of the upper classes were, for the most part, given an education that would prepare them for their intended life of leisure rather than one that might promote intellectual development. Although the information is incomplete, one or both of these factors are apparent in the lives of many of the women discussed in this book. Being the eldest daughter (or particularly, an only child) also seems to have favoured the pursuit of a university education.[91] Daughters of clergymen seemed to be over-represented among these early cohorts.[92] In addition, to go to college was one of the few avenues (nursing and missionary work being among the others) for a woman to escape the family home without the necessity of marriage.[93]

Going away to college was an exhilarating experience for this first generation. As one student remarked, she and her fellow co-eds were happy "... in the glorious conviction that at last, at last, we were afloat on a stream that had a real destination, even though we hardly knew what that destination was."[94]

For the second generation, there were different factors. In particular, by the end of the First World War, with so many young men having been killed, marriage prospects were minimal. Muriel Glyn-Jones recounted her own experience:

> My father decided to discuss my future with me. He said he would like to see me happily married, but after what had gone on in Flanders for the past four years that could only be doubtful. He would never be able to leave enough money for me to

> live on, but he was prepared to spend any money I could use for my education. It was later decided that I should go to Royal Holloway College ...[95]

As we will see in later chapters, attitudes from the men students ranged from outright hostility through amused tolerance. A student at Sheffield University in 1905 expressed her views poetically:

> 'Twas always said, in the long ago,
> That maids to College should never go;
> The very idea made the cheek grow pale
> and caused the stoutest of hearts to quail
> 'Tis all very well for men to praise
> The women of past and bygone days;
> 'Tis their opinion of us that's wrong
> at least so think I, and end my song.[96]

At co-educational facilities, women students were often constrained as to where they were allowed to go. In several institutions, women had to enter lecture rooms through a different door from men, and it was common for the lecture room itself to have a separate ladies row or section. Yet to these women students of the Victorian and Edwardian eras, the slights and insults were a small price to pay for the excitement of being in the first assault on the bastions of learning: to be in those "hallowed halls," where studying philosophy or physics was a joy, an end in itself.

By our standards, the activities of women students were severely circumscribed by the university rules, such as the need for chaperones. However, compared to the societal restrictions at home, universities were a haven of freedom for women. The historian J. F. C. Harrison described this feeling:

> For middle class girls the opportunity to have a room of one's own, to be able to organize one's life free from patriarchal

dominance, to have cocoa, tea or coffee parties unsupervised, to discuss what one liked with friends, to play games of hockey, and cycle around town — all this was immensely liberating, despite many restrictions and controls imposed by the college authorities.[97]

The Choice of University

For the 898 women chemists who became Associates or Fellows of the Royal Institute of Chemistry and/or Fellows of the Chemical Society between 1880–1949, we found information on the college or university attended for 841 of them.[98] The institutions where 10 or more women chemistry students obtained their undergraduate degree during that period are shown in Table 1.3.

As can be seen from Table 1.3, Bedford College, London — a small women-only college (see Chap. 4) — produced a disproportionately large number of women chemists. Royal Holloway College (RHC), too, is overrepresented for its size. For the constituent colleges of Cambridge, over 60% attended Newnham; while at Oxford, Somerville and St. Hugh's were the most popular. Queen Elizabeth College was founded as King's College, Ladies' Department, in Kensington, the graduates being awarded degrees through King's College (see Chap. 3). Likewise, the University of Newcastle was formerly a satellite college of the University of Durham; thus, women graduates from Newcastle were listed as having Durham degrees (see Chap. 5).

As students could obtain external degrees from the University of London, we find Battersea Polytechnic and Nottingham in the list, both offering London external B.Sc. (Chemistry) degrees during this time (as did Exeter, Northern Polytechnic, and others). In addition to those women chemists identifiable with a particular college, an additional 39 women had London chemistry degrees with no indication of their affiliated institution.

The total for all of the Welsh constituent university colleges (Aberystwyth, Bangor, and Cardiff; see Chap. 7) is given here.

Table 1.3. Most common universities/colleges attended for their undergraduate degree by British women chemists, 1880–1949.

University/College	Number of students
Bedford College, London	100
University College, London	57
Glasgow	48
Cambridge	45
Manchester	38
Royal Holloway College, London	36
Oxford	28
Birmingham	27
Queen Elizabeth College, London	24
Imperial College, London	23
Liverpool	22
Leeds	21
Queen Mary College, London	19
Edinburgh	19
Birkbeck College, London	15
Wales (combined)	15
Newcastle	13
Battersea Polytechnic	12
Aberdeen	10
St. Andrews	10
Nottingham	10

For Scotland (see Chap. 7), Glasgow was the overwhelming choice of women chemists. In part, this was probably due to the existence of a separate women's college, Queen Margaret College, in the early years; while the total for Edinburgh includes Heriot-Watt College (later Heriot-Watt University).

School–University Links

The most apparent link is between MHSG and the University of Manchester: nearly all MHSG graduates who became Associates or Fellows of the Royal Institute of Chemistry and/or

Fellows of the Chemical Society went to their neighbouring University; as a former MHSG student, Mary McNicol, commented in 1902:

> Owens College [later, Manchester] has always seemed to me the natural place at which [Manchester] High School girls should study after having finished their school course, and, judging by the number of Old Girls there, other people think so too. No High School girl needs ever feel lonely at Owens, for she will always come across girls she has known at school. A considerable number will be found in the Arts Classes, and several in the Science Department.[99]

Though the link between KEVI and Newnham College was mentioned earlier, in fact, only about 30% followed that route. Nearly all NLCS graduates entered either Bedford or RHC, with about half going to each. On the other hand, women chemistry students from South Hampstead High School for Girls all went to Bedford, according to our data set, as did a significant proportion of CLC graduates planning to follow a chemistry career.

Admission to university meant that another barrier to women had been overcome. But there was yet another one to face: admission to professional societies; and for certain societies, this proved a lengthier battle — the subject of Chap. 2.

Commentary

The advancement of women in chemistry would not have been possible without the development of academic schools for girls, particularly those emphasising a strong science component. The reader of subsequent chapters will observe that a very high proportion of the women chemists came from NLCS, CLC, KEVI, MHSG, or the GPDSC schools. Thus, it is important to give recognition to the pioneers who made it possible — particularly Buss, Beale, and Burstall — and to the forgotten institution from which they came: Queen's College, Harley Street (see Chap. 4).

Freund and Hartle were among those who fought for girls to take mainstream (men's) chemistry rather than domestic science. This route gave academically gifted girls access to university-level studies rather than diverting them to "women's science." On the other hand, the argument of Smithells and Stern — that making chemistry relevant to girls' lives would have broadened the appeal and usefulness of the subject — is as valid today as it was then. However, the majority of the activists insisted that for the good of future generations, it was necessary to have the same secondary chemistry as that for boys to enable their most talented girls to enter university chemistry degrees.

Notes

1. Bryant, M. (1979). Chapter 2: The agents. In *The Unexpected Revolution: A Study in the History of the Education of Women and Girls in the Nineteenth Century*, University of London Institute of Education, pp. 60–75.
2. Avery, G. (1967). Introduction to Shore, E., self-education. In Avery, G. (ed.), *School Remembered: An Anthology Edited, and with an Introduction by Gillian Avery*, Victor Gollancz, London, p. 184.
3. (a) McDermid, J. (1989). Conservative feminism and female education in the eighteenth century. *History of Education* **18**(4): 309–322; (b) Hunt, F. (1987). Divided aims: The educational implications of opposing ideologies in girls' secondary schooling, 1850–1940. In Hunt, F. (ed.), *Lessons for Life: The Schooling of Girls and Women, 1850–1940*, Oxford University Press, Oxford, pp. 3–21; (c) Jordan, E. (1991). "Making good wives and mothers?": The transformation of middle-class girls' education in nineteenth century Britain. *History of Education Quarterly* **31**: 439–462.
4. From Sewell's *Women and the Times We Live in*, cited in: Purvis, J. (1991). *A History of Women's Education in England*. Open University Press, Milton Keynes, pp. 111–112.

5. Becker, L. E. (1869). On the study of science by women. *Contemporary Review* **10**: 386–404.
6. Cited in: Gomersall, M. (1988). Ideals and realities: The education of working-class girls, 1800–1870. *History of Education* **17**(1): 43.
7. Allen, K. and MacKinnon, A. (1998). "Allowed and expected to be educated and intelligent": The Education of Quaker Girls in nineteenth century England. *History of Education* **27**(4): 391.
8. Sturge, H. W. and Clark, T. (1931). *The Mount School–York: 1785 to 1814, 1831–1931*. J. M. Dent, London, pp. 109–110.
9. Leach, C. (2006). Religion and rationality: Quaker women and science education 1790–1850. *History of Education* **35**(1): 69–90.
10. Stephen, L. (2004). Allen, William (1770–1843). *Oxford Dictionary of National Biography*, Oxford University Press, http://www.oxforddnb.com/view/article/392, accessed 27 Nov 2007.
11. Note 9, Leach, p. 76.
12. Cited in: Note 9, Leach, p. 77.
13. Delamont, S. (1993). Distant dangers and forgotten standards: Pollution control strategies in the British Girls' School, 1860–1920. *Women's History Review* **2**(2): 233–251.
14. Note 13, Delamont, p. 234.
15. Richardson, J. (1974). The great revolution: Women's education in Victorian times. *History Today* **24**: 420–427.
16. Glenday, N. and Price, M. (1974). *Reluctant Revolutionaries: A Century of Headmistresses 1874–1974*. Pitman Publishing, London, p. 23.
17. Scrimgeour, M. A. (ed.) (1950). *North London Collegiate School 1850–1950*. Oxford University Press, Oxford.
18. Steinbach, S. (2004). *Women in England 1760–1914: A Social History*. Palgrave Macmillan, New York, pp. 174–175.
19. Hughes, M. V. (1946). *A London Family 1870–1900: A Trilogy*. Reprinted 1991, Oxford University Press, London, p. 184.
20. Watson, N. (2000). *And Their Works Do Follow Them: The Story of North London Collegiate School*. James and James, London, p. 16.
21. Hill, E. M. (Nov 1899–Oct 1900). The Frances Mary Buss' Schools. *The Girl's Realm* **2**: 595.
22. Anon. (July 1895). In Memoriam: Grace Heath. *Our Magazine: North London Collegiate School for Girls* **20**: 60. See also: Eyre, J. V.

(1958). *Henry Edward Armstrong, 1848–1937. The Doyen of British Chemists and Pioneer of Technical Education.* Butterworths Scientific Publications, London, p. 272.

23. Hahn, G. (November 1897). A visit to Holloway College. *Our Magazine: North London Collegiate School for Girls* **17**: 107–110.
24. "Notes by one who was there." (December 1912). Science tea. *Our Magazine: North London Collegiate School for Girls* (n.v.): 89–90. Copies of *The Searchlight, NLCS Student Magazine for Science,* survive for 1911–1912, 1912–1913, and 1913–1914 (NLCS Archives, RS 4iv). See also: *A Short History of Science at North London Collegiate School*, unpublished, NLCS Archives.
25. (a) Kamm, J. (1971). *Indicative Past: A Hundred Years of the Girls' Public Day School Trust.* Allen and Unwin, London; (b) Goodman, J. F. (2004). Girls' Public Day School Company (*act.* 1872–1905). *Oxford Dictionary of National Biography*, Oxford University Press, http://www.oxforddnb.com/view/article/94164, accessed 17 Dec 2007. A "rival" network of Anglican church-organised schools was organised, the Church Schools Company; however, these schools did not place any significant emphasis on science teaching, see: Bell, E. M. (1958). *A History of the Church Schools Company: 1883–1958.* S.P.C.K., London.
26. Reynolds, K. M. (1950). VI: The school and its place in girls' education. In Scrimgeour, M. A. (ed.), *North London Collegiate School 1850–1950*, Oxford University Press, Oxford, pp. 114–115.
27. Anon. (June 1900). The jubilee of the Frances Mary Buss Schools, 4th April 1900. *The Magazine of the Manchester High School* 62–64.
28. Clarke, A. K. (1953). *A History of Cheltenham Ladies' College.* Faber & Faber, London.
29. (February 1854). *Governors Report.* The Ladies' College, Cheltenham.
30. Raikes, E. (1908). *Dorothea Beale of Cheltenham.* Archibald Constable & Co., London.
31. Note. 30, Raikes, p. 87.
32. Raikes, Mrs. (Nov 1899–Oct 1900). The Ladies' College, Cheltenham. *The Girl's Realm* **2**: 121.
33. Taylor, M. (1905). The new science wing, The Ladies' College, Cheltenham. *School World* **7**: 222.
34. Kamm, J. (1958). *How Different from Us: A Biography of Miss Buss and Miss Beale.* Bodley Head, London, p. 237.

35. Cited in: Dyhouse, C. (1976). Social Darwinistic ideas and the development of women's education in England, 1880–1920. *History of Education* **5**(1): 54.
36. Vardy, W. I. (1928). *King Edward VI High School for Girls Birmingham 1883–1925*. Ernest Benn, London.
37. Note 36, Vardy, p. 26.
38. Waterhouse, R. (n.d.). *King Edward VI High School for Girls 1883–1983* (no pub.), p. 23.
39. Note 36, Vardy, p. 32.
40. Note 36, Vardy, p. 26.
41. Anon. (n. d.). *Opening Ceremony of the New Building of King Edward VI High School*, p. 13.
42. Parker, J. E. (2001). Lydia Becker's "School for Sciencc": A challenge to domesticity. *Women's History Review* **10**(4): 643.
43. Fletcher, M. (1939). *O, Call Back Yesterday*. Shakespeare Head Press, Oxford, p. 47.
44. Burstall, S. A. (1933). *Retrospect and Prospect: Sixty Years of Women's Education*. Longmans, Green, London. See also: Burstall, S. A. (1911). *The Story of the Manchester High School for Girls*. Manchester University Press, Manchester.
45. McNicol, M. (July 1901). A school journey to Cambridge. *The Magazine of the Manchester High School* 48–49.
46. Pedersen, J. S. (1975). Schoolmistresses and headmistresses: Elites and education in nineteenth century England. *Journal of British Studies* **15**(1): 135–162. See also: Burstall, S. A. (April 1924). Memories. *Magazine of the Manchester High School* 15–17.
47. Marcet's book, *Conversations on Chemistry: In Which the Elements of That Science are Familiarly Explained and Illustrated by Experiments*, was first published in 1806. Designed with women readers in mind, it used conversations between a teacher and two female students to convey the information. The 18th and last edition was in 1853. See: Rayner-Canham, M. F. and Rayner-Canham, G. W. (1998). *Women in Chemistry: Their Changing Roles from Alchemical Times to the Mid-Twentieth Century*. American Chemical Society and the Chemical Heritage Foundation, Philadelphia, pp. 32–35.
48. Avery, G. (1991). *The Best Type of Girl: A History of Girls' Independent Schools*. Andre Deutsch, p. 61.

49. de Brereton Evans, C. (1898). The teaching of chemistry. In Beale, D., Soulsby, L. H. M. and Dove, J. F. (eds.), *Work and Play in Girls' Schools by Three Head Mistresses*, Longmans, Green & Co., London, pp. 310–311.
50. Bungay, J. (ed.) (1982). *Redland School, 1882–1982* (no pub.).
51. Bain, P. (1984). *St. Swithun's: A Centenary History*. Phillimore, Chichester.
52. Note 51, Bain, p. 10.
53. Goodman, J. (1997). Constructing contradiction: The power and powerlessness of women in the giving and taking of evidence in the Bryce Commission, 1895. *History of Education* **26**: 299.
54. Stuart, J. E. (1914). *The Education of Catholic Girls*. Longmans, Green & Co., London, pp. 120–121.
55. (1895). *Bryce Commission*, Vol. IX, pp. 428–429, cited in: Manthorpe, C. (1993). Science education in the public schools for girls in the late nineteenth century. In Walford, G. (ed.), *The Private Schooling of Girls: Past and Present*, Woburn Press, London, p. 70.
56. Howarth, J. (1985). Public school, safety-nets and educational ladders: The classifications of girls' secondary schools, 1880–1914. *Oxford Review of Education* **11**(1): 59–71.
57. Cited in: Note 25(a), Kamm, p. 77.
58. Phillips, G. W. (1980). *Smile, Bow and Pass On: A Biography of an Avante-Garde Headmistress, Miss Iris M. Brooks, M.A.(Cantab.) Malvern Girls' College*. St. Michaels Abbey Press, Farnborough, Hants.
59. Spencer, S. (2000). Advice and ambition in a girls' public day school: The case of Sutton High School, 1884–1924. *Women's History Review* **9**(1): 75–91.
60. Manthorpe, C. (1986). Science or domestic science? The struggle to define an appropriate science education for girls in early 20th century England. *History of Education* **15**: 195–213.
61. Flintham, A. J. (1977). The contributions of Arthur Smithells, FRS, to science education. *History of Education* **6**(3): 195–208; Flintham, A. J. (1975). Chemistry or Cookery? *Housecraft* **48**: 105–106.
62. Board of Education (1911). *Memorandum on the Teaching of Housecraft in Girls' Secondary Schools*, London, p. 26, cited in: Note 60, Manthorpe, p. 199.

63. Freund, I. (1911). Domestic Science — A Protest. *The Englishwoman* **10**: 147–163, 279–296. See also: The Editor (1911). Occasional notes. *Journal of Education* **33**: 451–452.
64. Hall, L. and Grünbaum, I. (1911). Correspondence: Housecraft in training colleges. *Journal of Education* **33**: 678–679; Hall, L. and Grünbaum, I. (1912). *The Chemistry of Housecraft: A Primer of Practical Domestic Science*. Blackie, London.
65. Hartle, H. J. (1911). Correspondence: Housecraft in training colleges. *Journal of Education* **33**: 849–850.
66. Hartle, H. J. (1912). Letter to the editor. *Journal of Education* **34**: 465.
67. Seward quotation in Flintham, A. J., unpublished thesis, reproduced as footnote 49 in Notc 60, Manthorpe, p. 202.
68. Stern, R. (1912). Science in Girls' Schools. *School World* **14**: 460–461. This issue of *School World* had a whole section (pp. 452–465) on the contentious issue of the appropriate science for girls. Amongst the contributers were L. M. Faithfull, Ida Freund, Charlotte L. Laurie (science teacher, CLC), Jessie White, and Arthur Smithells.
69. Anon. (1911). Science notes: Sir William Tilden on chemistry teaching. *Journal of Education* **33**: 167.
70. Faithfull, L. M. (1911). Home science. In Spender, D. (ed.), *The Education Papers: Women's Quest for Equality in Britain, 1850–1912*, Routledge & Kegan Paul, New York, 1987, p. 326.
71. Fortey, I. C. (1914). Correspondence: Sexless science. *Journal of Education* **36**: 618, 620. Born in 1864, Isabel Fortey followed a career as a science teacher, first in Britain, then in India. She died on 16 October 1954. See: White, A. B. (ed.) (1979). *Newnham College Register, 1871–1971*, Vol. I, 1871–1923, 2nd ed. Newnham College, Cambridge, p. 82.
72. White, J. (1914). Correspondence: Science for girls. *Journal of Education* **36**: 698.
73. Pope, R. D. and Verbeke, M. G. (1976). Ladies educational organizations in England 1865–1885. *Paedagogica Historica* **16**(2): 336–361.
74. Mitchell, S. (1995). *The New Girl: Girls' Culture in England, 1880–1915*. Columbia University Press, New York, p. 89.

75. Dixon, D. (Winter 1998/1999). Deprived and oppressed: Victorian and Edwardian magazines for girls. *Library News* #25, www.bl.uk/collections/nl25.html, British Library, accessed 9 December 2006. Upper-working-class girls had their own magazine, *Girl's Best Friend,* that focused on health and beauty rather than career options, see: Webb, A. (2006). Constructing the gendered body: Girls, health, beauty, advice, and the *Girl's Best Friend*, 1898–1899. *Women's History Review* **15**(2): 253–275.
76. Marden, O. S. (1914). The woman of the future: Her new opportunities and responsibilities — What will she do with them? *Girl's Realm* **16**: 449–451.
77. MacLeod, R. and Moseley, R. (1979). Fathers and daughters: Reflections on women, science, and Victorian Cambridge. *History of Education* **8**: 321.
78. Wein, R. (1974). Womens colleges and domesticity, 1875–1918. *History of Education Quarterly* **14**: 31–48.
79. Conway, J. (1972). Stereotypes of femininity in a theory of sexual evolution. In Vicinus, M. (ed.), *Suffer and Be Still: Women in the Victorian Age*, Indiana University Press, Bloomington, Indiana, p. 141.
80. Zschoche, S. (1989). Dr. Clarke revisited: Science, true womanhood, and female collegiate education. *History of Education Quarterly* **29**: 545–569; Alaya, F. (1977). Victorian science and the "genius" of woman. *Journal of the History of Ideas* **38**: 261–280.
81. Darwin, C. (2004). *The Descent of Man, and Selection in Relation to Sex*. Penguin Classics, New York, p. 629.
82. Hubbard, R. (1979). Feminism in academia: Its problematic and problems. In Briscoe, A. M. and Pfafflin, S. M. (eds.), *Expanding the Role of Women in the Sciences*, New York Academy of Sciences, New York, p. 251. See also Burstyn, J. N. (1973). Education and sex: The medical case against higher education for women in England, 1870–1900. *Proceedings of the American Philosophical Society* **117**: 79–89.
83. Purvis, J. (1995). Student life. In Purvis, J. (ed.), *Women's History: Britain, 1850–1945*, St. Martin's Press, New York, pp. 191–192; Blake, C. (1990). *The Charge of the Parasols: Women's Entry to the Medical Profession*. The Women's Press, London, pp. 164–165.

84. Sheffield, S. L.-M. (2004). The "empty-headed beauty" and the "sweet girl graduate": Women's science education in *Punch*, 1860–90. In Henson, L. *et al.* (eds.), *Culture and Science in the Nineteenth Century Media*, Ashgate, Aldershot, p. 19.
85. Tamboukou, M. (2000). Of other spaces: Women's colleges at the turn of the nineteenth century in the UK. *Gender, Place and Culture* **7**(3): 259.
86. Cited in: Burstyn, J. (1984). Educators' response to scientific and medical studies of women in England 1860–1900. In Acker, S. and Piper, D. W. (eds.), *Is Higher Education Fair to Women?*, SRHE & NFER-Nelson, Guildford, Surrey, p. 70.
87. Anon. (16 June 1870). The scientific education of women. *Nature* **2**: 117 118.
88. Sidgwick, Mrs. H. (1897). *University Education of Women: A Lecture Delivered at University College, Liverpool, in May 1896.* Macmillan and Bowes, Cambridge, p. 11.
89. Faithfull, L. M. (1900). *King's College Magazine, Ladies' Department* **4**(2): 6–10.
90. Howarth, J. and Curthoys, M. (June 1987). The political economy of women's higher education in late nineteenth and early twentieth-century Britain. *Historical Research* **60**: 208–231.
91. Sulloway has argued that birth order is of vital importance in determining traits; in particular, he argues that the first-born child has unique behavioural characteristics. See: Sulloway, F. J. (1996). *Born to Rebel: Birth Order, Family Dynamics and Creative Lives.* Random House of Canada, Toronto.
92. Wade, G. A. (1908–1909). Where do the cleverest girls come from? Is it from the rectory, the vicarage, the parsonage, and the manse? *Girl's Realm* **11**: 343–348.
93. Vicinus, M. (1985). *Independent Women: Work and Community for Single Women 1850–1920.* University of Chicago Press, Chicago, Illinois, p. 138.
94. Vicinus, M. (1982). One life to stand beside me: Emotional conflicts in first-generation college women in England. *Feminist Studies* **8**: 603–628.
95. Cited in: Bingham, C. (1987). *The History of Royal Holloway College 1886–1986.* Constable, London, p. 134.

96. "R.R.de L." (April 1905). The woman's side of the question. *The Blade: The Sheffield University Student's Magazine* (1): 4–5.
97. Harrison, J. F. C. (1990). *Late Victorian Britain 1875–1901*. Fontana Press, London, p. 171.
98. Rayner-Canham, M. F. and Rayner-Canham, G. W., to be published.
99. McNicol, M. (December 1902). College letters. *Magazine of the Manchester High School*, 115.

Chapter 2

The Professional Societies

These days we take for granted that scientific organisations are open to both men and women, but this was not always the case.[1] It is hard to realise that the admission of women chemists to chemical organisations was once a contentious issue. For example, in 1880, the American Chemical Society even held a formal Misogynist Dinner.[2]

During the 19th and early 20th centuries in the United Kingdom, there were a number of organisations that catered to the professional and social needs of chemists, the two aspects overlapping in the male club culture of the time.[3] What is particularly interesting is the very wide range of responses by the organisations when the issue of the admission of women arose. The Society for Analytical Chemists welcomed women from its early years (see below), though the first recorded women members only appeared in the 1920s. The Society for Chemical Industry seems never to have addressed the issue, though it too did not have any women members listed until the 1920s.[4]

Admission to the Royal Institute of Chemistry resulted from the accidental admission of a woman to the Institute's examinations in 1892, and the Biochemical Society, by vote of the existing members, first accepted women in 1913. The major holdout was the Chemical Society with its "Forty Years War," as Joan Mason, historian of chemistry, called it (see below); not until 1920 were women admitted to the Chemical Society. But the last of all was the Royal Society, which did not elect its first women Fellows until 1948, both of whom were chemists.

Each society treated the problem of the admission of women in a different way. In this chapter, we will focus particularly on the lives of those British women who led the fight for professional acceptance.[5] We will see that the paths of many of these women intersected and that, in fact, there must have been networking among them. The saga begins with the short-lived London Chemical Society.

The London Chemical Society

Events started promisingly for women. The London Chemical Society seemed to take pleasure in noting that women had participated in its events. At a pre-inaugural lecture on 7 October 1824, it was reported that: "Several ladies were present, taking a warm interest in all that was said, encouraging the lecturer by their smiles, and ensuring order and decorum by their presence."[6]

At the subsequent inaugural lecture, it was mentioned that among the 300 persons attending, there were "a great many ladies." The address was given by George Birkbeck,[7] who specifically welcomed the participation of women:

> It may not be out of place here to state, that chemistry is not only intended to be confined to *learned* men but not even to *men* exclusively. Hitherto, ladies have conferred the honour of their presence upon all our public proceedings; and we are extremely desirous, although it is not consistent with the present constitution of the Society, that they should hereafter become participators also, as members.[8]

Birkbeck continued by pointing out the contributions from the late 18th century of Elizabeth Fulhame[9] and Jane Marcet[10]:

> That they are well qualified for pursuing this branch of science, I may adduce, as evidence, the very able Essay, by Mrs. Fulhame, "On Combustion".... In further evidence, I may

> adduce the "Conversations on Chemistry," by Mrs. Marcet; which, as an interesting and instructive elementary work for the uninitiated, has never yet been equalled. And, lastly, for examples need not be multiplied, I may notice the elegant "Conversations on Mineralogy," by Miss Lowry.[8]

It is not noted whether the Society did, in fact, change its constitution to allow women to be formally admitted. Unfortunately, the London Chemical Society ceased to exist shortly afterwards.[11]

The Society for Analytical Chemistry

Women gained admittance to the Society for Public Analysts (later called the Society for Analytical Chemistry) without any problem. In 1879, five years after the founding of the organisation, the following comment was made in the society journal, *The Analyst*: "We are liberal enough to say that we would welcome to our ranks any lady who had the courage to brave several years' training in a laboratory ...."[12] However, we were unable to find any evidence that this invitation produced any results. It was not until the 1920s that significant numbers of women started to join the Society as a result of their entry into analytical positions in industry and government.[13]

Isabel Hadfield

In fact, the most prominent woman in analytical chemistry, Isabel Hodgson Hadfield,[14] did not join the Society until 1944. Born on 29 January 1893, her father a schoolmaster, Hadfield graduated from East London College (later Queen Mary College) in 1914. The following year, she obtained a Diploma of Education and became a Chemistry Mistress with the Birmingham Education Council.

During the First World War, the need for scientists overrode the traditions of society and women were pressed into scientific employment (see Chap. 12). Thus, in 1917, Hadfield joined the staff of the National Physical Laboratory. Studying chemical problems relating to aeronautics, she contributed many reports to the Fabrics Research Co-ordinating Committee of the Department of Scientific and Industrial Research. She retained her position at the end of the War, earning an M.Sc. in 1923.

Much of the analysis involved small samples and, as a result, Hadfield became a pioneer in the development of micro-analytical measurements.[15] In the 1930s, she was a founder member of the Microchemical Club. She retired with the rank of Principal Scientific Officer in 1953 in order to look after her elderly father. After his death, she went to live with a friend in Hampshire where she stayed until her own death in February 1965.

The Royal Institute of Chemistry

The entry of women into the Institute of Chemistry (later the Royal Institute of Chemistry) can best be regarded as "accidental." The Institute had been founded in 1877, and the successful sitting of an examination was a prerequisite for admission. In November 1888, the Council recorded a minute noting that they did not contemplate the admission of women candidates to the Examinations.[16]

Emily Lloyd

Nevertheless, it was only 4 years later that Emily Lloyd became the first woman Associate. Emily Jane Lloyd[17] was the daughter of Martin Lloyd, a nail manufacturer in Birmingham. Although there is very little information about her, and some of that is contradictory, we do know that she attended a private school in Leamington. She was admitted as a student to Mason Science College (later, the University of Birmingham) in 1883 at the age of 23, spending only 1 year there before transferring to University College, Aberystwyth (now the University of Wales,

Aberystwyth). She remained at Aberystwyth until 1887, gaining a pass in the University of London Intermediate examination in sciences. Lloyd is the only woman student mentioned in the history of the chemistry department of University College, Aberystwyth.[18]

At this point, Lloyd returned to Mason College, where she was awarded a B.Sc. from the University of London in 1892. In that same year, she applied under the name of E. Lloyd to sit the Associateship examination of the Institute of Chemistry. As the committee would have had no expectation that she was a woman, Lloyd was permitted to write the paper, which she duly passed. Once she had passed, the Institute had no means of denying her admission to the Society, and having admitted one woman, there was no feasible route of barring subsequent women applicants.[16]

Having gained her Associateship, Lloyd applied to the Institute to take the required examination to qualify as a public analyst. The admission of women candidates to this examination had not been contemplated by the Institute's council, but they had no excuse to refuse her. She was admitted to, and passed, this hurdle as well. The following year, she obtained an appointment as Science Mistress in a public school for girls at Uitenhage, Cape Colony (now South Africa). Lloyd remained there for 4 years and then returned to Wales, teaching at a school in Llanelly until about 1909, when she retired due to ill health. She died on 14 November 1912, aged 52 years. The first woman Fellow was to follow almost immediately after Lloyd. This was Lucy Everest Boole, whose life and work will be discussed in Chap. 4.

Rose Stern

The first woman Student Member of the Institute of Chemistry was Rose Stern.[19] Stern graduated in 1889 from King Edward VI High School for Girls, Birmingham (KEVI). Like Lloyd, Stern worked towards a B.Sc. (London) from Mason College, completing it in 1894.[20] At the time, there were few options open apart from marriage or school teaching for women science graduates,[21] so Stern obtained an appointment as Science Mistress at North London Collegiate School for Girls (NLCS).[22] As we described in the previous chapter, Stern provided the chemical inspiration for many young women at NLCS during the early decades of the 20th century. Following her retirement in 1920, she authored *A Short History of Chemistry*.[23] As a result of breaking a leg during the Second World War, she became less and less mobile, dying in October 1953.

The Biochemical Society

The Biochemical Club, as it was first called, was founded in 1911. At the first meeting, the second item on the agenda concerned the admission of women.[24] A letter had been received from "a lady" (possibly Ida Smedley or Harriette Chick; see below) requesting permission to become an original member. An amendment was therefore proposed to the rules that only men were eligible for membership. The amendment passed by a vote of 17 to 9. This vote was challenged, and at a Committee meeting the following year, the Club reversed its position, voting by 24 to 7 that women be admitted. In 1913, the Club held a meeting to elect new members and, of the seven elected, three were women: Ida Smedley, Harriette Chick, and Muriel Wheldale (see Chap. 8). Smedley was later (1927) to become the first woman Chairman of the Committee.

Ida Smedley (Mrs. Smedley Maclean)

Ida Smedley[25] was a key individual in the early advancement of women in chemistry. As we have seen above, she was one of the first women members of the Biochemical Society; but of more importance, as we will see shortly, she was one of the two women (Martha Whiteley — see Chap. 3 — being the other) who fought for decades for the admission of women to the Chemical Society.

Smedley was born on 14 June 1877 in Birmingham, the second daughter of William T. Smedley, a chartered accountant, businessman, and philanthropist, and Annie E. Duckworth, daughter of a Liverpool coffee merchant. Smedley had an idyllic upbringing:

> Ida grew up in a home where the two gifted and far seeing parents devoted themselves to the interests of the children, and one may add, to those of their children's friends. To some of us the house was a second home. Literature, theatricals, music and languages filled up the leisure hours. Independence of thought and action were encouraged, and the whole atmosphere of the home was decades ahead of its time.[26]

Smedley, like Stern, attended KEVI, and it was there that she met Beatrice Thomas (see Chap. 6), the two forming a friendship that endured throughout their lives. Smedley spent 3 years at Newnham College, Cambridge, graduating in 1899. She became a research student at the Central Technical College, London (later part of Imperial College, IC) with Henry Armstrong,[27] being awarded a D.Sc. by the University of London for her research on benzylaniline sulfonic acids. In 1903, she briefly returned to Newnham as a demonstrator in chemistry, resigning in 1904 to take up full-time research at the Davy–Faraday Laboratory of the Royal Institution, London.

Then in 1906, Smedley became the first woman appointed to the Chemistry Department at the University of Manchester, holding the rank of Assistant Lecturer. At Manchester, Smedley continued her research in organic chemistry long into the night, often until about 3 a.m., together with the other "night owls,"[28] Robert Robinson and D. L. Chapman.

Smedley embarked on a career change in 1910 that was to gain her more recognition. Awarded a Beit Research Fellowship, she returned to London to take up a position at the Lister Institute of Preventive Medicine, where she remained for the rest of her working life. Her field of study was that of fat metabolism and fat synthesis, and she became an expert in the subject.[29] The value of her research was recognised in 1913, when she received the Ellen Richards Prize[30] of the American Association of University Women for the woman making the most outstanding contribution of the year to scientific knowledge. In 1913, Smedley married Hugh Maclean, who became a Professor of Medicine at the University of London and at St. Thomas's Hospital, London; she had two children, a son (1914) and a daughter (1917).

During the First World War, Smedley worked for the Ministry of Munitions and for the Admiralty, her major contribution being the development of the large-scale production of acetone from starch by fermentation.[29] After the War, she returned to her work on fats. Her most influential research on fat metabolism came in 1938, when she followed up the classic work of G. O. Burr and M. M. Burr at the University of Berkeley, California, who had described stunted growth and dry scaly skin in young rats totally deprived of fat. Smedley, and her colleague, Margaret Hume, reproduced the results of the Burrs, and showed that it could be cured by either of the two unsaturated fatty acids — linoleic or linolenic acid. Then, she isolated and identified arachidonic acid and showed this fatty acid to be as active biologically as linoleic acid.[31] The three acids — linoleic, linolenic, and arachidonic — came to be known as "essential fatty acids" because without one or other of them in the diet, the characteristic signs of deficiency appeared. Of course, we are now aware of the major importance of these omega fatty acids in human diets. In addition to more than 50 research papers on fats, she authored a monograph, *The Metabolism of Fat*.[32]

Smedley was also active in the rights of women outside of chemistry. In particular, she played the leading role in forming the British Federation of University Women. As was noted at the time: "... the single gathering sufficed to reveal to several who were present (stimulated and led by Dr. Ida Smedley Maclean) the value and importance of forming an association of University women on a permanent basis ...."[33] In addition to being a founder, she served in various roles of the organisation, including President from 1929 to 1935.

In summing up the later years of Smedley's life, Mary Phillp (Mrs. Epps), a friend from KEVI and Newnham days, observed:

> Her marriage was a very happy one, and she showed as much skill in running her home, and bringing up her children, as she had done in other departments of her life. How she managed to hold three threads evenly in her hands, research, social work and home-life, was a constant wonder even to those who knew her best. When for a time serious illness fell upon a member of the family her devotion to the invalid, while all her other work was carried on, was little less than heroic. During the last two years of her life her health was failing; but true to her courageous nature she accomplished what she had set out to do, and in 1943 her last work was published.[26]

Smedley died on 2 March 1944, aged 67 years.

Harriette Chick

The second of the pioneer women to be admitted to the Biochemical Club was Harriette Chick.[34] Born on 6 January 1875, Chick was one of ten children (seven daughters) of Samuel Chick, a businessman in the lace trade, and Emma Hooley. All of the daughters attended Notting Hill High School, London, one of the Girls' Public Day School Company (GPDSC) schools (see Chap. 1).[35] Amazingly for the time, six of the seven daughters completed university degrees.

Harriette Chick attended classes at Bedford College and then entered University College, London (UCL), where she became an outstanding student in botany before shifting her focus to chemistry. Following the completion of her B.Sc., she turned her attention to bacteriology, undertaking research at the Hygienic Institutes of both the University of Vienna and the

University of Munich, before returning to Britain to work at the University of Liverpool.

In 1905, she applied for a Jenner Memorial Research Studentship at the Lister Institute. The application caused a furore as no women had previously been given this award:

> As soon as her application became known, two members of the scientific staff implored the Director not to commit the folly of appointing a woman to the staff. She was, nevertheless, appointed to the Studentship and soon accepted on terms of equality and friendship by the apprehensive males.[36]

Chick's early work at the Lister, undertaken with Charles Martin,[37] was on the chemical kinetics of the disinfection process. Of particular importance, she found that the temperature dependence of the disinfectant action did not follow the Arrhenius rate expression; instead, the rate increased by as much as seven- or eight-fold for a 10°C increase in temperature, thus showing that warm disinfectant solutions were far better for killing bacteria than cold solutions. This led to her being the co-developer of the Chick–Martin Test for the efficacy of a disinfectant.

Following the First World War, Chick was co-leader of a team studying the disease rickets. This research, showing the link between nutrition and the disease, and finding methods of treating and preventing it, resulted in the near-complete disappearance of this debilitating illness. She continued in the field of nutrition, particularly the role of vitamins, becoming one of the founder members of the Nutrition Society.[38] Despite formal retirement in 1945, she kept active in the nutrition field until her death on 9 July 1977, aged 103 years.

The Chemical Society

Though there was initial opposition to the admission of women to every chemical organisation, by the early part of the 20th century, there was only one major bastion to be breached — the Chemical Society. The fight for admission to this organisation was found to be tough and lengthy, spanning a period of 40 years. And the most interesting part is the women who spearheaded the attack during the long campaign.

The Chemical Society was founded in 1841, but it was not until 1880 that the question was raised of the admission of women. This convoluted saga has been described in detail by Joan Mason.[39] In the initial discussion, legal opinion was given that, under the Charter of the Society, women were admissible as Fellows. A motion was proposed in 1880 by A. G. Vernon Harcourt[40] to clarify the bye-laws, so that any reference to the masculine gender should be assumed to include the feminine gender.[41] The motion was rejected as "not expedient at this time." A similar proposal was put forward by William Ramsay[42] in 1888, but after lengthy discussion, the motion was withdrawn.[43]

The first attempt by a woman (possibly Emily Lloyd, see above) to enter the Society occurred in November 1892. The long controversy started innocuously, as the Minutes of the Council Meeting describes: "The Secretary having read a letter from Prof. Hartley suggesting the election of a lady as Associate, Prof. Ramsay gave notice that he would move that women be admitted Fellows of the Society."[44]

The motion, proposed by Ramsay and seconded by William Tilden,[45] came the following January.[46(a)] An amendment was then proposed by William Perkin, Sr.,[47] that it was not desirable at that time to amend the bye-laws for the purpose of admitting women. The amendment was defeated by seven votes to six, then curiously the motion itself was defeated by a margin of eight votes to seven. The Secretary commented:

> Does the Charter of the Society contemplate or permit of the admission of women as Fellows and if so what alteration, if any, is required in the Bye-Laws? The Council were advised that under the Charter women are admissible as Fellows, but

> when on three occasions (1880, 1888, and 1892) proposals were made to put it into effect no action was taken, the general feeling being that, although there was no objection in principle to the admission of women as Fellows, the case in their favour was not entirely established by any considerable number of applications for the Fellowship and that a change involving so radical an alteration in the policy of the Society, should be recommended by a maximum vote.[46(b)]

So things remained until 1904, when Marie Curie's name was put forward for election as a Foreign Fellow.[48] At the following meeting, discussion of her candidacy resulted in a motion to again request the opinion of legal council on the eligibility of women for admission as Ordinary Fellows and Foreign Members.[49] Presumably, either the opinion of 24 years earlier had been forgotten, or it was hoped that a new counsel would offer a different opinion. This was, in fact, the case. The new counsel, R. I. Parker, reported: "In my opinion married women are not eligible as Fellows of the Chemical Society and I think it extremely doubtful whether the Charter admits of the election of unmarried women as Fellows."[50] Under British common law of the time, a married woman was not an independent person and therefore could not be a Fellow in her own right. However, Parker deemed that as Honorary and Foreign members came with no duties and responsibilities, there was no impediment to the election of Curie. Curie was duly elected.[51]

The 1904 Petition

Emboldened by Curie's success, a memorial (petition) was presented to the Council in October 1904 by Ida Smedley and Martha Whiteley requesting admission of women to Fellowship.[39] Nineteen women chemists signed the petition (see Table 2.1), which stated: "We, the undersigned, representing women engaged in chemical work in this country desire to lay

Table 2.1. The signatories of the 1904 petition for admission of women to Fellowship in the Chemical Society (Note 52).

Lucy Boole [4]	Margaret Seward (Mrs. McKillop) [3]
Katherine Burke [3]	Ida Smedley [2]
Clare de Brereton Evans [4]	Alice Smith [7]
E. Eleanor Field [4]	Millicent Taylor [5]
Emily Fortey [5]	M. Beatrice Thomas [6]
Ida Freund [6]	Grace Toynbee (Mrs. Frankland) [11]
Mildred Gostling (Mrs. Mills) [11]	Martha Whiteley [3]
Hilda Hartle [13]	Sibyl Widdows [4]
Edith Humphrey [4]	Katherine Williams [5]
Dorothy Marshall [6]	

Note: Numbers in square brackets indicate the chapter in which their biography will be found.

before you an appeal for the admission of women to Fellowship in the Chemical Society."[52]

The petition authors then noted the increasing contributions of women chemists and the willingness of the Chemical Society to publish their results:

> Reference to the publications of the Chemical Society shows that during the last thirty years [1873–1903] the names of about 150 women of different nationalities have appeared there as authors or joint authors of some 300 papers. ... Seeing that the Chemical Society recognises the value of the contributions made by women to chemical knowledge by accepting their work for publication, we are encouraged to point out that their work would be greatly facilitated by free access to the chemical literature [in the Society's library] and by the right to attend the meetings of the Society.[52]

Following receipt of the petition, the Council, which at the time was women-friendly, unanimously adopted the proposal to alter the bye-laws, but the changes had to be approved by the body of the organisation. Of the over 2700 members, only 45 attended the Extraordinary General Meeting to approve the changes and, of them, 23 voted against. In his 1905 Presidential address, Tilden expressed his displeasure with the outcome:

> The number of women desiring admission is but small, and I fail to see any cogent reason, beside the legal one, for excluding them. Some of them are doing admirable scientific work, and all the memorialists are highly qualified. To deprive them of such advantages as attach to the Fellowship simply on the grounds that they are not men seems to be an unreasoning form of conservatism inconsistent with the principles of a Society which exists for the promotion of science. It seems to be unfortunate that when a subject so important is brought up for the judgement of a body numbering upwards of 2,700 members, the appointed meeting should be attended by no more than 45....[53]

It is the identity of these 19 women, and the factors that they had in common, that we will explore here. First of all, there were three individuals who played leading roles in the endeavour: the biochemist, Ida Smedley (see above); the organic chemist, Martha Whiteley (see Chap. 3); and the chemical educator, Ida Freund (see Chap. 6).[39]

One of the most interesting questions is: Why these 19? What did they have in common? There must have been networking in order to produce the signed petition. Whether it was solely among the women themselves or also through the correspondence of supportive male chemistry professors, we will never know for certain. The paths of many of the women signatories did cross. At the same time, several were researchers with sympathetic supervisors, and it may have been some of the supervisors who conveyed the news of the petition to the women in their research group.

Newnham College was the place where many of the women had met. Freund must have been a crucial link, as she was on staff there from 1887 to 1913. Those petitioners who would have been taught by Freund were Field, Gostling, Hartle, Marshall,

Smedley, and Thomas. Thomas became a close friend of Freund's and, in her later post at Girton College, she was initially an Assistant Demonstrator under Marshall. Some of the bonds between the women were initiated back in high school days. In particular, we know that Smedley and Thomas became friends at KEVI.[54]

Many of those in the "Newnham group" had a second commonality of Royal Holloway College (RHC). After Field was appointed at RHC, Thomas and Gostling held short-term demonstratorships there while Widdows' (see Chap. 4) sojourn at RHC overlapped with Thomas'. Widdows seems to be a key link into the next "circle" — that at the London School of Medicine for Women (LSMW). Widdows was at LSMW in 1904, as was Boole and Evans.

Fortey and Williams were at University College, Bristol, while Taylor undertook research at Bristol in her spare time. In fact, all three of them were there about the signing period of 1903–1904. But how did word reach the Bristol trio? Williams had been a collaborator of Ramsay when Ramsay had been at Bristol (see Chap. 5). As Ramsay was a fervent supporter of women's rights, it is quite possible that he conveyed news of the petition to Williams, who then passed it to her colleagues.

Hartle, who had been taught by Freund at Newnham, had spent 2 years at Birmingham as a researcher with Percy Frankland, spouse of Toynbee. Frankland was another champion of women chemists. In fact, it will always be an unanswered question whether it was the men, such as Ramsay and Frankland, who provided many of the links between women at different universities, as they would have had the professional and social network to do so.

Seward was Lecturer at RHC in 1887, one of her students being Whiteley. Perhaps their paths crossed again in 1904 when Seward (see Chap. 3) was Lecturer at King's College, Women's Department, in Kensington Square, while Whiteley had just taken up an appointment at nearby Royal College of Science in South Kensington.

This leaves Humphrey (Bedford College) and Smith (Owens College, Manchester). There seem to be no obvious links which would explain how news of the petition reached each of these women.

The Effect of the Petition

Tilden, the then-President of the Chemical Society, proposed another tack. He circulated a petition in support of women's admission, signed by 312 of the most distinguished Fellows of the Society.[55] Then in 1908, he co-sponsored with Henry Roscoe[56] a motion that there be a poll of members on the issue.[39] This motion passed, and a ballot was circulated accompanied by a list with six reasons to vote for admission and seven reasons to deny admission.

The reasons listed for supporting women's entry were as follows[57]:

1. That the petition has been signed by 312 Fellows (including 10 past Presidents, 12 Vice-Presidents, and 29 Members of Council, past and present) among whom are 33 Fellows of the Royal Society and the Professors of Chemistry or Heads of Chemical Departments of nearly all the most important Universities and Colleges in the country.
2. That a number of women are now devoting themselves to the science of Chemistry, the study of which is the chief object of the Society to promote.
3. The Chemical Societies of Berlin and America and the Institute of Chemistry admit women to full privileges of membership.
4. A small number of women chemists attend the meetings of the Society as visitors, and no inconvenience has arisen from their presence.
5. The Society numbers more than 2800 Fellows, while the number of women desiring admission at the present time is about 20; this number will probably increase as time goes on, but judging by the experience of other Societies, it is not

likely greatly to exceed that number in the present generation. ... consequently any fear that female influence might hereafter dominate the Council, or even that one woman might be elected to the Council except only in recognition of her scientific ability, is not worth consideration.

6. There is reason to believe that in the event of a decision in favour of applying for a supplemental Charter, the cost, or a large part of it, would be borne by the women chemists and their friends.

The arguments mounted by the opposition were as follows:

1. The expense, probably amounting to several hundred pounds, which would be incurred if a supplemental Charter is necessary, is not justified by the small number of new Fellows likely to seek admission.
2. It may be gravely doubted whether the deliberate encouragement of women to enter the chemical profession would not operate unfavourably on women themselves in view of the arduous nature of chemical work.
3. Although it is true, as urged by supporters of the petition, that the number of women seeking admission is small, it is also claimed as a reason for admission that the volume of chemical work contributed by women is increasing rapidly. ... but it is not stated, though equally true, that while the total number of Papers printed during that period exceeds 3400, only 23 are in the names of women alone.
4. Even assuming that these 23 contributions were independent of masculine inspiration, it may be questioned whether women have, as a group, shown marked aptitude for chemical pursuits, ...
5. Moreover, by being welcomed as guests to the Society, women have been able to enjoy that chemical atmosphere and intercourse which Fellowship of the Society involves.
6. As regards the admission of Madame Curie to Honorary Membership, it must be borne in mind that Honorary and

Foreign Members have no voting powers, and are not eligible for office.

7. Briefly stated, the position of those unfavourable to the admission of women is that, while gladly offering to those women who already have become chemists measures which would give them benefits derived from attendance at the meetings, they deem it inexpedient publicly to encourage women to adopt chemistry as a professional pursuit, since such a course would tempt them into a career in which they may ultimately not find employment in view of the already overcrowded state of the profession.

Reading the seven reasons for opposition of women's admission, it is difficult to identify seven separate arguments. In fact, a conclusion could be drawn that the contentions were simply spread over seven numbers to give the impression that there was an excess of negative arguments over positive ones.

In the pages of the journal *Nature*, the letter from the Secretaries of the Chemical Society listing the above arguments was preceded by an Editorial which expounded the opinions of the editors of that journal:

> It cannot be denied that women have contributed their fair share of original communications. Indeed, in proportion to their numbers they have shown themselves to be among the most active and successful of investigators. The society consents to publish their work which redounds to its credit. Why, then, should the drones who never have done, and never will do, a stroke of original work in their lives be preferred to them simply because they wear a distinctive dress and are privileged to grow a moustache?
>
> The women-chemists will doubtless smile at the futility of the adverse arguments which appear above the names of the two honorary secretaries of the society. They will have their own opinion concerning the arduous nature of chemical work, about which they know quite as much as those who profess so

> tender a solicitude for them. As to their chances of success in life, they have shown that they are quite able to hold their own, in spite of the alleged "overcrowded state of the profession." Overcrowded state of the profession, forsooth! With a delicious but wholly unconscious *naïvité*, the banging, barring and bolting people have herein revealed the true inwardness of their opposition. It is the argument of the weak-kneed-of persons whose *Zunftgeist* has warped their judgement and disturbed their mental balance. We trust the main body of the society will treat the argument with the contempt it merits.[58]

Of the 2900 Fellows of the Chemical Society, 1094 were in favour of the admission of women with full rights while 642 were opposed. One might have assumed that the battle had been won, but this was not to be the case.

At a special meeting of the Council on 3 December 1908, the motion was put forward by Tilden and seconded by Thomas E. Thorpe[59] that:

> The Council having referred the question of the admission of women to the Fellowship of the Society to the whole of the Fellows, and having ascertained that a majority of those who returned an answer are of the opinion that women should be admitted to the Fellowship on the same terms and with the same privileges as men, resolves to give effect to the wishes of the majority, and to take such action as they may be advised are necessary in order to permit of the inclusion of women as Fellows of the Society.[60]

The Minutes note that a discussion then ensued, but unfortunately no details are given. It would appear that Armstrong made very persuasive arguments against the motion. He proposed an amendment, seconded by Horace Brown[61]: "That in the opinion of this Council it is desirable that at any time, on the recommendation of three Fellows of the Society, women be accepted as Subscribers to the Society."[60] This new category would allow women to attend ordinary meetings, use the Society library, and receive the Society publications. However, women would not be able to attend extraordinary meetings, vote in Society elections, or hold offices in the Society. The amendment passed by 15 votes to 7. The passage of this reversal was prompted by the fear that the Armstrong-led minority would use legal means to block the proposed bye-law.

The rejection by the Council of the Society's own referendum resulted in a stinging editorial in the journal *Nature* in February 1909:

> ... No matter what the size of the majority in favour of the admission of women might be, a contumacious and recalcitrant element in the minority — a cabal of London chemists ... set themselves to thwart the wishes of the majority .... The whole business of the referendum was ... deliberately reduced to a fiasco.[62]

In a despairing attempt to right the wrong, at the Annual General Meeting of the Society in 1909, Edward Divers[63] proposed to move the adoption of a new bye-law on the admission of women.[39] However, the move was ruled out of order by reference to the legal opinion of 1898.

The Proponents and Opponents of Women's Admission

Throughout the struggle for the admission of women, there were men who fervently led the battle on behalf of the unfranchised women. Some of the male chemists knew one or more of the women personally; for example, Whiteley was a research student of Tilden. Another factor we will encounter in later chapters is that of very influential mothers. Again using Tilden as an example, his obituarist noted: "Their mother, however, was a woman of strong character and marked ability, devoting herself unreservedly to the care and instruction of her children during their early home-life...."[64]

Ramsay had proposed the 1893 motion for the admission of women that had been seconded by Tilden. In a historical review of the Chemical Society, Tom Moore and James Philip noted: "This controversy took place during the Presidency of Ramsay, and its result must have been particularly galling to him, as one of the earliest and most consistent advocates of the admission of women."[65]

A feature of all the women's supporters was their affable personality, and this was true of Ramsay as his biographer, Thaddeus Trenn, commented:

> Ramsay, the eternal optimist, whose motto was said to be "be kind," was above all a highly cultured gentleman. ... He was admired and beloved by almost all who knew him; and his boyish vigour and simple charm, unaffected by the many honours showered upon him, remained with him throughout his life.[66]

Yet a very different picture is presented by Margaret Tuke, Principal of Bedford College. Tuke was asked by J. B. S. Haldane in 1911 whether Bedford College (women) students went on to graduate school, for example, with Ramsay. Tuke replied: "They do not go to Professor Ramsay. He does not encourage women to research with him particularly. I think I am not mis-stating the fact that he rather discourages women in his laboratory for research purposes."[67] Certainly, in early years, Ramsay took on women students, both at Bristol and subsequently at UCL (see Chaps. 3 and 5). Whether he had found his first women students unsatisfactory, or whether he had a low opinion of the graduates of women's colleges, is impossible to say. Tuke's view is certainly at variance with impressions given by his colleagues.

Jocelyn Field Thorpe,[68] a later supporter of women in chemistry, also exhibited similar personality traits to Ramsay as

Thorpe's protégée and co-author of his obituary, Martha Whiteley, felt important to describe:

> Thorpe owed much of his success, both as a teacher and as a man of affairs, to his personal qualities, his joviality and charm of manner. He had an incorrigible faith in the goodness of human nature and refused to see anything but the best in the people with whom he came in contact.[69]

Against the many supporters, there was one figure who dominated the opposition: Armstrong. Moore and Philip commented: "It is agreed by all who have personal recollections of the Society from 1880 onwards that for many years Armstrong's influence was dominating."[70] They added that Armstrong's stubborn refusal to continence the admission of women was against the wishes of the majority of Fellows.

Armstrong was a traditionalist rather than a misogynist. In fact, he was the research supervisor of Ida Smedley from 1900 to 1904 (see above), while Clare de Brereton Evans (see Chap. 4) worked under him between 1895 and 1897, synthesising organonitrogen compounds and obtaining three publications. His opposition was not to women taking chemistry degrees, but rather to women becoming professional chemists instead of producing more "little chemists," as he stated himself:

> If there be any truth in the doctrine of hereditary genius, the very women who have shown their ability as chemists should be withdrawn from the temptation to become absorbed in the work, for fear of sacrificing their womanhood; they are those who should be regarded as chosen people, as destined to be the mothers of future chemists of ability.[71]

The 1909 Letter

In 1909, a report was circulated, claiming that the women petitioners were linked to the agitation for the political enfranchisement of women. Astonishing as it may seem today, 31 women chemists, including 14 of the original petitioners, felt it necessary to rebut the accusation that women chemists seeking admission as Fellows were associated with such radical elements of society. In a letter to *Chemical News*, the women noted that the sole bond between them was a common interest in chemistry:

> We, the undersigned women (actively engaged in chemical teaching and research), beg to ask for the hospitality of your columns in order to deny any such connection. The following facts, we venture to think, should prove the independence of the two movements:
>
> 1. Five years ago when some of us petitioned the Council of the Chemical Society to admit us to the Fellowship, the agitation in favour of "Woman Suffrage" was not prominently before the public.
> 2. The petition recently presented to the Council originated within the Chemical Society itself and was signed exclusively by Fellows of the Society.
>
> Moreover, we, as a body, have no knowledge of the political opinions and aspirations held by individual members; any such knowledge we should consider to be quite irrelevant, since the only link which unites us is a common interest in the science of chemistry. We are glad to take this opportunity of recording our thanks to those Fellows of the Chemical Society who have expressed themselves in favour of admitting women to the Fellowship of the Society. — We are, &c.[72]

The letter was followed by a statement from the same group of women concerning a "meeting of representative women chemists." In this declaration, the 312 Fellows were thanked for their support; in addition, women were urged not to become Subscribers on the grounds that it would prejudice their case for Fellowship of the Chemical Society (Table 2.2).

Table 2.2. The signatories of the 1909 letter to *Chemical News* (Note. 72) with their cited affiliations.

Heather H. Beveridge [7]	Carnegie Research Scholar, University of Edinburgh
Mary Boyle [4]	Lecturer and Demonstrator, Royal Holloway College
Katherine A. Burke (P) [3]	Assistant, Chemistry, University College, London
Frances Chick [2]	—
Louisa Cleaverley [2]	—
Margaret D. Dougal [13]	Indexer, Publications, The Chemical Society
Clare de Brereton Evans (P) [4]	Lecturer, Chemistry, London School of Medicine for Women
E. Eleanor Field (P) [4]	Senior Staff Lecturer, Royal Holloway College
Emily L. B. Forster [3]	Private Assistant, Prof. Huntington, King's College, London
Ida Freund (P) [6]	Staff Lecturer, Chemistry, Newnham College, Cambridge
Maud Gazdar [3]	Demonstrator, Chemistry, University College, London
Hilda J. Hartle (P) [13]	Lecturer, Chemistry, Homerton Training College, Cambridge
E. M. Hickmans [5]	—
Annie Homer [12]	Fellow, Newnham College, Cambridge
Ida F. Homfray [3]	—
E. S. Hooper [10]	Assistant Lecturer & Demonstrator, Portsmouth Municipal College
Edith Humphrey (P) [4]	Chemist to A. Sanderson & Sons
Zelda Kahan [2]	—
Norah E. Laycock [4]	Demonstrator, Chemistry, London School of Medicine for Women
Elison A. Macadam [3]	Private Assistant, Prof. Huntington, King's College, London
Effie G. Marsden [3]	—
Margaret McKillop (P) [3]	Lecturer, Chemistry, Women's Department, King's College
Agnes M. Moodie [7]	—
Nora Renouf [10]	Late Salter's Research Fellow, School of Pharmacy

(*Continued*)

Table 2.2. (*Continued*)

Ida Smedley (P) [2]	Assist. Lecturer & Senior Demonstrator, Victoria U., Manchester
Alice E. Smith (P) [7]	Assist. Lecturer & Senior Demonstrator, University College of North Wales, Bangor
Millicent Taylor (P) [5]	Lecturer, Chemistry, Ladies' College, Cheltenham
M. Beatrice Thomas (P) [6]	Lecturer, Girton College, Cambridge
M. A. Whiteley (P) [3]	Demonstrator, Chemistry, Royal College of Science, London
Sibyl T. Widdows (P) [4]	Head, Practical Chemistry Department, London School of Medicine for Women
Katherine I. Williams (P) [5]	—

Note: (P) indicates they were also a signatory of the 1904 petition. The numbers in square brackets indicate the chapter in which their biography will be found.

As with the original petition of 1904, there must have been links between these women enabling them to circulate and sign this document. For those signing the petition and the letter, one might infer that the social links established in 1904 had been maintained or resumed in 1909. For example, Humphrey, though in industry in 1909, had probably kept in contact with some of her former friends at the London Colleges. Nevertheless, it is difficult to find any contact points with such individuals as Beveridge at Edinburgh (see Chap. 7).

The Lesser-Known Signatories of the Letter

Most of the signatories of the letter continued with careers in chemistry or related areas, and their biographical details will be found in the most appropriate chapter of this compilation. However, there is little information on three of the signatories; thus, their brief biographies are compiled in this section.

Like her sister Harriette (see above), Frances Chick,[73] born in 1883, was educated at Notting Hill High School. She obtained a B.Sc. degree from UCL in 1908, and after graduation carried out postgraduate research on polymerisation reactions at University College with Norman T. M. Wilsmore.[74] Chick later moved on to medical and biochemical research, for a time working in the biochemical department of the Lister Institute. She married Sydney Herbert Wood and had one daughter, but unfortunately died in 1919 at the early age of 36.

Louisa Cleaverley had been a student at the East Ham Technical College, having a publication in 1907 with the chemist, Albert E. Dunstan.[75] In 1911, she married Dunstan and they had two children, Mary and Bernard. Cleaverley died in April 1963, and, deeply affected by her death, Dunstan died 8 months later.

Zelda Kahan[76] was educated at KEVI and then at Victoria University (later the University of Manchester). She entered the University of Leeds in 1900, graduating with a B.Sc. (Yorks) in 1905. The same year, she commenced research at UCL, where she authored three publications between 1906 and 1908. She was still at UCL in 1912, then she married Mr. Coates in 1914.

Admission at Last

For the 11 years of its existence, only 11 women availed themselves of Subscriber status, thus indicating a strong determination by most women that it was to be full Fellowship or nothing.[39] It was 1919 when everything changed: a Bill was moving through Parliament titled the Sex Disqualification (Removal) Act. Under this Act:

> A person shall not be disqualified by sex or marriage from the exercise of any public function, or from being appointed to or holding any civil or judicial office or post, or from entering or assuming or carrying on any civil profession or vocation, or for admission to any incorporated society (whether incorporated by Royal Charter or otherwise)....[39]

Though the Law did not receive Royal Assent until 23 December 1919, it would seem more than a coincidence that the sudden revival of the issue of Fellowship for women chemists occurred while the Bill was being debated. At an Extraordinary General Meeting of the Chemical Society on 8 May 1919, a resolution was proposed by James Philip[77] and seconded by J. W. Leather:[78] "That women should be admitted to the Society on the same terms as men."[79] The motion passed unanimously, and in 1920 the first women were admitted as Fellows.

Among the 21 women to be admitted at that auspicious first election[80] (Table 2.3) were four of the original petitioners: Smedley, Taylor, Whiteley, and Widdows. At subsequent meetings of the Society, three other petitioners — Burke, Humphrey, and Thomas — were elected.

The Lesser-Known Initial Members

Just as some of the signatories were low-profile individuals, so were some of the initial members. In this section, we will summarise what we know of them.

Eunice Annie Bucknell,[81] the daughter of Daniel Bucknell, a carpenter and joiner of Regent's Park, London, was born on 29 December 1888. She attended NLCS and entered Bedford College in 1907. Bucknell completed a B.Sc. there, and by 1920 she was an Analytical and Research Chemist at the South Metropolitan Gas Company.

Mary Cunningham[82] was born in Stamford Hill, London, in 1882. She completed a B.Sc. in 1907 and an M.Sc. in 1916 from UCL. In addition to becoming one of the first women members of the Chemical Society, she was also a member of the Society for Chemical Industry and of the Society of Dyers and Colourists. She had seven publications during the period 1908–1910 from the Borough Polytechnic Institute, London. Then, in 1918, she was sole author of two publications from the Chemistry Research Laboratories of the University of St. Andrews, from where she received a D.Sc. In 1920, Cunningham was appointed as a Research Chemist with the Fine Cotton Spinners and Doublers Association, Manchester.

Table 2.3. The first group of women to be elected as Fellows of the Chemical Society (Note 82) and their nominators.

Agnes Browne [3]	F. G. Donnan, J. N. Collie, B. D. Porritt, W. E. Garner.
Eunice A Bucknell [2]	E. V. Evans, H. Hollings, W. Barr, V. P. Hart.
Margaret Carlton [3]	H. B. Baker, G. Senter, J. C. Philip.
Mary Cunningham [2]	J. C. Irving, C. F. Cross, A. W. Crossley.
Ellen Field [11]	J. Walker, J. E. Mackenzie, A. R. Normand, H. G. Rule, G. Barger.
Mary Frances Hamer [13]	W. J. Pope, C. T. Heycock, W. H. Mills.
Elizabeth E. Holmes [2]	W. T. Burgess, R. N. Lennow, A. Teichfeld, J. D. Kettle, O. Silberrad.
Mary Johnson [6]	W. J. Pope, C. T. Heycock, W. H. Mills.
Hilda Mary Judd [13]	W. A. Tilden, M. O. Forster, J. C. Philip, J. Thorpe, A Stevenson, G. Kon.
May Sybil Leslie [5]	A. Smithalls, H. M. Dawson, W. Lowson.
Phyllis V. McKie [12]	K. Orton, J. N. Collie, F. G. Donnan.
Ida Maclean [2]	W. J. Pope, A. Lapworth, H. B. Dixon, R. Robinson, G. Barger.
Frances Micklethwait [3]	W. A. Tilden, G. T. Morgan, A. R. Ling, E. A. Cooper.
Nora Renouf [10]	A. W. Crossley, W. P. Wynne, G. Beilby.
Marion Crossland Soar [3]	C. K. Tinkler, K. Orton, T. M. Lowry.
Millicent Taylor [5]	J. W. McBain, M. Nierenstein, F. W. Rixon, F. Francis.
Gartha Thompson [2]	S. J. Lewis, A. Greeves, J. C. Philip, M. O. Forster.
Martha Whiteley [3]	W. A. Tilden, T. E. Thorpe, M. O. Forster, W. P. Wynne, G. T. Morgan, C. Jones, J. C. Philip, H. B. Baker, J. Thorpe.
Sibyl Widdows [4]	J. Thorpe, J. C. Philip, W. H. Mills, A. MacKenzie, G. Senter, J. A. Gardner.
Florence Mary Wood [2]	P. F. Frankland, H. McCombie, J. E. Coates, S. J. Lewis.
Olive Workman [2]	H. B. Baker, J. C. Philip, A. T. King, A. A. Eldridge, H. F. Harwood.

Note: The numbers in square brackets indicate the chapter in which their biography will be found.

In 1922 and 1923, she was granted two patents resulting from her work with the Association.

At the time of her admission to the Chemical Society, Edna Elizabeth Holmes[83] listed her occupation as a Chemical Assistant at the Lennox Foundry Company Research Laboratories, New Cross, London, where she had been for the previous 3 years. Nothing additional is known about her.

Olive Workman[84] obtained a B.Sc. from UCL in 1916, and an M.Sc. from the Royal College of Science, London, in 1919. In the Register of the Royal College of Science, she is noted as being a science mistress.

The other two women chemists mentioned here shared a common bond in their lives: they both worked with Samuel Judd Lewis.[85] Lewis was not part of the chemical academic establishment; instead, in 1909 he set up a private laboratory as a consulting and analytical chemist in Holborn. He became convinced that the future of analytical chemistry lay in spectroscopy, a view that was 20 years ahead of its time. Like so many of the women-supportive supervisors, his personality was empathetic, as his obituarist reported: "Dr. Lewis was courteous always, kindly and considerate to his staff, ..."[85]

Born in 1886, Gartha Thompson[86] attended KEVI. From there, she went to Wandsworth Technical College, and then to IC, obtaining a B.Sc. (London) in 1910. Her occupation was listed as private teacher from 1910 to 1912, then from 1912 to 1914 she was a chemist to the Polysulphin Company. For the following year, Thompson was secretary and assistant to the chief chemist of British Thomson-Houston Co. Ltd., Rugby; then, from 1915 she was senior assistant to Lewis.[85] In 1923, she had three sole-author publications listing an address of Rugby; thus, she may have returned to work at Rugby again. Thompson died on 4 November 1970, aged 84 years.

Florence Mary Wood[87] was educated at Bournemouth Collegiate School for Girls and the Municipal Technical College, Bournemouth. She obtained a B.Sc. in botany from the

University of London in 1911 and a B.Sc. in chemistry from the University of Birmingham in 1912. From 1912 to 1917, she was science mistress at Hampton School, Malvern, Jamaica, British West Indies. In 1918, she returned to England to join Lewis' research group; thus, it was possible that she overlapped with Gartha Thompson (see above). Wood received a Ph.D. from the University of London in 1924, and in the same year took up an appointment as Headmistress at Kensington High School for Girls, where she remained until 1929. After a short period at Twickenham County School, she was appointed biology mistress at Chiswick County School (now Chiswick Grammar School) in 1930, a post she held until her death on 18 November 1948, aged 61 years.

The Women Chemists' Dining Club

In addition to the Chemical Society, there had been a Chemistry Club since 1872.[3] This organisation provided the socialisation and bonding of the male chemists. Thus, the formation of the Women Chemists' Dining Club can be seen as providing an equivalent venue for women chemists.

The first meeting (only later was it called a Club) was organised by Martha Whiteley and Ida Smedley in November 1925 at the Lyceum Club in London.[88] The purpose of the Club was such that women chemists would have the opportunity to meet and to develop friendships (perhaps what we would now call "networking"). Three dinners were held per year with an occasional speaker, usually from outside the field of chemistry.

Though meetings of the Club were suspended during the Second World War, they resumed around 1946. Some social outings were arranged. For example, in the summer of 1948, 27 members visited the laboratories and colleges at Cambridge; in 1949, there was an excursion to the Royal Holloway College; while in 1951, there was an outing to Oxford with lunch at St. Hilda's College. During that period, the Club was run by two secretaries: Frances Hamer (see Chap. 13) and Ellie Knaggs (see Chap. 9).

The organisation appeared to be low-key. It was first mentioned in print in the journal *Chemistry and Industry* in 1952, at which time there were 66 members.[88] In the following issue, there was an editorial ruminating on the existence of the Club:

> It is with much interest that we learned a few weeks ago that women chemists in London had formed a Club. Most men are clubbable, one way or another, but we did not know this was true of women. We wonder if this formation of a Club for women chemists is another sign of female emancipation. We should be glad to think that they mellow over a bottle or two of fine wine. We commend claret — the Queen of wines. Presumably claret attained this title because of its beauty, its grace and its subtlety — admirable qualities which men have always associated with women.[89]

The article continued: "No doubt the ladies of this Club even smoke ..." though the editor recommends that the women consider snuff instead:

> If you give up smoking, ladies, we might permit you a little snuff. Think how beautiful snuff-boxes can be. How lovely they would look in your handbags: so easy to carry, so delightful to toy with. No more trouble filling petrol lighters: and the grace of the gesture, the poise of your arm, the curve of the extended fingers as you delicately administer a little snuff to your nostrils.[89]

Mary R. Truter,[90] former Professor of Crystallography, UCL, recalled: "My mother, Agnes Jackman, née Browne, took me to a few meetings [about 1946]. ... I do not know for how long the club continued. I suspect not very long because I do not remember Mother saying anything about it after I left home [in 1947]."[91]

The last mention of the Club was in 1953, when a meeting was held at Queen Elizabeth College, London, the former King's College of Household and Social Science (see Chap. 3). In the report, it was noted that: "After the meal, Mrs. A. Jackman, senior lecturer in the Chemistry Department and a keen member of the Club, conducted the party around the College."[92]

No records of the Club could be found in the Chemical Society holdings. The Club seems to have been very much run by the first generation of women members of the Chemical Society and it is quite possible that with their demise, particularly Whiteley's death in 1956, the Club ceased to be.

The Royal Society

The ultimate national accolade for a British scientist was, and is, election to the Royal Society. Hertha Ayrton,[93] a physicist, had been proposed in 1902, among the co-signatories being Tilden who had been so supportive of the admission of women chemists to the Chemical Society (see above). The Royal Society, like the Chemical Society, called upon the legal opinion of its lawyer, emphasising the non-status of married women of the time: "We are of the opinion that married women are not eligible as Fellows of the Royal Society. Whether the Charters admit of the election of unmarried women appears to us doubtful. ... A woman, if elected, would become disqualified by marriage...."[94] The legal opinion suggested a supplemental Charter if the Fellows so wished, but there did not seem any significant interest at the time for such a venture. Ayrton's case was denied.

In 1922 and again in 1925, Caroline Haslett, Secretary of the Women's Engineering Society, wrote to the Royal Society asking for their position concerning the admission of women following the passage of the 1919 Sex Disqualification (Removal) Act. In 1925, James Jeans, Secretary of the Royal Society, wrote to Haslett that there was now "general opinion" that women were eligible for admission, provided that "their scientific accomplishments were of the requisite standard."[95]

Yet despite this admission of women's eligibility, nothing happened for nearly 20 years. As Hilary Rose has commented:

> This extraordinary gap suggests at best a collective amnesia — or perhaps a repression of memory — within the Royal Society, in which the fact of legal eligibility and the political likelihood of success become conflated to become an unstated and legally false, but socially powerful, consensus that women were not admissible. ... But the pressure which had brought the reluctant admission that women were eligible had weakened. The interwar period ... meant that the feminist organizations were functioning at just tick-over.[96]

The decline of women in science during the 1930s will be addressed further in Chap. 13.

It was not until 1943 that matters came to a head, provoked by closing remarks in the British communist-leaning newspaper, the *Daily Worker*, by J. B. S. Haldane, following the election of six men of Indian nationality.[95(a)] In the article, he noted the "striking omission" of any women's names in the nomination list, and that there were "certainly half a dozen [women]" eligible. Lancelot Hogben wrote to Haldane on 30 July 1943 to ask Haldane for suitable names of women from the biologically related sciences.[95(a)] Haldane replied on 18 August 1943, supporting the nomination of biochemist Marjory Stephenson (see Chap. 8):

> I think the strongest claim is that of Dr Marjory Stephenson who was the first person in the world to do work on bacterial metabolism as exact as that on mammalian metabolism, and who has continued to do good work in this field, discovering, for example, a number of new enzymes, in particular those dealing with the production and consumption of hydrogen. Another possibility is Honor B. Fell whose work on tissue culture has certainly been pretty good, but I do not think her claim is as strong as that of Stephenson.[97]

While canvassing support for Stephenson, Charles Harrington, Director of the National Institute for Medical Research, asked Lawrence Bragg if he had any nominees of women from the physical sciences.[95(a)] Bragg, in turn, had written to William Astbury, asking whether he thought crystallographer Kathleen Lonsdale (see Chap. 9) was suitable. Astbury replied on 5 November 1943:

> I suppose the suggestion was bound to come sooner or later that women should be put up for the Royal Society, and once that is accepted I don't think you could find a woman candidate more likely than Mrs Lonsdale to be successful. I should put her at quite the best woman scientist that I know — but that probably is as far as I am prepared to go, because I must confess that I am one of those people that still maintain that there is a creative spark in the male that is absent from women, even though the latter do so often such marvellously conscientious and thorough work after the spark has been struck. ... And I think it is perhaps risky to argue that she may not be quite up to the best standards, because it seems to me that the standard in the Royal Society is pretty variable. It obviously contains people who soar into heights far beyond Mrs Lonsdale's ken, but it also contains people whom she could run rings around.[98]

The question was raised by the Society's Treasurer, Thomas Merton, that the election of a woman was a "break in the traditions of the Society." In response, Henry Dale, the previous Director of the National Institute for Medical Research, commented: "However much one might wrap it up, the question which we should put could only mean one thing — Do you or do you not wish the Council to continue a discrimination against women candidates which the Law has removed?"[99]

Dale proposed a motion to amend the Statutes of the Society as recommended by legal advice in 1922 to formally lift any restriction on the election of women. This move passed in a postal vote on 30 June 1944. A total of 336 Fellows had voted in favour, 3 others with qualifications, and 37 Fellows opposed the revisions. On 22 March 1945, Marjory Stephenson and Kathleen Lonsdale were the first women elected as Fellows of the Royal Society.

Commentary

The diversity in attitude between the different societies towards the admission of women is striking. It can be interpreted in terms of how much the society was a pure professional society and how much a "men's club." In particular, as the Royal Institute of Chemistry was the accrediting body, the role of the Chemical Society was much more towards the socialising end of the spectrum and therefore more hostile to women's admission. What comes through strongly is the importance of key individuals, particularly the pair of Ida Smedley and Martha Whiteley,[100] who had each established themselves as respected chemists. However, as women were excluded from the governing body, it was supportive male chemists, such as William Tilden and William Ramsay, who had to argue the women's case in the Council of the Society.

Notes

1. Noordenbos, G. (2002). Women in academies of sciences: From exclusion to exception. *Women's Studies International Forum* **25**: 127–137.
2. Kauffman, G. B. (1983). The misogynist dinner of the American Chemical Society. *Journal of College Science Teaching* **12**: 381–383.
3. Gay, H. and Gay, J. W. (1997). Brothers in science: Science and fraternal culture in nineteenth-century Britain. *History of Science* **35**: 425–453.
4. We thank I. Sheppard, Archivist, Society for Chemical Industry, for assistance in accessing SCI archives.
5. Part of this work has been published as: Rayner-Canham, M. F. and Rayner-Canham, G. W. (2003). Pounding on the doors: The

fight for acceptance of British women chemists. *Bulletin for the History of Chemistry* **28**(2): 110–119.

6. Anon. (1824). The London Chemical Society. *The Chemist* **2**: 56.
7. Lee, M. (2004). Birkbeck, G. (1776–1841). *Oxford Dictionary of National Biography*, Oxford University Press, http://www.oxforddnb.com/view/article/2454, accessed 25 Nov 2007.
8. Anon. (1824). The London Chemical Society. *The Chemist* **2**: 162–166.
9. Elizabeth Fulhame was a pioneer researcher in redox chemistry and the author of *An Essay on Combustion with a View to a New Art of Dying and Painting,* wherein the Phogistic and Antiphlogistic hypotheses are proved erroneous. See: Davenport, D. A. (2004). Fulhame, E. (fl. 1780–1794). *Oxford Dictionary of National Biography*, Oxford University Press, http://www.oxforddnb.com/view/article/39778, accessed 4 Nov 2004.
10. Jane Marcet was renowned for her book *Conversations on Chemistry,* in which the elements of that Science are familiarly explained and illustrated by experiments. The book was designed for the woman reader to enable her to understand principles of chemistry. See: Morse, E. J. (2004). Marcet, J. H. (1769–1858). *Oxford Dictionary of National Biography*, Oxford University Press, http://www.oxforddnb.com/view/article/18029, accessed 15 Nov 2007.
11. Brock, W. H. (1967). The London Chemical Society 1824. *Ambix* **14**: 133–139.
12. Chirnside, R. C. and Hamence, J. H. (1974). *The "Practicing Chemists": A History of the Society for Analytical Chemistry 1874–1974.* The Society for Analytical Chemistry, London, p. 87.
13. Horrocks, S. M. (2000). A promising pioneer profession? Women in industrial chemistry in inter-war Britain. *British Journal for the History of Science* **33**: 351–367.
14. (a) Butterworth, D. E. (1965). Obituary: Isobel Hodgson Hadfield. *Proceedings of the Society for Analytical Chemistry* **2**: 101; and (b) Queen Mary College, student records.
15. For example, see Hadfield, I. H. (1942). Two simple micro-burettes and an accurate wash-out pipette. *Journal of the Society for Chemical Industry* **61**: 45–50.

16. Pilcher, R. B. (1914). *The Institute of Chemistry of Great Britain: History of the Institute 1877–1914*. Institute of Chemistry, London, p. 113–114.
17. Anon. (1913). Obituary: Miss Emily Jane Lloyd. *Journal of the Institute of Chemistry, Part II* 32–33. The following are thanked for information on Lloyd: I. Salmon, Assistant Registrar, University of Wales, Aberystwyth; P. Bassett, Archivist, Special Collections, University of Birmingham; and N. Jeffs, Archivist, Special Collections, University of London Library.
18. James, T. C. and Davies, C. W. (1956). Schools of chemistry in Great Britain and Ireland–XXVII The University College of Wales, Aberystwyth. *Journal of the Royal Institute of Chemistry* **80**: 569.
19. (17 February 1893). Institute of Chemistry. *Minutes*.
20. The following are thanked for information on Stern: P. Bassett, Archivist, Special Collections, The University of Birmingham; and N. Jeffs, Archivist, Special Collections, University of London Library.
21. Gordon, A. M. (1895). The after careers of university educated women. *Nineteenth Century* **37**: 955–960.
22. Anon. (1954). Death. *North London Collegiate School Magazine* (3): 43.
23. Stern, R. (1924). *A Short History of Chemistry*. J. M. Dent & Co., London.
24. Goodwin, T. W. (1987). *History of the Biochemical Society 1911–1986*. The Biochemical Society, London, pp. 14–15.
25. Whiteley, M. A. (1946). Ida Smedley Maclean, 1877–1944. *Journal of the Chemical Society* 65.
26. "M.E. de R.E." [Phillp, M., Mrs. Epps] (January 1945). Ida Smedley Maclean. *Newnham College Roll Letter* 50–51.
27. Keeble, F. W. (1941). Henry Edward Armstrong, 1848–1937. *Obituary Notices of Fellows of the Royal Society* **3**: 229–245.
28. Todd, Lord, and Cornforth, J. W. (1976). Robert Robinson, 13 September 1886–8 February 1975. *Biographical Memoirs of Fellows of the Royal Society* **22**: 419.
29. Nunn, L. C. A. (22 July 1944). Obituary: Dr. Ida Smedley-Maclean. *Nature* **154**: 110.
30. Ellen Richards had been a leading American woman chemist. See: Rayner-Canham, M. F. and Rayner-Canham, G. W. (1998).

Women in Chemistry: Their Changing Roles from Alchemical Times to the Mid-Twentieth Century. Chemical Heritage Foundation, Washington, pp. 51–55. The Ellen Richards Prize was terminated in the 1930s as the organisers were convinced that women had finally established themselves in science and that the award was no longer necessary.

31. Chick, H., Hume, M. and Macfarlane, M. (1971). *War on Disease: A History of the Lister Institute*. Andre Deutsch, London, pp. 165–166.
32. Smedley-Maclean, I. (1943). *The Metabolism of Fat*. Methuen, London.
33. Tylecote, M. (1941). *The Education of Women at Manchester University*. Manchester University Press, p. 63.
34. Creese, M. R. S. (1998). *Ladies in the Laboratory? American and British Women in Science, 1800–1900*. Scarecrow Press, Lanham, Maryland, pp. 40, 149–150.
35. Sayers, J. E. (1973). *The Fountain Unsealed: A History of the Notting Hill and Ealing High School*. Broadwater Press, Welwyn Garden City, pp. 39–43.
36. Note 31, Chick, pp. 87–92.
37. Chick, Dame H. (1956). Charles James Martin, 1866–1955. *Obituary Notices of Fellows of the Royal Society* **2**: 172–208.
38. Sinclair, H. M. (2004). Chick, D. H. (1875–1977). *Oxford Dictionary of National Biography*, Oxford University Press, http://www.oxforddnb.com/view/article/30924, accessed 17 Nov 2007.
39. Mason, J. (1991). A forty years' war. *Chemistry in Britain* **27**: 233–238.
40. "H.B.D." (1920). Obituary notices of fellows deceased: A. G. Vernon Harcourt, 1834–1919. *Proceedings of the Royal Society of London, Series A* **93**: vii–xi.
41. (1880). Council of the Chemical Society. *Minutes of the Meetings* **IV**: 215, 219.
42. "J.N.C." (1917). Obituary notices of fellows deceased: Sir William Ramsay, 1852–1916. *Proceedings of the Royal Society of London, Series A* **93**: vlii–liv.
43. (1888). Council of the Chemical Society. *Minutes of the Meetings* **V**: 103, 219.

44. (17 November 1892). Council of the Chemical Society. *Minutes of the Meetings* **V**: 212, 226.
45. "J.C.P." (1928). Obituary notices of fellows deceased: Sir William Augustus Tilden, 1842–1926. *Proceedings of the Royal Society of London, Series A* **117**: i–v.
46. (a) (19 January 1893). Council of the Chemical Society. *Minutes of the Meetings*; and (b) Anon. (1893). *Proceedings of the Chemical Society* **9**: 84.
47. Meldola, R. (1908). Obituary notices: William Henry Perkin. *Journal of the Chemical Society, Transactions* **93**: 2214–2257.
48. (3 March 1904). Council of the Chemical Society. *Minutes of the Meetings.*
49. (23 March 1904). Council of the Chemical Society. *Minutes of the Meetings.*
50. Cited in Note 39, Mason, p. 234.
51. (20 April 1904). Council of the Chemical Society. *Minutes of the Meetings.*
52. (21 October 1904). Letter enclosed in Council of the Chemical Society. *Minutes of the Meetings.*
53. Tilden, W. A. (1905). Presidential address, delivered at the Annual General Meeting, 29 March 1905. *Journal of the Chemical Society, Transactions* **87**: 547–548.
54. Russell, D. S. (Michaelmas Term, 1954). In Memoriam: Mary Beatrice Thomas, 1873–1954. *Girton Review* 14–25.
55. Odling, W. *et al.* (1908). Letter from past Presidents communicating the Memorial to the Fellows. *Nature* **78**: 227–228.
56. "T.E.T." (1917). Obituary notices of fellows deceased: Sir Henry Roscoe, 1833–1915. *Proceedings of the Royal Society of London, Series A* **93**: i–xxi.
57. Reprinted (1908) in: Women and the Fellowship of the Chemical Society. *Nature* **78**: 226–227.
58. The editor (1908). Women and the Fellowship of the Chemical Society. *Nature* **78**: 226.
59. "A.E.H.T." (1925). Obituary notices of fellows deceased: Sir Thomas Edward Thorpe, 1845–1925. *Proceedings of the Royal Society of London, Series A* **109**: xviii–xxiv.
60. (3 December 1908). Extraordinary Meeting of the Council, Chemical Society. *Minutes* **VIII**.

61. "J.B.F." (1925). Obituary notices of Fellows deceased: Horace Tabberer Brown, 1848–1925. *Proceedings of the Royal Society of London, Series A* **109**: xxiv–xxvii.
62. The Editor. (1909). Women and the Fellowship of the Chemical Society. *Nature* **79**: 429–431.
63. "J.M." (1913). Obituary notices of fellows deceased: Edward Divers, 1837–1912. *Proceedings of the Royal Society of London, Series A* **88**: viii–x.
64. "M.O.F." (1927). Sir William Augustus Tilden. *Proceedings of the Chemical Society* 3190–3202.
65. Moore, T. S. and Philip, J. C. (1947). *The Chemical Society, 1841–1941: A Historical Review*. The Chemical Society, London, p. 97.
66. Trenn, T. J. (1975). William Ramsay. In Gillispie, C. C. (ed.), *Dictionary of Scientific Biography*, Vol. 11, Charles Scribner's Sons, New York, p. 278.
67. Dyhouse, C. (1995). *No Distinction of Sex? Women in British Universities, 1870–1939*. UCL Press, London, p. 144.
68. Ingold, C. K. (1941). Jocelyn Field Thorpe, 1872–1939. *Obituary Notices of Fellows of the Royal Society* **3**: 530–544.
69. Whiteley, M. A. and Kon, G. A. R. (1940). Obituary: Jocelyn Field Thorpe. *The Analyst* **65**: 483–484.
70. Note 65, Moore and Philip, pp. 78–79.
71. Nye, M. J. (1996). *Before Big Science: The Pursuit of Modern Chemistry and Physics 1800–1940*. Prentice Hall International, London, p. 17.
72. Beveridge, H. H. *et al.* (5 February 1909). Women and the Fellowship of the Chemical Society. *Chemical News* 70.
73. Note 34, Creese, p. 60, note 203.
74. Tattersall, G. (1941). Obituary notices: Norman Thomas Mortimer Wilsmore, 1868–1940. *Proceedings of the Chemical Society* 59–60.
75. Langton, H. M. (23 May 1964). A. E. Dunstan, D.Sc., F.R.I.C., 1878–1964. *Chemistry and Industry* 883–884.
76. University of Manchester: student records; University of Leeds: student records; Bedford College: files of the Chemistry Department; and *Calendars*: University College, London.
77. Egerton, A. C. (1942). James Charles Phillip, 1873–1941. *Obituary Notices of Fellows of the Royal Society* **4**: 51–62.

78. Anon. (1934). Obituary: John Walter Leather. *Journal and Proceedings of the Institute of Chemistry* **58**: 457–458.
79. (1919). Extraordinary General Meeting, Thursday, 8 May 1919. *Proceedings of the Chemical Society* 58–59.
80. (1920). Certificates of candidates for election at the Ballot to be held at the ordinary scientific meeting on Thursday, December 2nd. *Proceedings of the Chemical Society* 82–100.
81. Bedford College, student records; and (1920). Certificates of candidates for election at the Ballot to be held at the ordinary scientific meeting on Thursday, December 2nd. *Proceedings of the Chemical Society* 84.
82. University College, student records; and (1920). Certificates of candidates for election at the Ballot to be held at the ordinary scientific meeting on Thursday, December 2nd. *Proceedings of the Chemical Society* 85–86.
83. University College, student records; and (1920). Certificates of candidates for election at the Ballot to be held at the ordinary scientific meeting on Thursday, December 2nd. *Proceedings of the Chemical Society* 90.
84. (1920). Certificates of candidates for election at the Ballot to be held at the ordinary scientific meeting on Thursday, December 2nd. *Proceedings of the Chemical Society* 100; and (1951). *Register of Old Students and Staff of the Royal College of Science*, 6th ed. Royal College of Science Association, London.
85. Garton, F. W. J. (1960). Obituary notices: Samuel Judd Lewis, 1869–1959. *Proceedings of the Chemical Society* 156–157.
86. (1920). Certificates of candidates for election at the Ballot to be held at the ordinary scientific meeting on Thursday, December 2nd. *Proceedings of the Chemical Society* 98; and (1971). Personal news: Obituaries. *Chemistry in Britain* **7**: 35.
87. Anon. (1949). Obituary: Florence Mary Wood. *Journal of the Royal Institute of Chemistry* **73**: 62; and (1920). Certificates of candidates for election at the Ballot to be held at the ordinary scientific meeting on Thursday, December 2nd. *Proceedings of the Chemical Society* 99–100.
88. The Editor. (26 January 1952). For ladies only! *Chemistry and Industry* 71.
89. The Editor. (23 February 1952). A plea to the ladies. *Chemistry and Industry* 155.

90. Cruickshank, D. (June 2005). Mary Rosaleen Truter, 1925–2004. *Crystallography News* (93): 20.
91. Personal communication, letter, Mary R. Truter, 15 Oct 2001.
92. Anon. (9 May 1953). Meetings, notices, etc.: Women chemists' dining club. *Chemistry and Industry* 465.
93. Mason, J. (2004). Ayrton (Phoebe) Sarah (1854–1923). *Oxford Dictionary of National Biography*, Oxford University Press, http://www.oxforddnb.com/view/article/37136, accessed 2 Jan 2008.
94. Mason, J. (1991). Herta Ayrton (1854–1923) and the admission of women to the Royal Society of London. *Notes and Records of the Royal Society of London* **45**(2): 201–220.
95. (a) Mason, J. (1992). The admission of the first women to the Royal Society of London. *Notes and Records of the Royal Society of London* **46**(2): 279–300; see also (b) Mason, J. (1995). The women Fellows' jubilee. *Notes and Records of the Royal Society of London* **49**(1): 125–140.
96. Rose, H. (1994). *Love, Power and Knowledge: Towards a Feminist Transformation of the Science*. Indiana University Press, Bloomington, Indiana, pp. 115–135.
97. Note 95(a), Mason, p. 289.
98. Note 95(a), Mason, p. 290.
99. Note 95(a), Mason, p. 292.
100. As illustration, these are the only two women chemists of their era selected for entry in the *Oxford Dictionary of National Biography*; see Creese, M. R. S. (2004). Maclean, Ida Smedley (1877–1944). *Oxford Dictionary of National Biography*, Oxford University Press, http://www.oxforddnb.com/view/article/37720, accessed 28 Nov 2007; and Barrett, A. (2004). Whiteley, Martha Annie (1866–1956). *Oxford Dictionary of National Biography*, Oxford University Press, http://www.oxforddnb.com/view/article/46421, accessed 28 Nov 2007.

Chapter 3

The London Co-educational Colleges

The University of London became an umbrella organisation for a large number of constituent colleges. It was a unique institution in that women, once they were admitted, had a choice of both co-educational colleges and women-only colleges. In this chapter, we will summarise the struggle of women to gain admission to the University of London examinations, and then look in depth at four of the co-educational London colleges: University College, King's College, Imperial College, and East London College (Queen Mary College). In addition, we shall identify the important role that the polytechnics, particularly Battersea Polytechnic, played in the education of women chemists.

Admission of Women to the University of London Examinations

The University of London was founded in 1836 to administer the examinations for two rival colleges: University College and King's College. Then in 1856, its mandate was widened to allow candidates from around the world to sit its examinations and hence obtain a degree from the University of London. This could be accomplished without any residence requirement, unlike the other universities of the time. However, there was one stipulation — the candidate had to be male. Thus, the "women question" was simply whether women should be granted the right to take the University-set examinations and subsequently graduate with a University of London degree. It was to take 26 years from the first attempt by a woman to register until the final acceptance of women degree candidates.[1] In the space available here, we can only give the key features of the convoluted saga.

It was Jessie Meriton White[2] who, in 1856, first petitioned to be allowed to sit the University's examinations. White came from an affluent Nonconformist family and studied philosophy at the Sorbonne between 1852 and 1854. After visiting Garibaldi in Italy, she had decided to devote her life to the campaign for the unification of Italy. White returned to England in 1855 with the intent of gaining a medical degree so she could be more useful to the revolutionary cause. However, on 9 July 1856, the University Senate, on the advice of legal counsel, rejected White's application because they did not consider themselves "empowered to admit Females as candidates for degrees."[1]

The attempt by the second applicant, Elizabeth Garrett[3] (later Mrs. Garrett Anderson), though equally unsuccessful, followed a different path. Garrett had decided in 1860, at the age of 24, to become a doctor. In this aim, she was not only supported by her affluent father, but also by Emily Davies, Elizabeth Blackwell, Barbara Bodichon, and other members of the informal women's movement of the mid-19th century. The horizons for women had changed significantly during the late 1850s and early 1860s, in part because of the *English Woman's Journal*[4] founded by Bessie Parkes and Barbara Bodichon. Though the journal had a brief life from 1858 until 1864, it provided a vehicle for the development of feminist ideas and ideologies. Thus, whereas White represented herself, Garrett was at the forefront of a movement.

The case for admitting Garrett was put to the Senate meeting of 9 April 1862. The vote was close: by 7 votes to 6, the resolution was adopted that the Senate found "no reason to doubt the validity of Counsel's opinion given in the case of Miss Jessie Meriton White in 1856." Following this setback, Emily Davies, with the help of the Committee of the Society for the Employment of Women, circulated a petition for support from "persons of distinction and members of the University of London." The petition read: "I hereby express my opinion that it is desirable to obtain the admission of ladies to the Examinations of the University of London," and they collected the signatures of 13 peers, 44 members of Parliament, together with other prominent individuals. A draft Charter, which included the admission of women, was presented to the Senate on 7 May 1862. The vote on this issue split 20 for and 20 against (the majority of the opposition coming from the medical and legal professions), with the Chair having cast the tiebreaker for the status quo. Though Blackwell had lost again, the closeness of the vote and the advanced ages of many of the opponents led the women's movement to believe that success would soon be theirs.

The Senate of the University was not the only challenge. The Senate had been established in 1837 but, following agitation by graduates, Convocation was established by the Charter of 1858. Convocation consisted of all Doctors, all Bachelors of Law of two years' standing, and all Bachelors of Arts of three years' standing who paid a registration fee. Its consent was required for any changes in the University Charter. William Shaen,[5] one of the most forthright male proponents for women's rights of the period, declared:

> ... though the Senate, and not Convocation, is the governing body, still the Senate would not force female graduates upon the others, in opposition to the wishes of Convocation; and, on the other hand, there could be no doubt, that when Convocation makes up its mind that degrees shall be open to women, the Senate would be of the same opinion.[1]

Convocation was therefore seen as the key to women's admission.

As Shaen had predicted, it was Convocation that first declared itself in favour of permitting women to take degrees. This occurred on 12 May 1874. Senate followed suit on 1 July of the same year, but there seemed to be no enthusiasm for seeking the necessary statutory changes, and so exclusion remained the practice. The division of powers between Senate and Convocation of the University bedevilled progress on the issue, and it was not until 14 May 1878 that a new Supplemental Charter was finally approved admitting women to all degrees of the University. Finally, in 1880, the first four women graduated with B.A. degrees; the first two women B.Sc. recipients followed in 1881.

University College

University College London (UCL) had been founded in 1826 to provide a secular alternative to the religious universities of Oxford and Cambridge. In 1868, a Committee was established to provide classes for "ladies" parallel to those offered to men. As Gillian Sutherland has commented:

> This Committee spent a considerable amount of time on devices to keep male and female students separate, not only in the lecture rooms but in their comings and goings. Eventually it was agreed that lectures for men should begin and end on the hour, while those for women should begin and end on the half-hour. Practical considerations, such as the use of equipment, gradually brought a handful of shared classes. But there had to be a separate entrance to these; and in the autumn of 1870 a whole new door was knocked into the Chemistry Laboratory, the Ladies' Association having given a guarantee to cover the cost of blocking it off again, if the female demand for chemistry were to fall off.[6]

Emily Aston

When William Ramsay had been at Bristol College, he had had a woman research student, Katherine Williams (see Chap. 5).

Upon his arrival at UCL, he acquired another woman assistant, Emily Alicia Aston.[7] Aston was born in 1866, and between 1883 and 1885 she studied at both Queen's College, Harley Street (see Chap. 4), and Bedford College (see Chap. 4). Her first research was undertaken with Spencer Pickering[8] at Bedford College. She then spent a total of 14 years at UCL, receiving her B.Sc. in chemistry and geology in 1889. During her sojourn at UCL, she authored or co-authored a total of 12 publications, mostly with Ramsay.

Aston's versatility was remarkable, for she authored papers on topics in mineralogy, organic chemistry, inorganic chemistry, and physical chemistry. In Morris Travers's biography of Ramsay, it is noted that:

> Ramsay carried out further experimental investigations in the same direction with John Shields and with Miss Emily Aston, and made attempts to deduce from the results the degree of complexity of associating liquids.[9]

In the late 1890s, Aston spent some time at the Sorbonne in Paris where she worked with Paul Dutoit on electrolytic conductivity and molecular association, resulting in two publications. Then, she undertook research with Philippe Auguste Guye at the University of Geneva, Switzerland, on optical rotation. Four publications resulted from her work in Geneva.

Katherine Burke

Another researcher in Ramsay's group at UCL was Katherine Alice Burke.[10] Born in Surrey about 1875, Burke obtained her B.Sc. (London) degree from studies at Bedford College and later Birkbeck College, a small college emphasising technical and vocational subjects. Upon completion of a B.Sc. in 1899, she transferred to UCL to work in Ramsay's laboratory under Frederick Donnan.[11] Burke had two publications with Donnan and one with Edward Charles Cyril Baly.[12] In addition, she acted as a private research assistant to Ramsay, part of her task being to translate into English a book by the Danish chemist, Julius Thomsen, on systematic researches in thermochemistry. The translation appeared in print in 1905.

Perhaps learning from Ramsay of the 1904 petition for the admission of women to the Chemical Society, Burke was a co-signer of that, and of the 1909 letter to *Chemical News* (see Chap. 2). In 1906, she was appointed as Assistant in the Department of Chemistry, being promoted to Assistant Lecturer in 1921. Donnan noted in her obituary:

> In 1906 she was appointed a member of the Chemical Staff at University College, and from that time until her death on July 6th, 1924, she continued her teaching work, having charge of the practical laboratory work for students of the Intermediate Science class, and giving courses of lectures to more advanced students on the chemical aspects of radioactive transformations.[10(a)]

Other Women Researchers

Ida Frances Homfray,[7] too, was a researcher with Ramsay. She worked with him between about 1900 and 1910, authoring seven papers on surface properties, some, like Aston's, involving collaboration with Guye at the University of Geneva. During her time at UCL, she obtained a B.Sc. by research in 1905, followed by a traditional B.Sc. in 1910. Her research focused on the absorption of gases by charcoal, studies that were very relevant to the gas masks of the First World War. Homfray was another of the co-signers of the letter to *Chemical News* (see Chap. 2).

Two of the women researchers worked with the physical organic chemist, Edward Baly. Effie Gwendoline Marsden[13] was the most prolific woman researcher with Baly, co-authoring five papers with him between 1905 and 1910. Marsden had completed her secondary education at Kensington High School for

Girls (a GPDSC school; see Chap. 1) in 1899, then obtained a B.Sc. degree in chemistry from UCL before joining Baly's research group. Using absorption spectroscopy, she made major contributions to the study of keto-enol tautomerism. To illustrate the overlap of the women chemists, of Marsden's seven publications, one was also co-authored by Katherine Burke, while another was partly co-authored by Maud Gazdar (see below). All three were signatories of the 1909 letter to *Chemical News* (see Chap. 2), possibly having learned of it from Ramsay.

The other researcher with Baly, Maud Gazdar,[14] completed a B.Sc. at UCL in 1908, and then held the rank of Demonstrator from 1908 to 1911. While a Demonstrator, she worked with Baly and also with Samuel Smiles,[15] having a total of three publications. Gazdar then went to Trinity College, Dublin as a Research Assistant in the Biochemistry Department. Later, she acquired a Medical degree from the London School of Medicine for Women, and in 1918 was House Physician at the Royal Free Hospital, London (RFH). Gazdar remained at the RFH until 1924, being Clinical and Obstetric Assistant in the Skins Department from 1919 to 1920, and then Clinical Assistant in the Gynaecological Department. In 1924, she went into private practice in Bishop Stortford. She retired in 1929 to Godfreys Farm, Radwinter, Essex, and about this time she married.

King's College

King's College was founded shortly after University College in 1831. It was named for King George the Fourth and was to provide "an education in which the pursuit of knowledge and the practice of religion should be joined in indissoluble union"[16] — in contrast to the "godless" UCL to the north of it. However, financial problems forced King's to become an open institution by the end of the 19th century. King's was a male-only college, as Negley Harte described: "Female persons, with extremely rare exceptions on Saturday mornings [for laboratory work], were not admitted to the sacred precincts of King's itself until

1915."[17] Thus, prior to that date, women students at King's were to be found not on the Strand, but in Kensington.

The Women's Department of King's College

The foundation of the Women's Department (originally the Ladies' Department) of King's College was primarily the effort of Rev. Canon Alfred Barry.[18] In 1877, the Women's Education Union was in search of an institution willing to organise lectures for ladies in the west end of London. Barry proposed to King's College Council that classes for ladies would be offered under King's auspices. The lectures were to take place in Kensington, not the existing Strand campus, except chemistry and physics which were to be offered at the Strand on Saturday mornings using existing King's laboratories and equipment. The Council approved the former, but baulked at the latter, only agreeing in 1880 to women visiting the Strand for scientific laboratory work.

With the overwhelming success of the ladies lectures, a formal Women's Department of King's College was established in 1881, with official recognition in 1885. The Department expanded until 1910, when it was incorporated as a separate constituent College of the University of London: King's College for Women. As discussed by Christina Bremner in 1913, the type of student had changed from those of the early years: "In the early days a majority of students were drawn from the leisured classes, and sought culture largely for its own sake. ... The majority are now being prepared for work, to take up positions in life where their services are used and paid."[19]

As with the other women's colleges, the rules pertaining to residence life, and even academic life, were close to jail-like, as Neville Marsh commented: "Life in the lecture theatres and laboratories was as ordered as life in Hall. Degree students were expected to wear undergraduate gowns in lectures and the institutionalisation was completed by students having to wear coloured overalls in laboratories according to their course."[18] First-year students wore brown; second-year, orange; and third-year, navy; while laboratory assistants wore green. However, the limited number of possible colours led to the scheme later being abandoned.

If women were to be prepared for work, what sort of work and what sort of science? The dispute about the appropriate science for women that we described in Chap. 1 in the context of high schools was also debated for higher education. In 1908, a Home Science and Economics Department was opened, offering a 3-year programme giving a College Certificate in the subject.[20] Bremner contended that graduates from this programme would have excellent employment possibilities in hospitals, schools, and other public organisations, and, of course, such graduates would excel at scientific homemaking. She described the programme:

> To this end we have a new grouping of subjects useful to a new class of students, including biology, chemistry, physics, hygiene, physiology, economics, ethics, and psychology; the bearing of these on household work (cookery, laundry, housewifery, management of children and servants) is indicated, the practical arts are linked with the scientific principles on which they are based.[19]

Bremner then noted the chemistry component:

> Students of chemistry must learn to perform simple analyses, to study hydrocarbons, alcohols, acids, and so forth, so that in the final year they may deal effectively with water analysis, constituents and relative values of different foods, the chemical changes of ferments, preservation and deterioration of food, purity of milk, and so forth.[19]

She assailed those who argued that only “pure” or “men’s” chemistry should be taught to women students:

> It would be interesting to know precisely how far feminism and opposition to a Domestic Science course in a University

> coincide. I cannot think the lines of demarcation correspond perfectly, for I have known advanced feminists, and count myself amongst them, who for years have bitterly complained that so little of the money devoted to technical training has been spent on women, and also how very lacking in thoroughness have been many domestic science courses carried on all over the country.[19]

Bremner looked forward to the day — which would never come — when the University of London granted a B.Dom.Sci.

One of the most vocal supporters of the Home Science and Economics Scheme was Arthur Smithells[21] (see Chaps. 1 and 5). A report in the *King's College Magazine, Women's Department* noted his major role in the subject:

> At the second meeting [on the Scheme] Prof. Smithells, of Leeds University, was the chief speaker. He may be called the pioneer of this movement, as he was the first to start special classes in practical science for teachers of Domestic Economy and others in Leeds. He has been appointed Honorary Advisor to the Committee in charge of the King's College Scheme. … Prof. Smithells prefaced his remarks by reading, from the prospectus of a new Japanese University for Women, various paragraphs expressing with remarkable force and grace the idea he proceeded to elaborate, that women's training for domestic work may be conceived and planned with as high an ideal, as noble and inspiring a motive as any part of their higher education. He was led to see the importance of this form of training by being asked to assist in the future teachers of Cookery and other Domestic Arts in the ordinary branches of physical science.[22]

In 1915, the arts and science departments of King's College for Women were transferred to the Strand, at last making King's College a co-educational institution — in theory. In reality, the absorbed women students and staff had to adapt to functioning in a traditional male environment (see, for example, Rosalind Franklin's experiences at King's, Chap. 9).

The surviving portion, the Household and Social Science Department of King's College for Women, became completely independent on a new site at Camden Hill Road, Kensington, pioneering the role of nutritional science as a field of academic scholarship. Initially known as King's College of Household and Social Science, in 1953, it was renamed Queen Elizabeth College, lasting only until 1985, when it was amalgamated with King's College and Chelsea College.

Margaret Seward (Mrs. McKillop)

Margaret Seward[23] was the earliest chemist on staff at the Women's College, being there from 1896 to 1915. Seward, daughter of James Seward, Master at the Liverpool Institute, was born on 22 January 1864 and educated at Blackborne House, Liverpool. She entered Somerville College, Oxford, in 1881[24]; in 1884, she was the first Oxford woman student to be entered for the honour school of Mathematics. Seward then changed her focus to chemistry, and in 1885 became the pioneer woman to obtain a first class in the honour school of Natural Science. Upon graduation, Seward was immediately appointed Natural Sciences Tutor at Somerville, in addition to undertaking research with the Oxford chemist, W. H. Pendlebury. Two publications on chemical reactions resulted from her work, one of which was read to the Royal Society.

In 1887, Seward accepted a position as Lecturer in Chemistry at the Royal Holloway College (RHC). Martha Whiteley (see later), a student at RHC at the time, recalled that Seward was one of the six founding Lecturers at the College:

> Probably no one shouldered a heavier load than the Science Lecturer, Margaret Seward, for the College then possessed no science laboratories or equipment, and yet, during the four years during which she held that post, the first science building, comprising Chemistry, Physics, Botany and Zoology laboratories,

> was built and equipped under her direction; and, with occasional help in Physics and Zoology, she was responsible for all the science teaching in Chemistry, Physics and Zoology required to carry successfully the first group of science students through the Intermediate and Final B.Sc. Examinations.[25]

Resigning her position at RHC in 1891, Seward travelled to Singapore to marry John McKillop, a civil engineer. Seward, her husband, and son (Alasdair) returned to England in 1893, where she taught at the Girl's Grammar School, Bradford, and then at Roedean School. Seward was appointed to King's College, Women's Department, in 1896 to teach Elementary Science and Chemistry, the Chemistry Laboratory having opened in 1895.

Seward was a signatory of the 1904 petition for the admission of women to the Chemical Society and the 1909 letter to *Chemical News*. She was mentioned in an 1898 article on the College in *The Girl's Realm*:

> As regards science, there are two laboratories; the larger is for chemistry, the smaller serves various scientific courses: zoology, biology, botany, geology etc. In both the laboratories Princess Alice of Albany works, under the supervision of Mrs McKillop, one of our foremost women science-lecturers, and lately lecturer on Chemistry at the Royal Holloway College.[26]

Another member of royalty, Princess Margaret of Connaught, worked in the Chemical and Botanical Laboratories from 1899 to 1902.[27]

The year 1912 was the turning point for chemistry and for Seward herself. A report in the *King's College Magazine, Women's Department* suggested that the King's College administration wanted a male and/or a more research-oriented individual as Lecturer in Chemistry; as a result, Seward became marginalised:

> Mr. H. L. Smith has been appointed full-time lecturer in Chemistry, and as at present there is a very definite majority

> of Home Science over B.Sc. students in the Chemical Laboratory, he has taken charge of it, with Miss Masters as Demonstrator. Mrs. McKillop has thus been set free to undertake in addition to her tutorial work, the organization of the Library, a business which has been pressing for a little time.[28]

The Home Science Committee terminated Seward's position as of Spring 1914, perhaps in preparation for the transfer of the science programmes to the existing male-faculty departments of King's on the Strand. In 1915, Seward was teaching economics at Bradford Girl's Grammar School and giving food lectures to old girls. She then worked in the Ministry of Food, being awarded an M.B.E. in 1919 for her wartime studies on nutrition and human health, which included authoring a book: *Food Values*.[29] From 1920 until her sudden death on 29 May 1929, she acted as Librarian of the Sociological Society at Lepay House.

Chemistry at the King's College of Household and Social Science

The Chemistry Department of the newly formed King's College of Household and Social Science initially fell under the overall direction of Herbert Jackson[30] of King's College, The Strand.[18] Following Seward's departure in 1914, the staff based at Kensington Square were Henry Llewellyn Smith and an Assistant Lecturer, Helen Masters.

In 1915, just before the Women's College moved to Camden Hill Road, King's College decided a Reader in Chemistry should be appointed. The administration declared that: "… other things being equal, a woman should be appointed to the Readership in Chemistry and that under present conditions it would be of great value to the Department to secure the services of a woman with the high scientific standing and personality of Dr. Ida [Smedley] Maclean."[31] The statement is remarkable, for not only does it set an almost unobtainable benchmark — Smedley being an "exceptional" woman — but it also defines the expected personality of the successful candidate. Despite the wishes of the College, no woman having the academic and psychological attributes of Smedley presented herself. Instead, Kenneth Charles Tinkler was appointed.

Helen Masters

Helen Masters[32] had entered King's College, Women's Department, as a student in 1906. After completing a B.Sc. in Applied Chemistry in 1909, she took a year of postgraduate studies in Applied Chemistry, Domestic Arts, and Practical Chemistry. For the 1910–1911 year, Masters was Demonstrator in Physics at the Cheltenham Ladies' College (CLC); then, in 1911, she was appointed Demonstrator in Applied Chemistry back at King's in Kensington. In addition to teaching, Masters undertook a range of research related to food chemistry, resulting in at least eight publications. The first two papers were with Smith; four on her own, two of which were reports on the solubilisation of lead from lead-glazed casseroles; and one each with Phyllis Garbutt[33] and Marjory Maughan, two research students at the College.

Masters and Tinkler worked together, as Agnes Browne (Mrs. Jackman) noted in Tinkler's obituary: "In collaboration with Miss Helen Masters he [Tinkler] created a course in Applied Chemistry unique in its scope and to the present day regarded as a model in Home Science departments throughout the world."[34] The content of the course was published in 1926 as a two-volume set, *Applied Chemistry*.[35] This text became the standard reference work on analytical procedures for chemistry related to the home and it was still being reprinted in 1948, while the first American edition was published in 1950.

Becoming more and more interested in the chemistry of cooking, Masters spent the summer of 1924 visiting Household Science Departments in Canada and the United States. Her growing fascination with the subject led her to resign her position and take up an appointment as Head of Domestic Science at Battersea Polytechnic. Phyllis Garbutt followed Masters to Battersea, becoming an Assistant Science Teacher.

Marion Soar

In 1917, Tinkler had appointed Marion Soar[36] as a second Assistant Lecturer and Demonstrator. Marion Crossland Soar, daughter of Henry James Soar, was born on 3 March 1895 and attended County School for Girls, Bromley, Kent. She entered University College of North Wales, Bangor, in 1913 and graduated with a B.Sc. in Chemistry in 1917. Upon graduation, she accepted the offer of a position as Assistant Lecturer in Chemistry at King's College of Household and Social Science. Soar was among the cohort of the first women to be admitted to the Chemical Society (see Chap. 2).

In addition to teaching, Soar also undertook research with Tinkler, her most famous publication being in the *Biochemical Journal* on the formation of ferrous sulphide in eggs during cooking. Soar resigned in 1921, accepting a position as Lecturer in Chemistry at Battersea Polytechnic. It was possibly Soar who encouraged Masters and Garbutt to move to Battersea.

Agnes Browne (Mrs. Jackman)

Agnes Browne[37] succeeded Marion Soar in 1921. Browne, the youngest of seven children of John and Jane (née Eakin) Browne, a farmer in Londonderry, Northern Ireland, was born on 31 August 1896. She entered the Royal College of Science, Dublin, in 1915, and graduated with a degree from the University of London in 1919. After graduation, she remained there for a year doing research in organic chemistry, supported by a grant from the Department of Scientific and Industrial Research (DSIR), under the supervision of A. G. G. Leonard[38] and W. E. Adeney.[39] In 1920, they arranged for Browne to do research with Donnan at University College, London.

Browne was appointed to King's College of Household and Social Science as Assistant Lecturer in Chemistry in December of 1921. Browne married D. N. Jackman in 1924, and she resigned her position in 1925 on the birth of her daughter. However, she was asked to return in October 1926 on a temporary basis because Tinkler was ill and Masters had resigned to take up the position at Battersea, so there was no one left with the qualifications to teach the specialised course. Browne remained on annual appointment until February 1944 when, at last, the position was made full-time. While at the College, she co-authored a text on *The Principles of Domestic and Institutionalised Laundry Work*.[40] Browne retired in 1956 at the rank of Senior Lecturer in Chemistry and died on 6 November 1978.

Mary Thompson (Mrs. Clayton)

Overlapping with Browne, Mary Christina Thompson[41] was appointed in 1938 as Demonstrator and Assistant Lecturer. Mary, the daughter of William Thompson, a London accounts clerk, was educated at James Allen's Girls' School, Dulwich. She entered Bedford College in 1930, completing her B.Sc. in 1933, then continued to a Ph.D. in organic chemistry with Eustace Turner.[42]

Just before Christmas 1933, Thompson had an accident in one of the chemistry labs. She was distilling benzene using a Bunsen burner when the benzene caught fire. She extinguished the flames and then, as she told Turner, with her clothes on fire: "She threw herself on the floor 'as this was the proper thing to do', and, owing to her prompt action and that of Miss Cook and Miss Lockhart in wrapping her up in the fire blanket which was only a few feet away, Miss Thompson was saved from a much more serious accident."[43]

Following the accident, Thompson returned to her studies in May 1934, completing her Ph.D. in 1937. Thompson took a position at Roedean School, but left after one year as she was expected to teach biology and she did not know enough of that subject. It was then that Thompson obtained her position at King's, where she remained for many years. She married a Mr. Clayton in 1956, while in 1966 she was noted as living in Middlesbrough where she was employed as a teacher.

The Laboratory of Professor Huntington

Though at the time, women were not permitted to attend King's College, The Strand, as students, women chemists were hired as research assistants for King's leading metallurgist, Alfred Kirby Huntington.[44] His most prominent student was Elison Ann Macadam, a signatory of the 1909 letter to *Chemical News* (see Chap. 2). The sole biographical information on Macadam comes as a result of her subsequent marriage to the chemist Cecil Henry Desch.[45] Macadam was the second daughter of William Iveson Macadam, Professor of Chemistry at the College of Surgeons, Edinburgh. She had wished to study for a degree in chemistry at the University of Edinburgh, but at that time women were excluded. Curiously, despite the formal ban of women from the Strand campus, she was able to study chemistry with F. C. Thompson and Herbert Jackson in order to sit the Institute of Chemistry examinations. After she had successfully passed them,[46] Thompson and Jackson recommended her to Huntington, who hired her about 1902.

It was noted in McCance's obituary of Desch that: "Desch did not find it easy to work with Professor Huntington who was quick tempered and exacting."[45(a)] Macadam undertook accurate analysis of metal samples, which were then examined metallographically by Desch. In January 1909, Macadam and Desch were married. McCance quotes Huntington that: "Cecil Desch had robbed him of his best assistant."[45(a)] The Desch family moved shortly afterwards to Glasgow, where Desch had been appointed as Lecturer in metallurgical chemistry. Macadam had two daughters, and the only other reference to her was at a later appointment of Desch at the University of Sheffield, that:

"In promoting the social activities of the University, Mrs Desch took an active part."[45(a)]

Emily L. B. Forster was another of Huntingdon's research students. Little is known about her except that she was working with Huntington at the time she, too, signed the 1909 letter. Forster later became Lecturer at the Westminster College of Pharmacy (see Chap. 10). She authored two books: *How to Become a Woman Doctor* and *Analytical Chemistry as a Profession for Women*.[47]

Battersea Polytechnic

The role of women in polytechnics has been almost totally overlooked and yet they were an avenue for predominantly lower-social-class women to acquire chemical education, either pure or applied,[48] and for women to find teaching positions in chemistry.

The great polytechnic movement in London was designed to "promote the education of the poorer inhabitants of the metropolis by technical instruction, secondary education, art education, evening lectures, or otherwise, and generally to improve their physical, social and moral condition."[49] The first polytechnic, the Polytechnic at Regent Street (now the University of Westminster), dated back to 1839, and though many of the offerings were gender-specific, some women did enter the co-educational Department of Science (see Table 3.1).

Table 3.1. Department of Science, Chemistry aggregate examination results, Battersea Polytechnic Institute, 1897–1911 (from Note 49).

Courses	Total students	Female	Percent female
Theoretical inorganic	716	59	8.2
Practical inorganic	699	52	7.4
Theoretical organic	149	2	1.3
Practical organic	143	2	1.4
Chemistry total	1707	115	6.7

Battersea Polytechnic, in particular, attained considerable academic success, leading to an application for recognition as a School of the University of London in 1911.[49] However, the application was rejected and only in recent times did Battersea gain university status as the University of Surrey. Battersea Polytechnic was organised into six main departments: mechanical engineering and building trades, electrical engineering and physics, chemistry, women's subjects, art, and music.

In addition to the academic subjects, domestic science flourished in the polytechnics. The Domestic Science Training College at Battersea Polytechnic, renamed from the Department of Women's Subjects, had the largest number of students of any domestic science department in the country. There was significant chemistry content in most of the Domestic Science programmes; as an example, Fig. 3.1 provides the *Calendar* entry for the chemistry component of the Diploma in Advanced Cookery

Chemistry (Theoretical and Applied).—Air. Water. Chemical theory. Acids, alkalies and salts. Carbon and its oxides; fuels. Soaps. Textile fabrics. Water softeners. Sugars, starch, alcohol, acetic acid. Proteins. Fats. Vitamines. Yeasts, moulds and bacteria. Study of certain foods. Preservation and sterilisation of food stuffs. *The practical work* will be partly illustrative of the lectures, and partly experimental craft work, *i.e.* :—

Experimental Housewifery.—Study of metals, causes of tarnish, metal polishes and preservers, stainless cutlery. Study of woods, dry rot, furniture polishes, stains, paints and varnishes. French polish. Lacquers. Care of leather. Materials used in making floor coverings, and scientific reasons for methods of cleaning and preserving them. Household disinfection.

Experimental Laundrywork.—Comparative value of methods of softening water for laundry purposes. Study of detergents and their action on textile fabrics. Methods of testing fabrics, and the reactions of laundry reagents on them. Experimental removal of stains; bleaching and dyeing. Laundry blues. Microscopic and chemical examination of starches. Disinfection of clothing.

Experimental Cookery.—Examination of the chemical and physical natures of various food stuffs, *e.g.*, flour, fat, fish, meat, eggs, vegetables, pulses, milk. The effects of heat, and of different methods of cooking on these food stuffs. Study of yeast and its action in bread making. Examination of sugar substitutes. Experiments to attempt the solution of problems encountered in the kitchen.

Fig. 3.1. The chemistry (theoretical and practical) component of the Advanced Cookery and General Housecraft, with Cognate Chemistry Diploma Programme at Battersea Polytechnic, 1919–1920. (*Source*: *Battersea Polytechnic Calendar*).

and General Housecraft, with Cognate Chemistry. In 1948, the Department of Domestic Science became a separate entity — Battersea College of Domestic Science — and it is now known as Battersea College of Education.

The Domestic Science Training College was staffed very largely by women. It was to Battersea Polytechnic that Marion Soar, Helen Masters, and Phyllis Garbutt had moved from King's College for Women. In addition, the two chemistry positions — Chemistry & Physics, and Chemistry as Applied to Household Processes — were traditionally held by women. For example, in the 1919–1920 *Battersea Polytechnic Calendar*, the staff for both subjects were Claudia McPherson and Marjorie Sudds.

Arrowsmith's history of Battersea Polytechnic specifically mentions the hiring of Masters in 1926 as a key figure in the success of the Department:

> The Governors might well have feared that they could not possibly find anyone who could fill this important position [Head of Domestic Science] in the Polytechnic organization, but in Miss Helen Masters, B.Sc. (Lond.) they did. ... She had been on the staff of King's College for Women for 15 years, as a Lecturer in Chemistry in the Household and Social Science Department, and had published research on Cookery and Laundry work. She thus brought to the post at Battersea a training and experience, allied to a wide outlook on Domestic Science, which seemed to make her the right person to follow Miss Marsden. Miss Masters filled this post with great distinction and brought added lustre to the high reputation the Department already enjoyed.[49]

Masters resigned from her position in 1948.

After her arrival in the Department, Garbutt co-authored a book, *Food Wisdom*,[50] and in the 1940s she left to become the Principal of the Good Housekeeping Institute. She died in 1970.

Mary Corner

In addition to the Domestic Science programme, it was possible to take other degrees through Battersea Polytechnic. Mary Corner[51] was one of those to obtain a University of London B.Sc. in Chemistry at Battersea. She was born on 25 March 1899 and was educated at Beulah House High School, Balham, London. During the First World War, she worked in a pharmacy, entering Battersea Polytechnic in 1922 and graduating in 1927.

Corner obtained a position with the British Cotton Industry Research Association, Manchester, in 1928, working initially in the rayon department where she developed a fascination with microanalysis. As a result of her acquired background, she was promoted to Head of the Microanalytical Section; then in 1945, she obtained a similar post with the British Leather Manufacturers' Research Association. Two years later, Corner was invited to become Head of the newly formed Microanalytical Section of the Chemical Research Laboratory (later the National Chemical Laboratory).

As noted in her obituary by G. R. Davies,[51(a)] Corner had an "unfortunate accident" early in life and, "Burdened with a severe disability, she had, in addition, more than the usual share of suffering and trouble." In the 1930s, she became a founder member of the Microchemical Club (to be later joined by Isabel Hadfield; see Chap. 2). At the time of her death on 4 November 1962, she was Vice-Chairman of the Microchemistry Group of the Society for Analytical Chemistry.

East London College (Queen Mary College)

At the other end of the social spectrum from the Oxbridge Colleges (see Chap. 6) was East London College.[52] When it opened in 1888, it was titled the People's Palace Technical School, changing its name to East London Technical College in 1897. Then in 1915, the institution was incorporated in the

University of London as East London College, and renamed as Queen Mary College (QMC) in 1934.

Kathleen Balls

In 1915, John Theodore Hewitt,[53] the Head of Chemistry at the time, was commissioned in the Royal Engineers.[52] The only remaining chemist, F. G. Pope, was placed in charge and Kathleen Balls[54] was hired to assist Pope as Lecturer and Demonstrator in Chemistry, enabling courses to be run through the war period. Balls, the daughter of a Clapham carpenter, was born on 2 May 1890 and was educated at the City of London School for Girls. She entered East London Technical College in 1908 and completed a B.Sc. (Chemistry) degree in 1911.

At the outbreak of war, Balls was the science teacher at the County School, Enfield. With the greater urgency for university academic staff, she was released from her school-teaching duties and hired by the Chemistry Department of East London College.[55] Her appointment at the College stipulated the occupancy of the position as being for the length of hostilities only. Nevertheless, she actually continued at QMC into the interwar era, obtaining an M.Sc. in 1919. She submitted her resignation in 1924, but, though it was accepted, she was asked to continue as Lady Superintendent.[56]

Though she had lost academic status, Balls remained active in chemistry at QMC, working with James Riddick Partington,[57] Hewitt's successor as senior physical and inorganic chemist. Balls and Partington co-authored three research publications between 1922 and 1936. In addition, Balls, under her married name of Mrs. Stratton, and Partington co-authored a book on chemical calculations.[58]

Cecilie French

It was not until the aftermath of the Second World War that any more women chemists were appointed to the Chemistry

Department. The first of these was Cecilie Mary French,[59] who was born on 23 October 1915 in London. She was educated at Walthamstow Girls County High School and obtained her B.Sc. in chemistry from UCL in 1937. From 1938 to 1939, French was a Demonstrator in Chemistry at UCL and she also undertook research with Christopher Ingold[60] and Cecil Wilson, leading to her Ph.D. in 1940. That year, she took a position as a Research Chemist with Imperial Chemical Industries (ICI) Ltd., but later the same year accepted an appointment as Demonstrator and Assistant Lecturer in Physical and Inorganic Chemistry at Bedford College.

At this time, Bedford College had been evacuated to Cambridge to avoid the heavy bombing of London. In addition to her heavy teaching duties, French commenced research in magnetochemistry with James Spencer (see Chap. 4). Queen Mary College, too, had been evacuated to Cambridge, and thus began French's involvement with QMC. It started with her helping Partington with the preparation and revision of his books. Then, with the return of Bedford and QMC to London in 1944, she was offered a position of Lecturer in Inorganic and Physical Chemistry at QMC.

French stayed at QMC for the rest of her life, being promoted to Senior Lecturer in 1961. One of her obituarists, Dorothea Grove (see below), commented about French's teaching:

> Dr. French was an outstanding University teacher and will be remembered with gratitude by generations of undergraduates in whose welfare she took a keen personal interest. She proved her ability during the difficult post-war [Second World War] years when classes were swollen with ex-servicemen, and for a number of years she was also responsible for teaching the 1st M.B. students of the London Hospital.[59(a)]

Despite the heavy teaching load, French maintained an incredible research output. Until a Gouy balance was installed at QMC in 1951, French collaborated with Violet Trew at

Bedford College (see Chap. 4) on magnetochemistry. With the arrival of the balance, French began a collaboration with D. Harrison of QMC. French also undertook extensive research into the electrochemistry of nonaqueous solvents; and on the preparative side of chemistry, she synthesised novel boron compounds. During her years at QMC, 23 students undertook their graduate research under her supervision, resulting in close to 50 publications.

Unfortunately, French's health deteriorated, as Grove recounted:

> It was when Dr. French was barely forty that the first signs of encroaching illness came. With great courage and determination, as well as cheerfulness, she continued her work as usual despite partial blindness and increasing disability. At this stage her long-standing practice of lecturing without notes was invaluable to her. She learned Braille and characteristically gave her services to the National Institute for the Blind in advising them about the transliteration of scientific and mathematical symbols.[59(a)]

French travelled widely, lecturing in the Far East and in the United States, spending a year at Pennsylvania State University. She was invited to return to Penn. State in 1962 to give the Marie Curie Lecture, but sadly, she died on 6 August 1962 at Epsom.

Dorothea Grove

Dorothea Mona Grove[61] joined French at QMC in 1949. Grove was born on 3 April 1920, the daughter of Rev. P. R. Grove, and she was educated at Bolton School. She entered Somerville College, Oxford, in 1938, completing her B.Sc. in chemistry in 1943 and an M.A. in 1945. In 1945, Grove accepted a position as Demonstrator in Inorganic and Physical Chemistry at Bedford College, replacing French, and later becoming Assistant Lecturer.

Then in 1949, she, too, left Bedford for an appointment at QMC. Her initial rank was as Assistant Lecturer in the Chemistry Department, then later she was promoted to Lecturer when she received a Ph.D. from the University of London in 1950. Grove also served as Warden of the Women's Hall of Residence from 1954. She died on 2 August 1980.

Imperial College

In the late 19th century, the chemistry departments of the constituent Colleges of the University of London had been involved in the discourse on the teaching of chemistry as a pure or applied subject, or somewhere in between.[62] The antecedents of Imperial College (IC) certainly saw themselves as belonging to the "applied" camp.

The Chemistry Department of IC had two precursors.[63] The first was the Royal College of Chemistry, which had been established in Hanover Square, London, in 1845, the result of a private enterprise to found a college to aid industry.[64] In 1853, the Royal College of Chemistry affiliated with the Government School of Mines applied to the arts, effectively becoming its chemistry department. The name was changed to the Royal School of Mines in 1863, and new buildings were constructed in South Kensington in 1872.

Some of the courses, including chemistry, were merged with other science subjects to form the Normal School of Science (named after the École Normale in Paris). The Normal School of Science was renamed the Royal College of Science (RCS) in 1890. At first, the RCS saw itself primarily as a teacher training college in practical science and, as such, attracted women students including some from overseas.

The other part of the Chemistry Department originated with the Cowper Street Schools, which then became Finsbury Technical College. Finsbury College was intended as the first of a number of "feeder" colleges for the Central Institution (renamed the Central Technical College), but was almost certainly the only one founded. Needing a large new location, one was found in South Kensington adjacent to the Royal School of Mines and the Royal College of Science.

In 1907, the Royal School of Mines and the Royal College of Science were incorporated as constituent Colleges under the name of Imperial College. The same year, the Central Technical College was renamed the City and Guilds College, and it was then incorporated as a third constituent College of IC. Unique amongst the Colleges in London, IC had a very specific mandate: that for higher technological education and research, an area in which Britain lagged behind Germany.[65] Wishing to avoid the conflicts within the University of London, primarily between King's College and University College, it was only in 1929 that IC was finally cajoled into joining the University of London.

Women Chemistry Students

At IC Chemical Society meetings, the women chemistry students were expected to serve tea. In a 1904 issue of the College newspaper, *The Phoenix*, a male student reported back to his parents that: "You buy a six-penny ticket, and grab what you can get, whilst you ask one of the lady students, who are quite a jolly lot, for a cup of tea."[66]

However, attitudes towards women among some other male students were more negative. The following commentary in 1905 titled "Some Fallacies about US" shows how many males considered it more appropriate for women to study biology:

> In patronizing fashion masculine voices have enunciated that *they* are the students for Mathematics, Mechanics, Chemistry and Physics (that is, everything they consider worth knowing), but that we, *par excellence*, are of the right mettle for Biology, and that in attempting to work at any other science we must surely feel sadly out of our element. ... It seems scarcely in accordance with the theorem under discussion that neither Girton nor Newnham is overrun by ardent girl biologists, that

> Holloway and Bedford Colleges possess but a few lonely votaries of the science...[67]

The article concluded with a note that 14 of the 15 women students then at IC agreed totally with the sentiments expressed in the article.

This was the period of the suffragette movement, and an article in a 1907 issue of *The Phoenix* promoting the right of women to vote[68] provoked an ominous response:

> From the article [in the preceding issue] and also from a certain incident that took place in the College, I understand that all the lady students have turned suffragettes .... My real object in writing this article is to warn our own Little Band to be very careful and take great care of themselves. I think that perhaps they had better wait until they are older; for the present they must be good little girls and study their Chemistry.[69]

One of the early women students was Kathleen Mary Leeds.[70] Leeds was educated at Croydon High School (a GPDSC school) before entering the Royal College of Science in 1904 and graduating in 1908. In 1911, she was appointed Science Mistress at her former school, but died tragically in 1921.

Another pioneer was Lucy Alcock.[71] Alcock was a student from 1904 to 1907, and was then a Staff member for the year 1907/1908. A glowing account of a 1908 presentation by Alcock to the IC Chemical Society was reported in *The Phoenix*:

> Miss Alcock was then called upon to disclose the wonders of Colloids. Undismayed by the mysteries of a meaningless nomenclature, Miss Alcock laid bare all the secrets of science with a lucidity and clearness that would have done credit to Minerva herself. With a winning women's way we were conducted through the intricacies of one of the obscurest domains of chemistry, and the low murmur of pleasure which had been gradually growing through the meeting became a roar of

> delirious and deafening delight when the fair lecturer showed a colloidal solution of barium sulphate that had the consistency, appearance, and properties of condensed milk.[72]

Alcock moved to Canada in 1927; and in 1936, her address was listed as Kinsella, Alberta.[73]

Martha Whiteley

Much of the pressure for the admission of women to the Chemical Society came from the duo of Ida Smedley (see Chap. 2) and Whiteley. Martha Annie Whiteley,[74] born on 11 November 1866 at Hammersmith, London, was the second daughter of William Sedgwick Whiteley and Mary Bargh. Whiteley graduated from RHC with a B.Sc. in chemistry in 1890. Between 1891 and 1900, she was Science Mistress at Wimbledon High School (a GPDSC school); and for the next 2 years, science lecturer at St. Gabriel's Training College, Camberwell. During the 1898–1902 period, she was also undertaking research at the Royal College of Science.

In 1902, Whiteley was awarded a D.Sc. from the University of London; and the following year, she was invited to join the staff at Imperial College as an Assistant under William Tilden. Her appointment was reported in *The Phoenix*: "At the Prize Distribution, we had the very great pleasure of watching Dr. Whiteley take her place with the Staff, the first lady to occupy that position at the Royal College of Science."[75] Whiteley was promoted to Demonstrator in 1908 and, with the drafting of male scientists for war work in 1914, Whiteley was then appointed to the rank of Lecturer.

Retaining her position after the end of the First World War, in 1920, Whiteley was made an Assistant Professor (a senior post at Imperial, later designated as Reader). As T. S. Moore, one of her obituarists noted:

> A former R.H.C. [Royal Holloway College] student who went on to the Imperial College in 1920 writes: "She was in charge

of the undergraduates' organic laboratory while I was there. They were nearly all men and a very lively lot — being mainly ex-Service men, but she had them completely under control. She managed to turn them into fairly tidy, efficient practical workers."[74(d)]

Whiteley maintained an active research programme, authoring or co-authoring at least 15 publications. Creese, in her biography of Whiteley, commented:

> Although several British women from about Whiteley's time were productive researcher workers, most of them made their contributions as assistants to male chemists. ... Whiteley, however, was probably the only one who found a place as an independent worker in an established area of chemistry and remained active in research, teaching and technical writing throughout a long career at a major educational institution — a notable achievement for a woman chemist of her generation.[74(c)]

As will be described in Chap. 12, during the First World War, Whiteley became actively involved in war-related duties. Her contributions resulted in her being awarded the OBE in 1920.

In addition to spearheading, along with Smedley, the admission of women to the Chemical Society and co-founding the Women's' Dining Club of the Chemical Society, Whiteley was President of the Imperial College Women Students' Association from the organisation's inception in 1912. The women students of the time referred to her as the "Queen Bee."[63]

Whiteley co-authored a manual on organic chemical analysis,[76] but her greatest contribution was that on the *Dictionary of Applied Chemistry*.[77] She had helped extensively in the preparation of the second edition of (Edward) *Thorpe's Dictionary of Applied Chemistry*. Thus, when a third edition was contemplated, she was asked to co-edit it with Jocelyn Thorpe. She retired in 1934, but continued to work on the volumes as editor-in-chief following the death of her co-editor. This task was a labour of love that continued until 1954, when the last volume appeared. Whiteley was still editing and proofing other manuscripts until shortly before her death, just prior to her 90th birthday, on 24 May 1956.

Margaret Carlton

The second woman to teach in the Chemistry Department of IC was Margaret Carlton.[78] Carlton had commenced her education at Birkbeck College before transferring to Imperial College, where she completed her B.Sc. in 1919 and her Ph.D. in 1925.

Carlton became a research student with the Inorganic Chemistry Professor of the time, Herbert Brereton Baker.[79] As Hannah Gay noted: "Baker, like many of the early professors, had a gifted woman assistant. Margaret Carlton carried out much of the research in his laboratory and was acknowledged when Baker won the Davy Medal from the Royal Society."[63] Carlton held the rank of Assistant Lecturer for most of her career, being promoted to Lecturer only in 1946. Gay added: "This limited career progress was typical for women of her generation. While several of the women working at the college in this period were acknowledged as gifted scientists, they were not seen as serious candidates for professional advancement."[63] Carlton retired in 1960.

Frances Micklethwait

One of Whiteley's assistants (another being Edith Usherwood; see Chap. 11) was Frances Mary Gore Micklethwait.[80] Micklethwait was born on 7 March 1868, daughter of J. P. Micklethwait of Chepstow, Monmouthshire. After a private education, she attended the Swanley Horticultural College, where she gained her love of chemistry. In 1898, she transferred to the Royal College of Science, obtaining an Associateship in 1901.

Micklethwait joined the IC research group of Gilbert T. Morgan,[81] becoming one of the most prolific women authors of chemistry publications of her time, co-authoring at least 22 papers between 1902 and 1914. In his personal reminiscences, Morgan commented:

> In 1904 I was joined by Miss Micklethwait, a member of this [Chemical] Society, with whom I collaborated for nine years. During this period we succeeded in arousing the interest of many senior students, about 14 of whom were included in our joint publications. The work, which covered a wide field, included various studies of the diazo reaction, the preparation of organic arsenicals and antimonials, and the examination of certain coordination compounds of coumarin.[82]

At the outbreak of the First World War, Micklethwait came under the wing of Whiteley. For her war services, Micklethwait was awarded the MBE in 1918. George Kon,[83] one of Micklethwait's students during the First World War (and later Professor of Chemistry), commented that Micklethwait had high expectations: "I soon realized that my ideas of what work meant had to be overhauled."[83]

After the War, Micklethwait worked briefly in the research laboratory of Boots Pure Drug Company and then returned to Swanley Horticultural College, where she taught until 1921, being Principal for her last year there. From then until 1927, she compiled the Index for the second edition of *Thorpe's Dictionary of Applied Chemistry*, the series co-edited by her friend, Whiteley. She died on 25 March 1950, aged 83 years.

Helen Archbold (Mrs. Porter)

Helen Kemp Archbold,[84] born in 1899, was educated at Clifton High School for Girls, Bristol, and entered Bedford College in 1917. Graduating in 1921, she obtained one of the places reserved for women in Jocelyn Thorpe's organic chemistry research group at IC, as her biographer, Donald Northcote, noted:

> She [Helen] obtained one of the two or three places in Professor Thorpe's organic chemistry department that were made available to women by the persuasion of Dr. Martha Whiteley. .... under the supervision of the strict and assiduous Dr. Whiteley set the high standards of Helen's future work.[84]

During the large-scale importation of apples, whole cargoes in the refrigerated holds of ships were found to have turned brown and unfit to eat. A joint Cambridge–Imperial research team was set up in 1918 to investigate this problem. Archbold was invited to join the plant biochemistry part of the study in 1922. Over time, her research shifted from the chemical analysis of apples to the study of the metabolic processes of the maturing apple. In 1931, she was transferred to the plant physiology team at IC to work on the origin of the starch in barley. This study continued into the 1940s and laid the foundation for her subsequent research on polysaccharide synthesis.

The work with barley led to a lasting interest in the enzymes responsible for starch formation and breakdown. In 1947, she spent a year in St. Louis with the Nobel Laureates, Gerti and Carl Cori, the discoverers of the Cori cycle. Returning to IC, she received her first major grant, enabling her to set up a research group and equip a laboratory. The original intention was to study enzyme systems by conventional means, but in the 1950s, her group focused on the newly developed radioactive tracer methods. In particular, they were among the first to prepare radioactive biochemicals and use them to study the intermediate metabolism of plants. The work was so innovative that she was elected a Fellow of the Royal Society in 1956 and, in 1959, she became the first woman professor ever at Imperial College.

Archbold combined her academic brilliance with a taste for adventure and travel. Unfortunately, her personal life was dogged by tragedies. Both her first husband, William George

Porter, and her second husband, Arthur Huggett, died after only a few years of marriage. Archbold herself died in 1987 at the age of 88.

Commentary

At both University College and Imperial College, women chemists were readily accepted. It was Ramsay's group that was welcoming at UCL; however, the women remained as laboratory assistants. In those early years, many of the chemists at Imperial College were particularly supportive. For example, Tilden had been one of the women chemists' most vociferous advocates for admission to the Chemical Society (see Chap. 2). Of particular note, Whiteley was able to advance to the rank of Assistant Professor, a senior position at the time. Even more exceptional, Jocelyn Thorpe agreed (at Whiteley's suggestion) to assign a number of research places in his laboratory to women students.

Religious affiliation played a significant role in how women chemists were treated. As we saw in Chap. 1, Unitarians and Nonconformists were advocates of the equality of women. Here, we see that Anglican-affiliated King's College preferred to "banish" women students to Kensington so that they would not "contaminate" the male-exclusive College on the Strand. When co-education was imposed, the women-dominated Domestic Science programme was excluded.

Though many feminists were opposed to academic Domestic Science, particularly domestic chemistry, women chemists had more opportunities of employment as lecturers of domestic chemistry at King's College of Household and Social Science and at the polytechnics, such as Battersea Polytechnic, than in mainstream chemistry.

Finally, at the time of the First World War, small East London College more closely resembled the English provincial universities (see Chap. 5) in that a women chemist (Balls) was hired to do most of the teaching during that War. In general, as we show in later chapters, it appears that women chemists played more important roles at smaller institutions.

Notes

1. Willson, F. M. G. (2004). *The University of London, 1858–1900: The Politics of Senate and Convocation.* Boydell Press, Suffolk, pp. 85–144.
2. Daniels, E. A. (1972). *Jessie White Mario: Risorgimento Revolutionary.* Ohio University Press, Athens, Ohio.
3. Manton, J. (1965). *Elizabeth Garrett Anderson.* Methuen, London.
4. (a) Dredge, S. (2005). Opportunism and accommodation: The *English Woman's Journal* and the British mid-nineteenth century women's movement. *Women's Studies* **34**: 133–157; and (b) Rendell, J. (1987). "A moral engine": Feminism, liberalism and the *English Women's Journal.* In Rendell, J. (ed.), *Equal or Different: Women's Politics 1800–1914*, Basil Blackwell, Oxford, pp. 112–138.
5. William Shaen, the most vigorous male supporter of admission of women students to the University of London, was also a solicitor for the Girls' Public Day School Company; benefactor of Newnham College and Girton College, Cambridge; legal advisor to Somerville College, Oxford; and Chairman of the Board of Bedford College, London. A brief biography of Shaen was edited by his daughter: Shaen, M. J. (ed.) (1912). *William Shaen: A Brief Sketch.* Longmans, Green & Co., London.
6. Sutherland, G. (1990). The plainest principles of justice: The University of London and the higher education of women. In Thompson, F. M. L. (ed.), *The University of London and the World of Learning, 1836–1986*, The Hambledon Press, London, p. 39.
7. Creese, M. R. S. (1991). British women of the nineteenth and early twentieth centuries who contributed to research in the chemical sciences. *British Journal for the History of Science* **24**: 275–303.
8. "A.H." (1926). Obituary notices of Fellows deceased: Percival Spencer Umfreville Pickering, 1858–1920. *Proceedings of the Royal Society of London* **111**: viii–xii.
9. Travers, M. W. (1956). *A Life of Sir William Ramsay.* Edward Arnold, London, p. 96.

10. (a) Donnan, F. G. (1926). Obituary notices: Katherine A. Burke. *Journal of the Chemical Society* 3244; (b) Personal communication, letter, Keith Austin, Archivist, Senate House, University of London, 1998; and (c) (1925). University of London, University College, Report of the University College Committee, p. 22.
11. Freeth, F. A. (1957). Frederick George Donnan (1870–1956). *Biographical Memoirs of Fellows of the Royal Society* **3**: 23–39.
12. Donnan, F. G. (1948/1949). Edward Charles Cyril Baly, 1871–1948. *Obituary Notices of Fellows of the Royal Society* **6**: 7–21.
13. University College, London, student records.
14. *Calendars*, London School of Medicine for Women; and from Bedford College staff records.
15. Benett, G. M. (1953). Samuel Smiles, 1877–1953. *Obituary Notices of Fellows of the Royal Society* **8**: 583–600.
16. Hey, D. H. (1955). Schools of chemistry in Great Britain and Ireland–XVIII King's College, London. *Journal of the Royal Institute of Chemistry* **79**: 305–315.
17. Harte, N. (1979). *The Admission of Women to University College, London: A Centenary Lecture*. University College, London, p. 20.
18. Marsh, N. (1986). *The History of Queen Elizabeth College: One Hundred Years of Education in Kensington*. King's College, London.
19. Bremner, C. S. (January 1913). King's College for Women: The Department of Home Science and Economics. *Journal of Education* 72–74.
20. Anon. (1908). College notes. *King's College Magazine, Women's Department* (33): 3.
21. Raper, H. S. (1940). Arthur Smithells, 1860–1939. *Obituary Notices of Fellows of the Royal Society* **3**: 96–107.
22. Anon. (1908). The Home Science and Economics scheme. *King's College Magazine, Women's Department* (33): 8–9.
23. Note 7, Creese, p. 272.
24. Pottle, M. (2004). McKillop [née Seward], Margaret (1864–1929). *Oxford Dictionary of National Biography*, Oxford University Press, http://www.oxforddnb.com/view/article/51779, accessed 22 Dec 2007; personal communication, e-mail, P. Adams, Archivist, Somerville College, Oxford, 1998; personal communication, e-mail, B. Ager, Archivist, King's College, London, 1998; and *Register*, Somerville College, Oxford.

25. Whiteley, M. A. (November 1929). Margaret McKillop, M.B.E., M.A., 1864–1929. *College Letter, Royal Holloway College Association* 51–52.
26. Rawson, Mrs. S. (1898). Where London girls may study, I. King's College, Kensington Sq. *The Girl's Realm Annual* 1203.
27. Anon. (1905). College notes. *King's College Magazine, Women's Department* (24): 3.
28. Anon. (1912). College Notes. *King's College Magazine, Women's Department* (46): 4.
29. McKillop, M. (1916). *Food Values*. E.P. Dutton, New York; expanded and republished as: McKillop, M. (1922). *Food Values: What They Are, and How to Calculate Them*. Routledge, London.
30. Moore, H. (1938). Sir Herbert Jackson, 1863–1936. *Obituary Notices of Fellows of the Royal Society* **2**: 306–314.
31. Note 18, Marsh, p. 124.
32. Anon. (1910). College notes. *King's College Magazine, Women's Department* (38): 21 & (39): 17; Anon. (1910). Appointments. *King's College Magazine, Women's Department* (40): 31; and Anon. (1911). Appointments. *King's College Magazine, Women's Department* (43): 27.
33. (a) *Register*, Royal Institute of Chemistry, 1926 to 1948; and (b) Anon. (1971). Personal news: Obituaries. *Chemistry in Britain* **7**: 78.
34. Jackman, A. (1952). Charles Kenneth Tinkler 1881–1951. *Journal of the Chemical Society* 1191–1192.
35. Tinkler, C. K. and Masters, H. (1926). *Applied Chemistry: A Practical Handbook for Students of Household Science and Public Health. Volume I. Water, Detergents, Textiles, Fuels, Etc. Volume II. Foods*. The Technical Press, London.
36. (1920). Certificates of candidates for election at the Ballot to be held at the ordinary scientific meeting on Thursday, December 2nd. *Proceedings of the Chemical Society* 97; and University of Wales, Bangor, student record (with thanks to E. W. Thomas, Archivist).
37. (a) Truter, M. R. (1987–1988). One of the first women Fellows of the Chemical Society Agnes Jackman (née Browne). *Journal of the Chemical and Physical Society, University College* 12–13; and (b) personal communication, letter, M. R. Truter (Lady Cox, daughter of Jackman), 31 December 1999.

38. Thornton, H. D. (1967). Obituary: Alfred Godfrey Gordon Leonard, 1885–1966. *Chemistry in Britain* **3**: 224.
39. Anon. (1935). Obituary: Walter Ernest Adeney. *Journal and Proceedings of the Institute of Chemistry* **59**: 326.
40. Jackman, A. and Rogers, B. (1934). *The Principles of Domestic and Institutionalised Laundry Work*. E. Arnold & Co., London.
41. (a) (1933). Certificates of candidates for election at the Ballot to be held at the ordinary scientific meeting on Thursday, December 7th. *Proceedings of the Chemical Society* 72; (b) Bedford College, student records.
42. Ingold, C. (1968). Eustace Ebenezer Turner, 1893–1966. *Biographical Memoirs of Fellows of the Royal Society* **14**: 449–467.
43. Letter, E. E. Turner to Miss Monkhouse, 19 December 1933, Bedford College Archives (held at Royal Holloway College). Dorothy Ellen Cook completed a B.Sc. at Bedford in 1933 and a Ph.D. in 1935. Her first position was Assistant to the Research Chemist at the Shellac Bureau, London; then from 1936 to 1938, she was librarian at Imperial Chemical Industries. That year, she married Frank Stewart and ceased employment. Diana Lockhart obtained her B.Sc. and Ph.D. degrees the same years as Cook. In 1936, she was a Demonstrator in the Biochemical Department at King's College, London. Then in 1952, it was reported that she was married to Mr. Huish, had school-age children, and was attempting to find "work for interest."
44. Desch, C. H. (15 May 1920). A. K. Huntington. *Journal of the Society of Chemical Industry* 162.
45. (a) McCance, A. (1959). Cecil Henry Desch, 1874–1958. *Obituary Notices of Fellows of the Royal Society* **5**: 49–68; pp. 57, 60; and (b) Thompson, F. C. (1958). Prof. C. H. Desch, FRS. *Nature* **182**: 223–224.
46. (a) (1906). New associates. *Proceedings of the Royal Institute of Chemistry* 62; and (b) (1909). New Fellows. *Proceedings of the Royal Institute of Chemistry* (pt. i): 30.
47. Forster, E. L. B. (1920). *Analytical Chemistry as a Profession for Women*. Charles Griffin & Co., London.; and Forster, E. L. B. (1920). *How to Become a Woman Doctor*. Charles Griffin & Co Ltd., London.

48. Stevenson, J. (1997). "Among the qualifications of a good wife, a knowledge of cookery certainly is not the least desirable" (Quentin Hogg): Women and the curriculum at the Polytechnic at Regent Street, 1888–1913. *History of Education* **26**(3): 267–286.
49. Arrowsmith, H. (1966). *Pioneering in Education for the Technologies: The Story of Battersea College of Technology 1891–1962*. University of Surrey, Surrey.
50. Cottington, T. D. D. and Garbutt, P. (1928). *Food Wisdom*. Sir Isaac Pitman & Sons, London.
51. (a) Davies, G. R. (1963). Obituary: Mary Corner. *The Analyst* **88**: 155; (b) Anon. (1963). Obituary: Mary Corner. *Journal of the Royal Institute of Chemistry* **87**: 147.
52. Hickinbottom, W. J. (1956). Schools of chemistry in Great Britain and Ireland–XXVI Queen Mary College, London. *Journal of the Royal Institute of Chemistry* **80**: 457–465.
53. Turner, E. E. (1955). John Theodore Hewitt, 1868–1954. *Biographical Memoirs of Fellows of the Royal Society* **1**: 79–99.
54. Queen Mary College, London, staff and student records.
55. (1 Feb 1916). East London College Council. *Minutes*. A. Nye, Queen Mary College Library, is thanked for this information.
56. (29 May 1924). East London College Standing Committee. *Minutes*.
57. Partington is best known for his multi-volume series on the history of chemistry. See: Butler, F. H. C. (1966/1967). Obituary: James Riddick Partington. *British Journal for the History of Science* **3**: 70–72.
58. Partington, J. R. and Stratton, K. (1939). *Intermediate Chemical Calculations*. Macmillan, London.
59. (a) Grove, D. M. (1963). Cecilie Mary French, 1915–1962. *Proceedings of the Chemical Society* 30–31; and (b) Anon. (1963). Obituary: Cecilie Mary French. *Journal of the Royal Institute of Chemistry* **87**: 66.
60. Shoppee, C. W. (1972). Christopher Kelk Ingold, 1893–1970. *Biographical Memoirs of Fellows of the Royal Society* **18**: 348–411.
61. (a) (1944). List of applications for Fellowship. *Proceedings of the Chemical Society* 67; (b) Anon. (1980). Personal news: Deaths. *Chemistry in Britain* **16**: 629; and (c) *Register*. Somerville College, Oxford.

62. Sanderson, M. (1972). The University of London and industrial progress, 1880–1914. *Journal of Contemporary History* **7**: 243–262; and Donnelly, J. F. (1986). Representations of applied science: Academics and chemical industry in late nineteenth-century England. *Social Studies in Science* **16**: 195–234.
63. Gay, H. (2007). *The History of Imperial College London 1907–2007*. Imperial College Press, London.
64. (a) Hall, A. R. (1982). *Science for Industry: A Short History of the Imperial College of Science and Technology and Its Antecedents*. Imperial College Press, London; (b) Whitworth, A. (ed.) (1985). *A Centenary History: A History of the City and Guilds College, 1885 to 1985*. City and Guilds College of Imperial College of Science and Technology, London; and (c) Bud, R. and Roberts, G. K. (1984). *Science versus Practice: Chemistry in Victorian Britain*. Manchester University Press, Manchester.
65. Vernon, K. (2001). Calling the tune: British Universities and the State, 1880–1914. *History of Education* **30**(3): 251–271.
66. "D.M.L." (1905–1906). The letters of a self-made student to his parents. *The Phoenix* **18**: 13–14.
67. "C.E.A.S." [Speed, C.]. (1904–1905). Some fallacies about US. *The Phoenix* **17**: 32–35. By permission of the archives, Imperial College of Science, Technology and Medicine, London. Clarisse Speed completed an associateship of the Royal College of Science (A.R.C.S.) from 1902–1906, and then became a lecturer in music (extra-mural) at the University of Cambridge.
68. "Suffragist." (1907–1908). The question of the hour. *The Phoenix* **20**: 64–65.
69. "R.M.H." (1907–1908). Suffrage. *The Phoenix* **20**: 90. By permission of the archives, Imperial College of Science, Technology and Medicine, London.
70. Anon. (November 1921). In memoriam: Kathleen Mary Leeds. *The Phoenix* (new ser.) **7**(1): 6.
71. (1951). *Register of Old Students and Staff of the Royal College of Science*, 6th ed. Royal College of Science Association, London.
72. Carvel, H. (1908–1909). Chemical Society. *The Phoenix* **21**: 120. By permission of the archives, Imperial College of Science, Technology and Medicine, London.

73. Anon. (May 1927). Royal College of Science Association (old students and staff). *The Phoenix* (new ser.) **12**(5): 81.
74. (a) Eldridge, A. A. (1957). Martha Annie Whiteley, 1866–1956. *Proceedings of the Chemical Society* 182–183; (b) Owen, L. N. (1956). Dr. M. A. Whiteley, OBE. *Nature* (June 30) 177, 1202–1203; (c) Creese, M. R. S. (1997). Martha Annie Whiteley (1866–1956): Chemist and editor. *Bulletin for the History of Chemistry* **20**: 42–45; and (d) "T.S.M." [Moore, T. S.]. (December 1956). Martha Annie Whiteley, O.B.E., D.Sc., F.R.I.C. *College Letter, Royal Holloway College Association* 61.
75. "C.E.A.S." [Speed, C.]. (1904–1905). From our lady correspondent. *The Phoenix* **17**: 12.
76. Thorpe, J. F. and Whiteley, M. A. (1925). *A Student's Manual of Organic Chemical Analysis, Qualitative and Quantitative.* Longman Green & Co., London. 2nd ed. (1927).
77. *Dictionary of Applied Chemistry*, Supplement by J. F. Thorpe and M. A. Whiteley, 3 vols., Longmans, Green & Co., 1934 (vol. 1), 1935 (vol. 2), 1936 (vol. 3). The preface of the Glossary & Index acknowledges help and guidance of several, including Miss D. Jordan Lloyd, M.A., D.Sc., and Miss P. R. E. Lewkowitsch, ARCS, Ph.D.
78. (1951). *Register of Old Students and Staff of the Royal College of Science*, 6th ed. Royal College of Science Association, London.
79. Thorpe, J. F. (1935). Herbert Brereton Baker, 1862–1935. *Obituary Notices of Fellows of the Royal Society* **1**: 522–526.
80. (a) Burstall, F. H. (1952). Frances Mary Gore Micklethwait (1868–1950). *Journal of the Chemical Society* 2946–2947; and (b) Anon. (1950). Obituary: Frances Mary Gore Micklethwait. *Journal of the Royal Institute of Chemistry* **74**: 277.
81. Irvine, J. C. (1941). Gilbert Thomas Morgan, 1872–1940. *Obituary Notices of Fellows of the Royal Society* **3**: 354–362.
82. Morgan, G. T. (1939). Personal reminiscences of chemical research. *Chemistry and Industry* 665–673.
83. Linstead, R. P. (1952–1953). George Armand Robert Kon, 1892–1951. *Obituary Notices of Fellows of the Royal Society* **8**: 172.
84. Northcote, D. H. (1991). Helen Kemp Porter, 10 November 1899–7 December 1987. *Biographical Memoirs of Fellows of the Royal Society* **37**: 400–409.

Chapter 4

The London Women's Colleges

The account of the London co-educational colleges only provides part of the story of the education of women chemists in the capital. The women's colleges may have been small, but, as we showed in Chap. 1, they graduated a disproportionately large number of women chemists. The first colleges for women in London were set up independently of the University of London. The earliest of all, Queen's College, graduated many of the pioneer leaders in women's education; however, its role was taken over by Bedford College, later to be joined by Holloway College. One of the limited employment opportunities for women chemists was the London School of Medicine for Women; therefore, this institution is included here as well.

Queen's College, Harley Street

In the middle of the 19th century, the position of governess was the only employment option for unmarried middle-class women. Many of the governesses had little education themselves, so a group of lecturers of King's College, London, started a series of Lectures for Ladies in 1847. These classes were so successful that they provided the springboard for the establishment of Queen's College, the first institution for the comprehensive education of women.[1] The founder was Frederick Denison Maurice, a Christian Socialist. The Christian Socialists believed that religion and social progress went hand in hand, and, rare for the period, that women had rights, in particular to education. The founding of the college coincided with a campaign by Amelia Murray, one of Queen Victoria's ladies-in-waiting, for a college for women, and it was Murray who persuaded the Queen to allow her name to be used.

The college was run by men, but with Lady Visitors as Chaperones,[2] one of the chaperones being Jane Marcet (see Chap. 1). Though the initial focus was on the education of governesses, the aims grew rapidly broader as the 19th century progressed. As Shirley Gordon noted in her analysis of the Queen's College curricula from 1848 to 1868:

> The scientific studies at Queen's College were, for the period, amongst the most remarkable in the curriculum. At a time when it was very much an open question whether boys should study science, the committee at Queen's College included Natural Philosophy in the survey of academic knowledge which represented their timetable.[3]

The course on Natural Philosophy included a chemistry section covering the facts and classifications of chemistry, illustrated by experiments performed by the lecturer.

Among the many women graduates to become influential were the educators Dorothea Beale and Frances Buss (see Chap. 1) and medical doctor Sophia Jex-Blake (see Chap. 7). Though producing such pioneers, the College Council, dominated by absentee clerics and peers, refused to adapt the curricula so that students could aim for degree qualifications. Instead, students graduated by means of the college's own internal examinations. As Beale wrote to Buss: "Queen's College began the Women's Education Movement undoubtedly, but it grew conservative ..."[1] Over time, the offering of advanced courses dwindled and finally ceased in the latter part of the 19th century, and the college settled into the role of a secondary school. As a result, the torch for the higher education of women was passed to Bedford College.

Bedford College

Bedford College had evolved from a set of classes held in the house of Mrs. Reid in Bedford Square. Mrs. Reid was a Unitarian, and when she founded Bedford College, initially called the Ladies' College, Bedford Square, in 1849, she wanted women to be part of the governance of the College and for the College to be non-denominational.[4] The early history was full of turmoil: to Reid's disappointment, she was not flooded with women students, and lecturers from King's College — an Anglican institution — were forced by their Principal to resign from part-time teaching at non-conformist Bedford.[5] In 1856, merger talks with Queen's College floundered, probably in part because of differences in religious outlook between the two institutions.

Three factors proved to be the turning point: first, the formation of the academic girls' secondary schools under the Girls' Public Day School Company (GPDSC — see Chap. 1), providing cohorts of academically qualified and enthusiastic young women; second, a move to more spacious facilities on Baker Street (now the Sherlock Holmes Hotel); and third, the opening of the University of London examinations to women.[5] In 1880, the first women at Bedford graduated with external University of London degrees. However, the future of Bedford College was not assured until its inclusion in the University of London Act. One of the people who had strongly pressed the case of Bedford as a constituent College was a Bedford alumna, Sophie Bryant, by then Head Mistress of North London Collegiate School (NLCS).

The Founding of the Chemistry Department

The requirements of the University of London necessitated the construction of a new wing for chemistry and physics laboratories — the first university-level ones for women in London.[6] Though lectures in chemistry had been given as early as 1877, it was the hiring of Holland Crompton[7] in 1888 as Head that really marked the founding of the Department of Chemistry.[8] One of Crompton's first tasks was the design of the new laboratory. An article in *The Girl's Realm Annual* reported: "... Further up the stairs you come across the chemistry laboratory, where Mr. Crompton is supervising a large class. Each girl wears a smock-pinafore of coloured linen."[9] By 1900, having more students than it could handle, Bedford College moved again, this time to Regent's Park, a site which it was to occupy from 1913 until the College's merger with Royal Holloway College in 1985.

In 1906, Crompton was joined by a Demonstrator, James F. Spencer,[10] a physical chemist. Spencer had several women research assistants, including Mary Crewdson (see below), Margaret Le Pla (see Chap. 13), and Eleanor Marguerite Stokes.[11] Stokes, born on 27 October 1886, was the daughter of Albert Edward Stokes, warehouseman. She had been educated at Friern Barnet High School and Dame Alice Owen's Girls' School, London, and entered Bedford College in 1905. After completing a B.Sc. degree in chemistry and obtaining a Teacher's diploma in 1908, she accepted a position as assistant science mistress at Highbury Hill High School.

The department was split in 1919, Crompton becoming Head of the Department of Organic Chemistry, while Spencer became Head of the Department of Inorganic and Physical Chemistry. Having suffered from bad health, Crompton retired in 1927, to be replaced by Eustace E. Turner[12] in 1928. It was these three individuals who provided the core of the chemistry department through the first part of the 20th century.

Though men held the senior ranks in chemistry at Bedford College, women played key roles as support staff. The first woman appointee occurred in 1898. This was Barbara 'Ally' Tchaykovsky,[13] who held the rank of Assistant Lecturer in Chemistry. Tchaykovsky was born on 26 September 1875 in New York, the eldest child of Nicholas Tchaykovsky, a Russian university professor who had fled the Tsarist regime, the family moving to London shortly after her birth. Tchaykovsky was educated at NLCS and then completed a B.Sc. in chemistry at

Bedford College in 1897. She was hired by Crompton the following year.

After working two years at Bedford, Tchaykovsky was offered a Reid Fellowship, which she used to study medicine at the London School of Medicine for Women, completing an M.B. and B.Sc. (Medicine) in 1906. In 1909, she obtained an M.D. from the University of London, being the recipient of the University Gold Medal in State Medicine.[13(b)]

For all of her working life, Tchaykovsky held a part-time position as School Medical Officer with the London County Council, refusing offers of a full-time position, as she contended it would interfere with her many social causes. During the First World War, she became active with the East London Federation of Suffragettes.[14] Tchaykovsky was forced to retire from the position of School Medical Officer at the age of 65, but kept active with her voluntary work until her death in Watford on 4 February 1956, aged 80 years.

Women Staff of the 1920s

The incumbents of the Demonstrator and Lecturer positions changed periodically, one being Mary Sumner Crewdson.[15] Crewdson was born on 23 April 1889, daughter of M. Crewdson, a Wesleyan Methodist minister. She was educated at NLCS and entered Bedford College in 1907. Completing her B.Sc. in 1910, she was a science teacher at Saltburn High School for Girls between 1910 and 1918. Returning to Bedford in January 1919, she was appointed as Demonstrator in inorganic and physical chemistry, following Spencer's promotion to Professor. Crewdson became Assistant Lecturer and then Junior Lecturer, both in 1920, and finally Lecturer in 1922.

During this time, Crewdson worked with Spencer on an M.Sc., which she completed in 1923. In 1926, her nomination for recognition as Teacher of the University of London was turned down on the grounds of insufficient research. She resigned due to ill health before her appointment was terminated, and that same year became Warden of Northcutt House and then Lindsell Hall, both residences of the College. Retiring in 1954, Crewdson died on 18 February 1966.

Two other Demonstrators during this period were Nellie Walker (see Chap. 7), who was at Bedford from 1915 to 1918 before she returned north to Dundee, and the peripatetic Phyllis McKie from 1921 to 1925 (see Chap. 12).

The most tragic story was that of Ivy Rogers.[16] Ivy Winifred Elizabeth Rogers was born on 14 February 1900, the only daughter of E. W. Rogers, a public works contractor. From Notting Hill High School (a GPDSC school), Rogers came to Bedford in 1918 as a junior laboratory assistant, being promoted to senior laboratory assistant in 1925. In her spare time, she studied for a Special B.Sc. in chemistry, completing it in 1928. It was Spencer who conveyed the news to the chemical community in 1934:

> Miss Rogers died as a result of an accident on June 20th [1934] in the Chemistry laboratories of Bedford College. ... Bedford College, by her tragic and untimely death, has lost a very willing, faithful and efficient servant and those who knew her have lost a loyal and staunch friend.[16]

The exact nature of the accident was not noted in the department files.

By 1921, the Bedford Chemistry Department had grown to a substantial size, with Crompton having two Demonstrators, while Spencer had a Junior Lecturer and two Demonstrators. One of the Demonstrators with Spencer was Ruth Drummond.[15] Drummond, daughter of a minister, attended South Hampstead High School (a GPDSC school) and then NLCS before entering Bedford in 1915. For four years after getting her degree, she was a Demonstrator in the Chemistry Department at Bedford, after which she acted as temporary Demonstrator and Warden when help was needed. Drummond then went to the University of Birmingham as a Warden at the women's residence, Ashbourne

Hall. By 1938, she was listed as being an H. M. Inspector of Factories. She died in December 1975.

A later Demonstrator was Margaret Elsie Snowden Appleyard.[17] Appleyard was born on 7 January 1908 in Huddersfield and was educated at Greenhead High School, Huddersfield. She entered Newnham College in 1927, completing the degree requirements in 1931, and then undertaking research in photochemistry with Ronald Norrish[18] during the following year. In 1933, she was appointed as Demonstrator at Bedford, while also undertaking research with Spencer. She returned to Cambridge in 1936 to work on an M.A. degree. She left Cambridge again in 1937 to join the Asphalte and Bitumen Company, Preston.

The following year, Appleyard crossed the Atlantic to accept a one-year appointment at Mount Holyoke College, Massachusetts. In 1939, she was hired by Wellesley College, Massachusetts, where she taught until 1942, interrupted by periods of illness. In 1942, Appleyard took up a position at Williams College, also in Massachusetts, where she was one of the first women to serve on the faculty. It was here that she died tragically at the early age of 36, her death being ascribed to a sudden heart attack.

However, the long-time staff members at Bedford were Mary Lesslie, an organic chemist, who started in 1927, and Violet Trew, a physical chemist, in 1930. The pair were the anchors of the Bedford chemistry department for the next 40 years (see Fig. 4.1).

Mary Lesslie

Mary Stephen Lesslie,[19] daughter of Andrew J. W. Lesslie of Dundee, was born in 1901 and educated at Morgan Academy, Dundee. She was only 16 years old when she entered University College, Dundee (now the University of Dundee), graduating with an M.A. (St. Andrews) in 1922, and a B.Sc. (St. Andrews) in 1924. She then completed a Ph.D. (St. Andrews) with Alexander McKenzie (see Chap. 7) on stereochemistry in 1927, her results being published in two papers.

Lesslie immediately left Scotland to take up an appointment as Demonstrator at Bedford College. She carried with her from Dundee "a trunk containing not only her personal effects and books but a vacuum desiccator, the first seen in the Department."[6] A significant proportion of the organic chemistry teaching fell on Lesslie's shoulders, particularly in her first year when Crompton was in failing health. With the arrival of Turner, the load was eased slightly and, in addition, Lesslie and Turner formed a research partnership that blossomed over the following decades. During the period 1928–1956, Lesslie co-authored 21 research publications with Turner, nearly all on organic stereochemistry involving the resolution of isomers.

Lesslie's initial appointment at Bedford terminated in 1931 and was nonrenewable. With her research flourishing, she continued to work unpaid during the day at Bedford, taking evening employment elsewhere in order to survive. Fortunately, in 1932, she obtained a re-appointment as a Junior Lecturer, and subsequently rose to the rank of Senior Lecturer in 1947. Appointed a Recognised Teacher of the University of London, Lesslie was awarded a D.Sc. (London) in 1950.

During the Second World War, the staff and students of Bedford were evacuated to the University of Cambridge. Teaching duties became even more onerous for Lesslie, as the degrees had to be completed within two years. In addition, she and Turner, together with Margaret Jamison (later Mrs. Harris), undertook a program of syntheses of pure hydrocarbon isomers, mostly using Grignard reactions, for the Ministry of Aircraft production.[6] Lesslie's specialty was getting recalcitrant Grignard reactions to work. In her obituary, this focus of her life was noted: "As ever, it was in the laboratory that the Organic staff gathered for coffee: on one bench there would be a boiling kettle, and on the bench opposite a bubbling Grignard reaction, and something very pure and clean crystallising in slow perfection."[19] The infrared spectra of each isomer was then obtained, much of the analysis being undertaken by Delia Simpson (see Chap. 6). The purpose was to take samples of fuel from captured aircraft and, comparing the infrared spectra with Lesslie's reference samples, endeavour to identify the fuel source by determining isomer ratios.

In 1954, Lesslie was appointed as Dean of Lindsell Hall, following Crewdson's retirement. Her new duties were in addition to those of lecturing, running laboratory sessions, and supervising a Ph.D. student. During vacations, Lesslie would hurry north to Dundee to care for her elderly widowed father and to see her nephews and nieces. She was periodically asked by students why she herself had never married, and her response was always: "It was not for want of asking!"[19] Her attitude is quite typical for the period in that she saw herself as dedicating her life to the cause of chemistry and to her students.

Lesslie retired to a small house in Dundee in 1968 where she survived for another 20 years, dying on 24 February 1987 at age 86. Her obituarist quoted Lesslie's former Ph.D. student, Yvonne Bernstein, as saying:

> Her knowledge was encyclopædic and she shared it with her students unstintingly. She taught me patience when my instinct was to take short cuts or give up too easily. She had high principles as befitted her Calvinistic upbringing but combined these with tolerance and a marvellous sense of humour.[19]

Violet Trew

Violet Corona Gwynne Trew,[15] born on 26 June 1902 to A. N. Trew (mother), was educated first at St. Olaf's, Beckenham, and then at James Allen's Girls' School, Dulwich. She completed her B.Sc. at Bedford in 1926 and a Ph.D. in 1928, becoming Bedford's first internal doctoral degree in chemistry.

After two years of part-time positions, Trew was appointed as Junior Lecturer at Bedford in 1930. In 1933, she was promoted to Lecturer, and then to Senior Lecturer in 1949. The teaching of inorganic and physical chemistry largely fell to her.

When Margaret Jamison arrived at Bedford as an undergraduate, she commented:

> On our first morning in 1932 we assembled in the inorganic laboratory, each to be allotted a cupboard (personal territory) and given, by Dr V. C. G. Trew, a tube containing a powdered mixture whose composition we had to identify. We set about a regular series of tests; qualitative investigations such as fusion on a charcoal block using a blowpipe to get maximum heat, or bending platinum wire and forming a borax bead whose colour, after fusing with our mystery mixture, could (with other diagnostic tricks) lead to the identification of four or five components.[6]

The physical chemistry lab was also under Trew's command, the experiments being performed on Saturday mornings.

The 1930s were a period when spiritualism and all manners of occult sects thrived. Trew became involved with the Theosophical Society and the occult chemistry of Annie Besant.[20] Besant and her group believed that in a semitrance, it was possible to make oneself infinitesimally small and visit atoms on a voyage of discovery. In fact, she and her group reported the shapes of each atom as they saw them. As Jamison commented:

> Dr. Trew had another interest which proved pastorally helpful later. She was an early member of the Theosophical Society (she even inherited Annie Besant's typewriter). When cults began to capture the minds of innocent students who were diverted from their work, Miss Trew was able to produce a useful folder of relevant information.[6]

Trew's research area was magnetochemistry, leading to 16 publications. The early research papers were co-authored with Spencer, while much of her later work was published with her as sole or senior author. In 1937, Trew was granted the honour

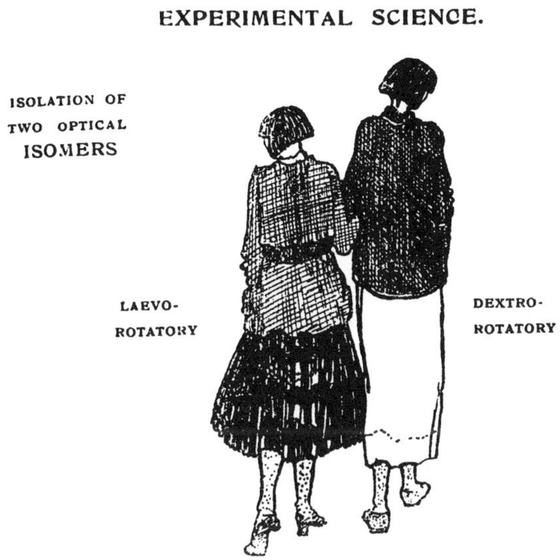

Fig. 4.1. An engraving from *The Bedford College Union Magazine* (#2, March 1922, p. 15), which may be of Lesslie and Trew.

of Recognised Teacher of the University of London; and then, in 1955, she was awarded a D.Sc. (London). Trew retired in 1969, and died in 1995 at the age of 92.

The Next Generation of Women Staff

The two stalwarts of the next generation, both alumnae of Bedford, were Margaret Manderson Jamieson (Mrs. Harris) and Dorothy Muriel Hall (Mrs. Hargreaves).[6] Margaret Jamieson[21] was born on 3 February 1914, the daughter of a Head Master with the London County Council. She attended Putney High School (a GPDSC school) and entered Bedford in 1932, completing her B.Sc. in 1935 and a Ph.D. with Turner in 1937. In 1938, Jamieson was hired as a Demonstrator, replacing Ishbel Campbell (see Chap. 5), who had a brief sojourn at Bedford from 1936 to 1938. Over the following decades, Jamieson was promoted first to Assistant Lecturer, then Lecturer, and finally Reader. She co-authored at least 30 research publications, the early papers with Turner, and the later papers with various different co-authors including Lesslie and Hall. She also co-authored the textbook *Organic Chemistry* with Turner in 1952.[22] In 1945, she married Rev. H. L. Harris, who taught at Harrow School, and had two children. Jamieson retired in 1979, and died on 26 January 2002, aged 87 years.

Dorothy Hall[15] was born on 20 March 1920 in Walthamstow. Daughter of F. H. Hall, a company director, she was educated at Wanstead County High School. She followed the same path as Jamieson, with a B.Sc. at Bedford in 1941, and completed a Ph.D. with Turner in 1944. Again, like Jamieson, she was promoted first to Assistant Lecturer, then Lecturer, and finally to a Readership in 1957. Hall was author or co-author of 32 publications, mostly with Turner, and a few with Lesslie or Jamieson as co-authors. She married Michael Hargreaves on 21 December 1957 and had one son. Hall retired in the mid-1980s.

Women Chemistry Students

Life at Bedford in the early decades was unlike the universities of today. A dress code was enforced; for example, about 1915, during weekdays students were required to wear "day dress," a jacket-suit or a dress and jacket in a dark colour or, in summer, a dark skirt and light blouse (ties were commonly worn). Dinner was a specific ceremonial occasion, as a student in 1915, Margaret McDonald, recalled:

> Staff and students forgathered in the Common Room and in order of precedence each chose a partner, senior offering an arm to juniors. Any fresher left over bought up the rear of the procession, sometimes with relief at not having to make conversational efforts.[23]

The early volumes of the *Bedford College Magazine* provide an insight into the enthusiasm for chemistry that prevailed at the institution. In particular, some of the articles highlight the differences from today's chemistry. During those early years, danger was regarded as an intrinsic part of laboratory life, as the following commentary from an 1898 issue shows:

> A certain amount of excitement was caused one afternoon by the fact that one student was suddenly seen to be in flames. However, she lay down quite calmly, and was immediately knelt upon by her nearest neighbours, so that all danger was over before most people knew what had happened and they only caught a glimpse of her as she lay 'smiling and smouldering' on the floor. We consider the behaviour of those concerned a credit to the College and to the cause of women's education.[24]

Explosions also seemed commonplace, according to a subsequent remark in 1900: "At present, one's life is a series of adventures in the Chemistry Laboratory, for bits of flying glass are as plentiful as the smuts on the window-sill."[25] Drinking and eating in the lab were taken for granted: "B.Sc.'s on the whole, seem to enjoy life in all places and at all times, be it at their work, or having tea-fights in the Chemistry laboratory, when beakers take the place of cups, and flasks do for kettles. Some say that tea-cakes cooked on asbestos have a very savoury taste."[25]

Chemical spillage was also prevalent: "The final chemistry students have been performing original research on the action of saturated solution of caustic potash on their books and garments. It has been found that the result is the destruction of the parts acted on."[26] And the unique appearance of chemistry students was observed: "This science [chemistry] indeed leaves a hall-mark upon its devotees. They can usually be distinguished by the moth-eaten appearance of their clothes and the peculiar colour of their hands."[26]

The practical work involved both volumetric and gravimetric analysis, as an anonymous author complained:

> Practical work has been confined chiefly to volumetric analysis, which has, alas! almost landed some of us in Colney Hatch

> [Lunatic Asylum]. Weight pincers are frequently used by distracted students for the purpose of extracting chemicals from bottles. We must put this absent-mindedness to the exciting times we live in.[27]

Qualitative work was accomplished in part by Fresenius's analysis scheme using hydrogen sulphide, or sulphuretted hydrogen, as it was then called. In the annual report for 1906, the purchase of a new Kipp's apparatus was announced:

> We have, I think, somewhat diminished the odour of sulphuretted hydrogen by the substitution of a new generating apparatus instead of the old one, which always seemed to be out of order; though, poor thing, with so many calls on its attention, it is perhaps no wonder that it broke down, when old age made it feeble.[28]

Chemistry had the largest enrolment among the sciences.[29] In addition to the classes and laboratory sessions, there were weekly Chemical Society meetings. A student by the pseudonym of "C=M" was obviously critical of a 5 p.m. meeting held in 1905, at which 18 papers were presented:

> The victims of the evening invariably retire from the laboratory to remove traces of the day's labour from their hands five minutes before five. ... There is no lack of information in the papers, but the authors generally consider it incumbent upon them to deliver it at such a rate that the audience is left almost too exhausted to applaud.[30]

Edith Humphrey

Among the Bedford College chemistry alumnae was the forgotten pioneer of co-ordination chemistry, Edith Ellen Humphrey.[31] Born in 1875, she attended NLCS, and it was there that Humphrey first took chemistry: "At the North London we did quite a bit of it. We had a good teacher who had no degree but who was very good."[31(b)] Her father encouraged all of his children, even the girls, to obtain a good education, and Humphrey entered Bedford College in 1893. Humphrey continued with chemistry, as she considered it to be "where things were happening."

Following the completion of her B.Sc. degree in 1897, she applied and was accepted to do a Ph.D. at the University of Zürich. Zürich first admitted women in the mid-1860s, and it had become a haven for women students from all over Europe.[32] Humphrey wrote a long essay on her life at Zürich, which was published in the *Bedford College Magazine*.[33] For the research, students had to provide their own apparatus and chemicals. Her doctoral program was four years in length, as opposed to the English system of the time, which allowed completion of Ph.D. in as little as two years.

It was her thesis work with Alfred Werner[34] that was to enter her in the annals of the history of chemistry. Among the compounds she made was *cis*-bis(ethylenediamine)dinitrocobalt(III) bromide. This was the very first synthesis of a chiral octahedral cobalt complex, though at the time the significance of her synthesis was overlooked.[31(a)] Werner was so impressed with her work that, for her last year, he took her on as his personal assistant, the first women to be chosen for this prestigious post.[35] More important for the impoverished Humphrey, she at last had some income in very expensive Switzerland.

Her Ph.D. finished in 1901, it was recommended that Humphrey continue her studies with Wilhelm Ostwald[36] in Germany. There was a problem, as Humphrey herself commented: "But they wouldn't have me in Germany. They said I could go to lectures but not practicals because the men wouldn't do any work."[31(a)]

Thwarted in her plans, Humphrey returned to England, spending the rest of her working life as a research chemist with the company Arthur Sanderson & Sons, who specialised in such products as fabrics and wallpapers. It was during her time with

Sanderson that she became a signatory of the 1904 petition for admission of women to the Chemical Society and of the 1909 letter to *Chemical News* (see Chap. 2). She died in 1977 at the age of 102. The chiral crystals, synthesised by Humphrey, are now in Burlington House, Piccadilly, London, as they were donated in 1991 to the Royal Society of Chemistry on the occasion of its 150th anniversary by the Swiss Committee on Chemistry.

Rosalind Henley (Mrs. Pitt-Rivers)

Rosalind (Ros) Venetia Henley,[37] born on 4 March 1908, developed her love of chemistry at the age of 12, when she was given a chemistry set by an uncle:

> This present proved to be such a success that she, along with her cousin Ed, ... was given the stables ... to be used as a laboratory. The inhabitants of the estate were not only alarmed by the smells that emanated from there but remember at least one explosion that came quite close to a disaster.[37(a)]

Her secondary education was at Notting Hill High School (a GPDSC school), followed by Malvern Girls' College. However, her science teacher pronounced that: "I fear Rosalind will never make a chemist."[37(a)]

At Bedford, Henley completed a B.Sc. in 1930 and an M.Sc. in 1931. That year, she married George Lane-Fox Pitt-Rivers, and the following year, her son, Anthony, was born. The marriage was not a happy one, as George Pitt-Rivers became increasingly eccentric and developed a fanatical enthusiasm for fascist politics. In 1937, Henley left the family home and an acrimonious divorce ensued. Pitt-Rivers' loss was science's gain. Henley commenced a Ph.D. at University College Medical School on methyl glucosaminides and their hydrolysis by snail enzymes.

Henley's research supervisor was Charles Harington,[38] and he proved to be just the sort of mentor that Henley needed.

Following the completion of her Ph.D. in 1939, Henley accompanied Harington on his move to the National Institute for Medical Research (NIMR) in Hampstead. Her field of specialisation was iodoproteins, and her research on them over the rest of her career was to garner her worldwide fame. The Second World War disrupted her research, as she was seconded to various war-work projects. The last of these, helping the survivors at the liberation of a concentration camp, proved so traumatic that she became a heavy smoker, the probable cause of her death 25 years later. In 1952, she isolated the thyroid hormone, triiodothyronine. The discovery of this new and highly unusual hormone contributed largely to her election as Fellow of the Royal Society only two years later.

Henley continued working on iodine-containing biomolecules until her retirement in 1972, co-authoring 61 papers and two books: *The Thyroid Hormones*[39] and *The Chemistry of Thyroid Diseases*.[40] Her research style had always been one for small groups, her own never containing more than four people at a time. This practice of "small science" enabled her to continue to be active at the research bench throughout her career — her favourite place to be. Her health started to deteriorate in 1985, and she died in 1990.

Royal Holloway College

Whereas Bedford College was located in the middle of London, Holloway College, later Royal Holloway College (RHC), was built in 95 acres of Surrey countryside at Egham, though with easy rail access to London. It was Jane, wife of Thomas Holloway, who suggested that part of Holloway's fortune be used to construct a college for women which would be "... founded on those studies and sciences which the experience of modern times has shown to be the most valuable, and best adapted to meet the intellectual and social requirements of the students."[41] The College, modelled on the Chateau at Chambord in the Loire Valley, was formally opened in 1886.

Whereas Bedford was partially a day college, RHC was entirely residential. Bedfordians thought of themselves as sophisticated "girls about town" and considered their "sisters" at RHC to be mere "girls," protected in a boarding-school atmosphere.[42] In reality, the students at both Colleges came from a similar range of middle-class and lower-middle-class backgrounds.

Though much closer to London, Royal Holloway's original connections were with the University of Oxford. It was at a Conference in December 1897 that 52 educators, including 22 women, debated the future of the College.[43] Three options were put before them: to make Holloway College an independent degree-granting institution; to initiate a Women's University of which Holloway would be a part; or to merge into the University of London. The first path was quickly rejected as too risky; the second, after much debate, was dismissed on the grounds that it would be a "rash experiment." This left the third option. With Bedford as a role model, the educators overwhelmingly opted for incorporation with the University of London.

E. Eleanor Field

The Department of Chemistry had been founded in 1891 by M. W. Robertson, but it was Elizabeth Eleanor Field[44] who dominated the department for two decades. Field was educated at Newnham College, graduating in 1887. From 1889 to 1890 she was an Assistant Demonstrator in Chemistry at Newnham, and from 1891 to 1893 held a Bathurst studentship, carrying out research under Matthew Moncrieff Pattison Muir.[45] After leaving Cambridge, she held a post as assistant mistress for two years (1893–1895) at Liverpool College for Girls, and then became Lecturer and Head of Chemistry at RHC.[46] Women Heads were titled "Senior Staff Lecturer," while the male faculty of the same rank were titled "Professor."

One of the signatories of the Petition for the admission of women to the Chemical Society (see Chap. 2), Field remained at RHC for 19 years. Her RHC obituary noted: "She was not herself a great conversationalist; her thought perhaps moved too slowly for sparkling repartee, but to listen to her slowly expressed but lucid exposition of some problem in Chemistry, or even to hear her tell a story, was an intense pleasure and a liberal education."[46(a)]

After her retirement in 1913, Field lived near the College until 1922, when she "... elected to follow the fortunes of her cousin and ward to [Brno] Czechoslovakia."[46(a)] It was there that she died on 17 November 1932.

Mary Boyle

Mary Boyle,[47] like Field, another petitioner for admission to the Chemical Society, was also a stalwart of the Chemistry Department at RHC. She was born in 1874 and entered the Royal College of Science in 1898. Boyle took the full three year course for chemistry before transferring to RHC in 1901, where she gained her B.Sc. after one year of study. She remained at RHC, firstly as a Demonstrator, then Assistant Lecturer in 1906, rising to Staff Lecturer.

In 1910, only four years after being appointed as Lecturer, Boyle received her D.Sc., the subject of her research being the iodosulphonic acids of benzene. One of her obituaries noted:

> This extensive research, the work of many years, was published in four large instalments from 1910 to 1919 [all under her name alone]. Five of the six theoretically possible diiodosulphonic acids, three of the six theoretically possible triiodosulphonic acids, and one tetraiodosulphonic acid, were first prepared by her, and incidentally she prepared a large number of new nitro- and amino-sulphonic acids. The orientation of these substances was worked out, their important derivatives were accurately characterised, and the electrolytic behaviour of the more important members was investigated. Our knowledge of this group of substances still rests mainly on Dr. Boyle's work. After 1920, the expansion of the teaching

> work in the Chemistry Department left her little time for research ...[47(c)]

Boyle retired in 1933, moving to Leeds to be near her family, particularly her many nephews and nieces and their children, and she became active in community service. She died in November 1944.

The Succession Controversy

Up until Field's resignation, the Chemistry Department had always been in the hands of women chemists. However, her upgraded replacement position of a University Professorship of Chemistry, tenable at the RHC, was advertised in the 15 March 1913 issue of *The Times* as open to men and women on equivalent terms. Sibyl Widdows (see below) sent a letter to many of the former RHC students expressing her opposition:

> I and several other Old Students feel rather strongly that this post which has been in the hands of women ever since the opening of the College, and which has been markedly efficiently run, ought not to pass out of their hands without some strong reason. Of course we do not want any woman to be given the post, but we think that if a woman of sufficient standing and ability applies it should be offered to her. It is becoming a very serious thing for science women the way in which the science posts in women's Colleges are gradually being placed in the hands of men when there are quite good and efficient women to fill them. It means that a woman can never obtain those opportunities for research and association with other scientists which are so necessary for their work.[48]

Accompanying this letter was a petition, pointing out that of the nine members of the Selection Committee for the post, only one was a woman. Among the members of the Vigilance Committee elected by the Royal Holloway College Association to promote the hiring of a woman Professor was Margaret Seward (Mrs. McKillop) (see Chap. 3).

A counterpetition was then sent by other RHC graduates, arguing that so long as there was a reasonable proportion of women on the staff at RHC, it was undesirable to reserve any specific post for a woman. It also argued that to require a woman to be appointed implied that there were no women candidates capable of matching those of male chemists. In response to the original petition, 30 expressed the opinion that no gender preference should be given; 207 considered that preference should be given to a woman, of whom 156 added the caveat, "subject to equal qualifications"; while 48 contended that the position had to be given to a woman.

The Governors of the College continued with their original plan, and George Barger[49] was appointed as the first Professor of Chemistry at RHC. Barger held the position for only one year before being appointed as Research Chemist under the wartime Medical Research Committee. Nevertheless, he had time to co-author two research publications with Field, one in 1912 and another in 1915. Barger was succeeded as Head by T. S. Moore,[50] who held the post from 1914 until 1946. One of Moore's obituaries was written by a former student, Rosa Augustin, and she added: "Teaching in a woman's college and with a daughter of his own (as well as a son), T.S. was immensely interested in the education of women and was delighted when we did well in our careers."[50(a)]

Millicent Plant

In 1933, Moore was joined by Millicent Mary Theodosia Plant.[51] Plant had received her B.Sc. from the University of Birmingham, where she had worked with Norman Haworth[52] on carbohydrates. While Moore taught the inorganic and physical courses, Plant took over responsibility for teaching organic chemistry. These two constituted the entire chemistry staff at RHC until the end of the Second World War.

Plant embarked on a research programme at RHC, her major publication being on aldol condensation products. However, her interests and influence spread far beyond the chemistry laboratory, as an article on her retirement noted:

> There are few sides of college life which did not feel the shaping pressure of her serious interest, for she thought of life in a resident college as a whole and was concerned for the total well-being of the students. We remember her against many backgrounds as well as the familiar one of the Chemistry Laboratory; in Choral, in Chapel; on the tennis-courts and in the swimming-bath; at Staff Meetings; on her potato-patch at Highfield — this unsorted and far from exhaustive list suggests the variousness of her gifts and interests.[51]

In 1947, Plant married Alfred Georg, resigned from her position, and moved with him to Geneva. It would seem possible that she returned to England, as in 1954 she co-authored a publication on an algal polysaccharide under the name of M. M. T. Georg-Plant.

Women Chemistry Students

The chemistry laboratories, designed by the chemist Vernon Harcourt, did not open until 1889. Martha Whiteley (see Chap. 3), a chemistry student at RHC at the time, noted: "... there were only four science students, and no laboratories. They used three rooms in the North Tower, which to this day have sinks and taps in them, and did Chemistry, Physics or Botany according to the way the wind blew, because one chimney always smoked."[53]

Student life at RHC was very regimented in the early years, as one student, a Miss Dabis, recounted in 1888:

> The order of the day is as follows:- A bell is rung at 7 a.m., when a servant brings around hot water. At 7.55 a short service is held in the chapel, which is absolutely compulsory, and

> is begun most punctually... Breakfast follows at 8.20 in the large dining hall, where three long tables are spread with good and plentiful provisions. Lectures begin at 9, and last till 1, five minutes being allowed after each lecture, and at 11 there is an interval of ten minutes for letters and lunch. Luncheon at 1 o'clock is a slightly informal meal, to which people straggle in when they please, and dispense with servants. The afternoons are given up to recreation, and are generally spent at the tennis ground, which consists of three asphalt and ten grass courts. At 4 o'clock comes tea, which the students generally partake of in small parties, each one bringing her own particular little tray, with teapot, &c., from the pantry. They then work from 4.45 to 6.45, dine at 7, and, after evening prayers, seek "Nature's sweet restorer, balmy sleep."[54]

As at Bedford, dinner was an important ritual, though at RHC it was organised slightly differently, as Marion Pick recalled from the 1902–1907 period:

> At the beginning of each term we filled our dinner books with the names of our partners for every night of the term; the ceremonial for this changed as the years went on and the last I heard was the First Years were ranged in the Picture Gallery underneath the 'Marriage Market.' Each night at 6.55 p.m., we assembled in lines down the Library in order of our years and pair by pair, subsequently joined the procession led by the Principal or Senior Resident, which passed from the Drawing Room. .... The staff with their partners took the heads of the tables and a student marshal then saw the rank and file filled the tables in due order. After grace, an ample domestic staff waited on us and our business was to converse, which we did.[55]

London School of Medicine for Women

With women barred from admission to British medical schools, Sophia Jex-Blake, Elizabeth Garrett Anderson, Emily Blackwell,

Elizabeth Blackwell, and Thomas Henry Huxley established the London School of Medicine for Women (LSMW) in 1874.[56] The School began in a small house on Henrietta Street (now Handel Street) with sympathetic male doctors teaching the women. The radical lawyer William Shaen (see Chap. 3) agreed to act as solicitor for the School. Shaen is another of the forgotten heroes of women's educational rights, having also assisted in the foundation of Bedford College and having been connected with the campaign for women's entry to Cambridge University.

In 1877, the Royal Free Hospital was persuaded to admit female students to its wards, enabling the women to obtain clinical practice. The arrangement between the School and the Hospital worked so well that in 1898 the two were merged to become the London (Royal Free Hospital) School of Medicine for Women. By 1914, the school had 300 students, and in its first 40 years the LSMW graduated 600 women, 60% of the total women doctors in Britain. In the 1920s, the LSMW again became a haven for women medical students as more and more of the other British medical schools either closed their doors completely to women or made life intolerable for them.[57]

In a letter to the magazine *Time and Tide* in 1922, Winifred C. Cullis and Sibyl T. Widdows reminded readers of the role of London (Royal Free Hospital) School of Medicine for Women:

> This hospital [The Royal Free Hospital], it should never be forgotten, was the very first to open its doors to women; it opened them not when it was a popular thing to do during a great European war, but when every other teaching hospital closed its doors and kept them closed until nearly forty years later. This hospital at the present, as in the past, is doing more for the medical education of women than any other teaching hospital.[58]

Lucy Boole

The courses at the LSMW included theoretical and practical chemistry, and the Department of Chemistry was staffed by a series of talented women chemists. The first woman to hold a chemistry position at the LSMW was Lucy Everest Boole,[59] a signatory of the petition of 1904 for the admission of women chemists to the Chemical Society. Boole, born on 5 August 1862 in Cork, was one of five children (all daughters) of the famous mathematician, George Boole, and Mary Everest. Each daughter was highly talented, in particular, Alicia (Alice) Boole (Stott), who became a distinguished mathematician.[60] Despite little formal education in her early years, Lucy Boole was a student at the School of the Pharmaceutical Society from 1883 to 1888 (see Chap. 10). After completion of the examinations, she became a researcher at the Pharmaceutical Society Laboratory.[61]

In 1891, Boole was appointed as a Demonstrator in Chemistry at the LSMW under Charles William Heaton.[62] Very shortly after, due to Heaton's ill health, she was called upon to take over his duties; and on his resignation in 1893, she was formally appointed Lecturer. Louisa Martindale, a student at the LSMW, recalled:

> The chemistry laboratory, presided over by Miss Lucy Boole was a poor one. We much revered Miss Boole on account of her clever mother, a mathematician who had 'proved the divinity of Christ by mathematics', and of her sister who had written the *Gadfly* and then married a Nihilist.[63]

The following year, Boole was elected as the first woman Fellow of the Institute of Chemistry. Unfortunately, in 1897, deteriorating health resulted in her submitting her resignation; however, to keep her, the Council of the School divided the duties of the position and appointed Boole as Teacher of Practical Chemistry. She held this position at the time of her death on 5 December 1904 at the age of 42.[64]

Clare de Brereton Evans

To take up the remainder of Boole's load, Clare de Brereton Evans[65] was appointed as a Lecturer in Chemistry. Evans had been educated at Cheltenham Ladies College (CLC) and had also obtained a B.Sc. (London) in 1889 while at CLC. Following graduation, she had undertaken research at the Central Technical College with Armstrong, from which, in 1897, she was to be the first woman granted a D.Sc. degree, her research being on aromatic amines. Evans was another signatory of the 1904 petition for admission of women to the Chemical Society and of the 1909 letter to *Chemical News* (see Chap. 2).

During her time at LSMW, Evans undertook part-time research at University College (UCL) as part of William Ramsay's group. She only held the Lectureship at LSMW until 1912, resigning so that she could devote her time completely to research at UCL.[66] One of her publications[67] describes her attempt to separate an unidentified element from iron residues supplied by Ramsay; and of note, she thanked Katherine Burke (see Chap. 3) for the analytical studies, hence providing another link among the signatories.

Sibyl Widdows

It was Sibyl Taite Widdows[68] who was to dominate the Chemistry Department at LSMW for 40 years. She was yet another signatory of the petition for admission of women to the Chemical Society and the 1909 letter to *Chemical News* (see Chap. 2). Widdows, born on 27 May 1876, had been educated at Dulwich High School for Girls (a GPDSC school), and then obtained a degree in chemistry at RHC in 1900. In 1901, she was appointed as Demonstrator in Chemistry at LSMW. Over the years, she progressed to the rank of Lecturer, during which time she authored or co-authored at least 12 research papers. Interestingly, her second paper involved research with Mills, spouse of Mildred Gostling (see Chap. 11), while her third paper was co-authored with Ida Smedley (see Chap. 2). Widdows was placed in joint charge of the Chemical Department in 1904, and became Head in 1935.

In a lengthy obituary, her successor, Phyllis Sanderson, noted how Widdows had been "one of the last of the remarkable women who staffed the School during the first forty years of this century, an uphill and critical period in the history of this medical school."[69] Sanderson added: "As so many of her contemporaries, she was an ardent feminist and willingly sacrificed her own career as a chemist for the cause most dear to her heart, the training of women doctors at Hunter Street, the only training ground in Medicine open to women in England at that time."[69]

Life in the Chemistry Department at LSMW was also described by Sanderson:

> Of miniature stature, alert and sprightly, Miss Widdows possessed such vitality and drive that it seemed a store of dynamite must be housed within her small frame. As with all who have a gift for it, she loved teaching and did so with untiring verve, never despairing even of the lowest of her flock. ... Practical classes, certainly no playtime, held an element of excitement (possibly mixed with terror) that kept everyone on their toes; for S.T.W. would systematically work her way down the laboratory, visiting student after student to ensure that each in turn was fully understanding what they were doing. Suddenly a loud scream of dismay would ring out and all would shudder, knowing full well that some unfortunate student had uttered an appalling chemical howler or had committed some dangerous crime such as heating an inflammable liquid with a naked flame. Near neighbours of the offender would immediately rush off to recharge their washbottles or busy themselves at the fume-cupboard hoping (in vain) to escape the deadly searching questions so soon to reach them. Just as frequently there were roars of laughter at the odd joke or cries of triumph as she found one of her flock had at last understood some basic chemical principle.[69]

Widdows retired in 1942 and died on 4 January 1960.

Phyllis Sanderson

Phyllis Mary Sanderson[70] also spent 40 years teaching at the LSMW. Sanderson, daughter of a medical doctor, was born in Hove, Sussex, in 1901, and completed her B.Sc. in chemistry at UCL in 1924. After a year's postgraduate research at the Children's Hospital, St. Vincent's Square, which resulted in her first publication, she was appointed as Demonstrator at LSMW in 1925.

During the 1930s, Sanderson investigated industrial dusts, including chemical aspects of silicosis in miners, at Imperial College, research for which she was awarded a Diploma of Imperial College. During the years 1936 and 1937, she co-authored at least 11 publications on industrial dusts with Vincent Briscoe,[71] Janet Brown (Mrs. Matthews), and P. F. Holte; and then she returned to UCL for her Ph.D., which was granted in 1939.

During her time at the LSMW, she rose through the ranks to Senior Demonstrator in 1933, Assistant Lecturer in 1934, and Lecturer in 1946. She undertook some research in heterocyclic chemistry, but her interests changed to research in the history of science, as her obituarist commented:

> Dr Sanderson loved to delve into the history of chemistry and scientific thought in general. It was typical of her sense of justice that in one of these studies she should have rescued from oblivion a hitherto obscure 18th Century scientist William Cruickshank, by re-establishing his claims to several important discoveries that had been erroneously ascribed to another investigator. Not the least of the results of these efforts was her familiarity with all the great libraries of London. Her lively tales of the peculiarities of their arrangement and procedure, and the idiosyncrasies of both librarians and readers were a source of much amusement to her friends.[70(b)]

Sanderson died on 7 September 1965.

Other Women Chemists

During Evans' Lectureship, the post of Demonstrator was held by Norah Ellen Laycock.[72] Laycock obtained her B.Sc. from RHC, graduating in 1901. She was a Demonstrator in Chemistry in 1906, the position she stated as signatory of the letter in 1909 to *Chemical News* (see Chap. 2). In 1916, her appointment became a joint one between the Chemistry and Biology Departments, and the following year, solely as a Biology Demonstrator. She remained in this position for 25 years before resigning to take up the appointment of Headmistress of Mayertorne School, Wendover, Buckinghamshire.

Laycock was succeeded in 1916 as Demonstrator in Chemistry by May Williams.[73] Williams, born on 7 May 1886, was the daughter of Ralph Williams, minister of Maida Vale, London. She was educated at Notting Hill High School (a GPDSC school), and then obtained a B.Sc. in Chemistry at RHC. Williams was first appointed to the rank of Demonstrator, and then was promoted to Senior Demonstrator in 1920 and to Assistant Lecturer in 1921, the year she received her M.Sc. degree. She retired in 1946.

Another graduate of UCL, Anne Ratcliffe,[74] followed Sanderson in her appointments. Ratcliffe was born in 1896 in London, and completed her B.Sc. degree in chemistry in 1924. Her appointment as Demonstrator in Chemistry at the LSMW commenced in 1929, with her promotions being a few years behind those of Sanderson: Senior Demonstrator in 1940 (a year after obtaining her M.Sc. degree); Assistant Lecturer in 1945; Lecturer in 1947; and finally, Senior Lecturer in 1949.

Commentary

So much has been forgotten of the early history of the London women's colleges. The initial role of Queen's College has been overlooked, in part by its relegation to high school status by its blinkered and conservative administrators. With Bedford now absorbed into Royal Holloway, and chemistry no longer taught in the combined College, the past glories of their many chemistry graduates have vanished from the collective memory.[75] The women's colleges also acted as one of the few prospects of academic employment for women chemistry graduates in the early decades of the 20th century.

Just as the memory of the women chemists of Bedford and Royal Holloway Colleges have been forgotten, so has the important role of the London School of Medicine for Women prior to its absorption into the Royal Free Hospital. Not only did it provide a reliable conduit for the education of women doctors, but its chemistry department employed a series of talented women, a significant proportion of the chemistry graduates of RHC.

Notes

1. Kaye, E. (1972). *A History of Queen's College, London 1848–1972.* Chatto & Windus, London.
2. Grylls, R. G. (1948). *Queen's College 1848–1948.* Routledge & Sons, London, p. 16.
3. Gordon, S. C. (1955). Studies at Queen's College, Harley Street, 1848–1868. *British Journal of Educational Studies* **3**(2): 144–155.
4. Watts, R. E. (1980). The unitarian contribution to the development of female education, 1790–1850. *History of Education* **9**(4): 273–286.
5. Pakenham-Walsh, M. (2001). Bedford College 1849–1985. In Crook J. M. (ed.), *Bedford College: Memories of 150 Years,* Royal Holloway and Bedford College, pp. 13–45.
6. Harris, M. M. (2001). Chemistry. In Crook, J. M. (ed.), *Bedford College: Memories of 150 Years,* Royal Holloway and Bedford College, pp. 81–94.
7. Spencer, J. F. (1932). Obituary notices: Holland Crompton. *Journal of the Chemical Society* 2987–2988.
8. Turner, E. E. (1953). Schools of chemistry in Great Britain and Ireland–XVII Bedford College, London. *Journal of the Royal Institute of Chemistry* **79**: 236–238.
9. Rawson, S. (1899). Where London girls may study: I. Bedford College. *The Girl's Realm Annual* 925–929.

10. Anon. (1951). Obituary. James Frederick Spencer. *Journal of the Royal Institute of Chemistry* **75**: 127.
11. We thank Vicky Holmes, Archivist, Royal Holloway College, for information on Stokes.
12. Harris, M. M. (1966). Obituary: Eustace Ebenezer Turner (1893–1966). *Chemistry and Industry* 1953–1955.
13. (a) Anon. (3 March 1956). *British Medical Journal* 524; (b) Anon. (n.d.). Pioneer of the Child Welfare Service: Death of Dr. Barbara Tchaykovsky. Unidentified newspaper clipping, North London Collegiate School Archives.
14. Winslow, B. (1996). *Sylvia Pankhurst: Sexual Politics and Political Activism*. UCL Press, London, p. 77, 85; see also: (March 1915). Appeal for Help. *Our Magazine: North London Collegiate School for Girls*, p. 16.
15. Bedford College, Staff and Student Records.
16. Spencer, J. F. (1934). Ivy Winifred Elizabeth Rogers, 1900–1934. *Journal of the Chemical Society* 2016.
17. (a) Anon. (1945). Obituary: Margaret Elsie Snowden Appleyard. *Journal of the Royal Institute of Chemistry* **69**: 86; (b) White, A. B. (ed.) (1981). *Newnham College Register, 1871–1971*, Vol. II, 1924–1950, 2nd ed. Newnham College, Cambridge, p. 37.
18. Dainton, F. and Thrush, B. A. (1981). Ronald George Wreyford Norrish, 9 November 1897–7 June 1978. *Biographical Memoirs of Fellows of the Royal Society* **27**: 379–424.
19. Anon. (May 1988). Obituary: Mary Lesslie. *Royal Holloway and Bedford New College Association, College Journal* 17–19.
20. Butler, A. (1991). An extraordinary excursion into atomic structure. *Chemistry in Britain* **27**(1): 40–42.
21. Personal communication, letter, Margaret Jamieson (Mrs. Harris), 19 March 2000.
22. Turner, E. E. and Harris, M. M. (1952). *Organic Chemistry*. Longmans, Green & Co., London.
23. McDonald, M., cited in: Bentley, L. (1991). *Educating Women: A Pictorial History of Bedford College, University of London, 1849–1985*. RHC and Bedford New College, London, p. 40.
24. "H.A.B." (June 1898). Science. *Bedford College London Magazine* (36): 16.
25. Anon. (June 1900). *Bedford College London Magazine* (42): 3.

26. Anon. (June 1903) *Bedford College London Magazine* (51): 5.
27. Anon. (June 1914). College letters: Science. *Bedford College London Magazine* (84): 3.
28. "μχ" (June 1906). *Bedford College London Magazine* (60): 4.
29. Anon. (July 1918). Science letter. *Bedford College London Magazine* (93): 3.
30. "C=M." (March 1905). *Bedford College London Magazine* (56): 11.
31. (a) Bernal, I. (1999). Edith Humphrey. *The Chemical Intelligencer* **5**(2): 28–31; (b) Brandon, R. (11 September 1975). Going to meet Mendeleev. *New Scientist* 593.
32. Bridges, F. (1890). Coeducation in Swiss universities. *Popular Science Monthly* **38**: 524.
33. Humphrey, E. (June 1900). The University of Zurich. *Bedford College London Magazine* (42): 25.
34. Kauffman, G. B. (1966). *Alfred Werner: Founder of Coordination Chemistry*. Springer-Verlag, Berlin.
35. Anon. (1901). College notes. *Bedford College London Magazine* (44): 18.
36. Fleck, G. (1993). Wilhelm Ostwald 1853–1932. In Laylin, J. K. (ed.), *Nobel Laureates in Chemistry, 1901–1992*, American Chemical Society, Washington, DC, pp. 61–68.
37. (a) Tata, J. R. (1994). Rosalind Venetia Pitt-Rivers. *Biographical Memoirs of Fellows of the Royal Society* **39**: 327; (b) Tata, J. R. (1990). Obituary: Rosalind Pitt-Rivers. *The Independent*; (c) "J.H." (2000). Pitt-Rivers, Rosalind Venetia (Henley) (1907–1990). In Ogilvie, M. and Harvey, J. (eds.), *The Biographical Dictionary of Women in Science*, Routledge, New York, pp. 1028–1030.
38. Himsworth, H. and Pitt-Rivers, R. (1972). Charles Robert Harington, 1897–1972. *Biographical Memoirs of Fellows of the Royal Society* **18**: 266–308.
39. Pitt-Rivers, R. and Tata, J. R. (1959). *The Thyroid Hormones*. Pergamon Press, London.
40. Pitt-Rivers, R. and Tata, J. R. (1960). *The Chemistry of Thyroid Diseases*. Chas. C. Thomas, Springfield, Illinois.
41. Finch, A. (1963). Royal Holloway College. *Chemistry and Industry* 1132–1135.
42. Mackinnon, A. (1990). Male heads on female shoulders? New questions for the history of women's higher education. *History of Education Review* **19**(2): 36–47.

43. Bingham, C. (1987). *The History of Royal Holloway College 1886–1986*. Constable, London.
44. White, A. B. (ed.) (1979). *Newnham College Register, 1871–1971*, Vol. I, 1871–1923, 2nd ed. Newnham College, Cambridge, p. 83; also, personal communication, letter, S. Badham, Archivist, Royal Holloway College, 1998.
45. Morrell, R. R. (1932). Obituary notices: M. M. Pattison Muir 1848–1931. *Journal of the Chemical Society* 1330–1334.
46. (a) Anon. (November 1932). Eleanor Field (1893–1913). *College Letter, Royal Holloway College Association* 57–58; (b) Fortey, I. C. (1933). Elizabeth Eleanor Field. *Newnham College Roll Letter* 52.
47. (a) Moore, T. S. and Whiteley, M. A. (1945). Obituary: Mary Boyle, 1874–1944. *Journal of the Chemical Society* 719; (b) Anon. (1933). IV. College news. 1. Staff news. *College Letter, Royal Holloway College Association* 15; and (c) Anon. (November 1944). Dr. Boyle. *College Letter, Royal Holloway College Association* 52–54.
48. Widdows, S. (July 1913). *College Letter, Royal Holloway College* 14–26.
49. Dale, H. H. (1941). George Barger 1878–1939. *Obituary Notices of Fellows of the Royal Society* **3**: 63–82.
50. (a) Augustin, R. (1967). T. S. Moore 1881–1966. *Chemistry in Britain* 494; (b) Bourne, E. J. (1967). Obituaries: Professor T. S. Moore. *Nature* **214**: 1063.
51. Anon. (1947). Resignation: Millicent Plant. *College Letter, Royal Holloway College Association* 16.
52. Hirst, E. L. (1951). Walter Norman Haworth, 1883–1950. *Obituary Notices of Fellows of the Royal Society* **7**: 372–404.
53. Whiteley, M., cited in: Salt, C. and Bennett, L. (eds.) (1986). *College Lives: And Oral Panorama Celebrating the Past, Present and Future of the Royal Holloway and Bedford New College*, Royal Holloway and Bedford Colleges, London, p. 17.
54. Dabis, A. (June 1888). Holloway College. *Magazine of the Manchester High School* 236–238.
55. Pick, M., cited in: Salt, C. and Bennett, L. (eds.) (1986). *College Lives: And Oral Panorama Celebrating the Past, Present and Future of the Royal Holloway and Bedford New College*, Royal Holloway and Bedford Colleges, London, p. 31.
56. (a) Bell, E. M. (1953). *Storming the Citadel: The Rise of the Woman Doctor*. Constable, London; (b) Blake, C. (1990). *The*

Charge of the Parasols: Women's Entry to the Medical Profession. The Women's Press, London, pp. 167–171.
57. Dyhouse, C. (1998). Women students and the London Medical Schools, 1914–1939: The anatomy of a masculine culture. *Gender and History* **10**(1): 110–132.
58. Cullis, W. C. and Widdows, S. T. (17 March 1922). Correspondence: Deeds not words. *Time and Tide*, in the Archives of the Royal Free Hospital, S. T. Widdows file.
59. MacHale, D. (1985). *George Boole: His Life and His Work*. Boole Press, Dublin.
60. Coxeter, H. S. M. (1987). Alicia Boole Stott (1860–1940). In Grinstein, L. S. and Campbell, P. J. (eds.), *Women of Mathematics: A Biobibliographic Sourcebook*, Greenwood Press, Westport, Connecticut, pp. 220–224.
61. (1905). Obituary: Lucy Everest Boole. *Proceedings of the Institute of Chemistry, Part II* **29**: 26.
62. "S.A.V." (1894). Obituary notices: Charles William Heaton. *Journal of the Chemical Society, Transactions* 386–388.
63. Martindale, L. (1951). *A Woman Surgeon*. Victor Gallancz, London, p. 32.
64. Anon. (1905). *London School of Medicine for Women Magazine* 454–455.
65. Anon. (1898). *London School of Medicine for Women Magazine* 17; Personal communication, letter, Keith Austin, Archivist, Senate House, University of London. General Register, volume III, 1901.
66. Anon. (1912). *London School of Medicine for Women Magazine* **8**(15): 77.
67. Evans, C. de B. (1908). Traces of a new tin group element in thorianite. *Journal of the Chemical Society, Transactions* **93**: 666.
68. Anon. (1960). Sibyl Taite Widdows. *Journal of the Royal Institute of Chemistry* **84**: 233; and letter from Senate House, University of London. Keith Austin, Archivist (1998).
69. Sanderson, P. M. (1960). Obituary: Sibyl Widdows. *Royal Free Hospital Journal* **23**: 21–22.
70. (a) *Royal Free Hospital School of Medicine for Women Report*, 1964/1965, RFH Archives; (b) Anon. (1966). *Royal Free Hospital Journal* **27**(83): 190–191.
71. Anon. (1961). Obituary: Henry Vincent Aird Briscoe. *Journal of the Royal Institute of Chemistry* **85**: 425.

72. Anon. (1929). *London School of Medicine for Women Report*. RFH Archives.
73. (a) *Royal Free Hospital School of Medicine for Women, University of London Report*, 1945/1946; (b) Royal Holloway College, student records.
74. University College, Student Records; *Calendars*, London School of Medicine for Women.
75. Rayner-Canham, M. F. and Rayner-Canham, G. W. (2006). The pioneering women chemists of Bedford College, London. *Education in Chemistry* **43**(3): 77–79.

Chapter 5

English Provincial Universities

In England, up until 1850, there were four universities: the historic Oxford and Cambridge, Durham, and London. The six English provincial, or civic, universities which came into existence over the next 60 years (Birmingham, Bristol, Leeds, Liverpool, Manchester, and Sheffield) were to change the availability and range of university education tremendously.[1] The provincial universities saw themselves as serving their specific cities and regions; thus, most of the students commuted from home, unlike the Oxbridge Colleges which were residential. Each institution started from College status and was initially only able to prepare students for University of London degrees.

From the opening of the doors of each provincial college, women were eligible for degrees, and they did not suffer most of the restrictions of their Oxbridge sisters (see Chap. 6). However, women's presence at university was not always accepted, as we will see below.

Pre-First World War, the distribution of the backgrounds of women students at provincial universities also differed substantially from those at Oxbridge, as Julie Gibert reported:

> The upper and professional classes who represented over half of the students at Oxford's women's colleges represented less than one-fifth of the female student body at the civic universities, while working-class occupational groups, entirely unrepresented at both Oxford and London, accounted for 8% of the

students at the civics. A third notable discrepancy is the much higher proportion of 'semi-professional' rank students at the civic universities. At the civic universities 27% of students fell in this group, while at Oxford and London the figures were 4% and 7%, respectively.[2]

Armstrong College (University of Newcastle)

The University of Durham was founded in 1832, and it was run very much as an arts college on the Oxbridge model.[3] A School of Physical Science had been opened in Durham in 1865, but it closed in 1871.[4] With the founding of the Newcastle Chemical Society (later the Society of Chemical Industry, Newcastle Section) in 1869, pressure mounted for a science college in that city; as a result, the Durham College of Physical Science was opened in Newcastle in 1871. Three years later, the school was affiliated with the University of Durham as Armstrong College. Armstrong College was merged with the Newcastle College of Medicine and renamed King's College of the University of Durham in 1937. King's was given independent status as the University of Newcastle-upon-Tyne in 1963.

Chemistry was not taught on the Durham campus until 1926, and even then it laboured under disadvantages, as C. W. Gibby recalled: "Research in the science subjects received no help from industry nor from Government Departments.... The attitude of Palace Green [the arts, theology, and education faculties] towards the Labs. was not hostile, it was just one of blank incomprehension."[5]

The first woman on staff at Armstrong College was Charlotte Bean Schofield[6] (Mrs. Cole), who was Assistant Lecturer and Demonstrator in Chemistry from 1919 to 1920. Schofield obtained a B.Sc. (Durham) in 1915 and an M.Sc. (Durham) in 1917 in chemistry from Armstrong College. After teaching for 1 year at Armstrong College, she entered medical school, moving to South Africa after graduation. She died on 1 August 1972.

Grace Leitch

Replacing Schofield was Grace Cumming Leitch,[7] who provided the backbone of much of the teaching in the first part of the 20th century. Leitch was born on 14 July 1889 at Cupar, Fife, and was educated at Bell Baxter School in Cupar. She obtained a B.Sc. from St. Andrews in chemistry in 1913, followed by a Ph.D. in 1919, her research work on the structure of sugars being undertaken with Norman Haworth.[8] During the First World War, Leitch had been appointed Junior Lecturer, and over the same period she also undertook work on mustard gas. However, the war had a more direct impact on Leitch in that her fiancé was killed in France just before the Armistice.

When Haworth left St. Andrews and took up a position at Armstrong College in 1920, Leitch accompanied him, herself being appointed Lecturer in organic chemistry. She continued her research on sugars, co-authoring three papers with Haworth and being the sole author of another. When Haworth subsequently moved to Birmingham, Leitch joined the research group of George Clemo,[9] co-authoring three papers on alkaloids. However, as she was responsible for a substantial part of the teaching load, there was little time for research.

In 1926, Leitch became the subwarden of the women's residence, Easton Hall. She became seriously ill in the mid-1930s, and though she never fully recovered, she continued to hold both Lectureship and subwarden appointments until her health deteriorated again in July 1941. She died on 12 March 1942 in her hometown of Cupar. As was noted by her friend and mentor, Haworth:

> Among scientific workers she had a wide acquaintance, fostered by her regular attendance at the annual British Association meetings. Few could have failed to appreciate her gaiety of spirit, warm enthusiasm and loyalty, her capacity for friendship, and her vigorous personality; for it was these qualities which commended her to her students and colleagues, by whom she will be greatly missed.[7(a)]

Catherine Mallen (Mrs. Elmes)

When Grace Leitch arrived at Newcastle, Catherine Eleanor Mallen[10] (later Mrs. Elms) had already taken up her appointment as Assistant Lecturer and Demonstrator. Mallen and Leitch became close friends, and it was Mallen who wrote one of the obituaries of Leitch.[7(b)] Mallen was born on 22 April 1896 at Sunderland, where she attended the Bede School for Girls. She entered Newnham College in 1915, completing the chemistry degree requirements in 1919 and obtained the appointment at Armstrong College in the same year. Mallen was promoted to Lecturer in 1939, a rank which she held until her retirement in 1960. In addition to teaching, Mallen authored two research publications — one alone, and one with the Australian chemist Thomas Iredale, who was a Lecturer in Chemistry at Newcastle from 1925 until 1927.

Mallen had married Ralph S. Elms, also a Cambridge graduate and a Lecturer in English at Newcastle, in August 1943. The Elms retired to Wensleydale, where Mallen became active with the Wensleydale Society until her death in 1984. As her obituarist, Stella Buckley, explained:

> This [the Wensleydale Society] had evolved from a series of lectures on local history and had become a most successful institution providing lectures by well-known specialists on the archaeology, geology, botany and history of the district. Catherine was no mean botanist and greatly enjoyed our walks and expeditions until well over eighty.[10(a)]

Life for Women Students

Though the first women students did not appear at Durham itself until 1896, the College of Physical Science, Newcastle, admitted women to lectures and laboratories from its opening in 1871. Women were able to register for the qualification of Associate of Science (A.Sc.), but Durham barred them from admission to a bachelor's degree. As Marilyn Hird noted: "When one enterprising young lady completed the qualifying examinations and presented herself, without success, for admission to the degree of B.Sc., the whole question was raised."[11] Finally, in 1895, a Supplemental Charter was approved that gave the university the power to confer degrees on women.

The suffrage movement was an active issue at Armstrong College. In 1906, a very heated debate among the women students resulted in a vote of 29 in favour of women's suffrage and 55 against.[12] As the author of the report noted: "Some of the suffragettes did not look very happy when the result was announced," however, "several converts to the cause were made."[12]

Friction between the male and female students arose in 1912, which prompted a retort in the student magazine, *The Northerner*: "Certainly if we were to take the College as an example, there can be little doubt that man is a less industrious creature than woman. Who do the most work in college, men or women? Ask the lecturers, their opinion would be unanimous — the women."[13]

Owens College (Victoria University of Manchester)

In his Will, John Owens, a Manchester merchant, left a substantial sum for the founding of a university-level institution and, as a result, Owens College was established in 1850–1851. At the time, the primary concern was that the College be non-sectarian, and no thought was given to the possibility of women wanting to attend.[14] By 1875, pressure was building for the admission of women. Manchester High School for Girls (MHSG) had opened its doors in 1874; thus, it could no longer be argued that women students were ill-equipped to face a university education. Heated debate ensued in the correspondence columns of the *Manchester Examiner*, but as Mabel Tylecote observed: "The men poo-poohed, often with little taste, the claims of women as a sort of bad joke. The women often began with serious arguments and then ended by losing their tempers."[14]

A defining step was taken in 1877 with the opening of the Manchester and Salford College for Women at 223 Brunswick Street, near the Owens campus.[15] Most of the teaching at the College for Women was given by professors and lecturers of Owens College, though there was no legal connection between the two institutions. The Women's College was absorbed into Owens College in 1882 as the Department for Women. Initially, women students were only permitted onto the Owens campus in preparation for their final examinations, and it was not until 1892 that women were first allowed to attend physics and chemistry instruction and laboratory work towards a B.Sc.[16]

In 1880, Victoria University was formed, comprising Owens College, Manchester; University College, Liverpool; and Yorkshire College, Leeds.[17] The university was designed to encompass, in time, all of the provincial colleges; but the Federation was shattered in 1900 when Mason College, Birmingham, refused to join, insisting on the right to its own Royal Charter as a unitary and autonomous university for Birmingham and the Midlands. The members of the Victoria Federation then decided to go their separate ways, and three individual universities — Manchester, Liverpool, and Leeds — were formed.

In addition to the Victoria University of Manchester, there was the Manchester School of Technology, which had been founded in 1824 as the Manchester Mechanics Institute, and which later became the University of Manchester Institute of Science and Technology.

Life for Women Students

Even the Owens College Library was initially "off-limits," as a woman student in 1901 later recalled:

> It would have been the height of impropriety [for a woman student] to enter the library and demand a book in the hardened

> manner now usual ... No, we had to "fill up a voucher", and a dear little maid-of-all-work, aged about 13, went to the library with it. If we were not quite sure of the volume required, she might have to make the journey ten times, but it was never suggested that she should be chaperoned.[18]

As at many co-educational colleges, it was understood that certain rows in the lecture room were unofficially reserved for women students. A student, J. Harold Bailey, described in 1892 how this fact had not been imparted to a new male arrival:

> A tale is told of an Arts man commencing his courses late in the session, who came in some little time before the lecture began and took his seat in the middle of the row usually occupied by the ladies. Presently the ladies trooped in as a body, but on catching sight of the intruder they shrank back in dismay, and crowded round the doorway undecided as to the proper thing to do under the circumstances. Just as the Professor entered the room one of the men students, thinking that the fun had gone far enough, went up to the innocent offender and explained matters, whereupon he meekly took another seat amidst the suppressed titters of his fellow men and the ladies were able to occupy their accustomed seats without running the risk of being contaminated by having a man sit amongst them.[19]

Leonore Kletz (Mrs. Kletz Pearson)

Two early women chemists contributed to the Chemistry Department at Manchester, the first of whom was Leonore Kletz[20] (later Mrs. Pearson). Kletz was born on 18 June 1891, her father being a house furnisher, and she attended North Manchester High School and Pendleton High School. She completed a B.Sc. in chemistry at Manchester University in 1912, and an M.Sc. in 1913 in organic chemistry. Kletz worked for a year with Arthur Lapworth[21] as Schlunk Research Assistant before being appointed Assistant Lecturer and Demonstrator in the Chemistry Department.[22] It is not clear for how long she held the position, but her seven research publications (all under the name of L. K. Pearson) are spread from 1915 to 1927. It was noted in her obituary that: "For several years she had given up the practice of chemistry but had recently renewed her activities in this direction."[20(a)] Kletz died on 23 July 1947, aged 57 years.

May Badger (Mrs. Craven)

May Badger,[23] the other pioneer, had a more varied career. She was born on 18 February 1887 and attended Ardwick Higher Grade School. She entered the Faculty of Technology of Victoria University,[24] being awarded a B.Sc. Tech. (Applied Chem. Hons. Div.) in 1907. Receiving a postgraduate scholarship, she worked with William Pope[25] towards a M.Sc. Tech., which she completed in 1908. After graduation, Badger became a Chemist with the Pilkington Tile and Pottery Co., carrying out research on glass and pottery, after which, in 1911, she became Chemist to the Clifton and Kersley Coal Co.

In 1916, Badger returned to the University of Manchester Faculty of Technology as Senior Demonstrator in Chemistry. She continued her interest in coal chemistry, co-authoring (as M. B. Craven) with Frank Sinnatt[26] a bulletin on the heat content of coal in 1921, and with Sinnatt and others a book titled *Coal and Allied Subjects* in 1922.[27] Badger remained at the University of Manchester until her retirement in 1952, having risen to be Head of the Inorganic Chemistry Laboratories. She died on 24 November 1953, her obituarist noting:

> Mrs Craven had a flair for teaching the methods of inorganic analysis and her enthusiastic approach was always stimulating to her students. She will be affectionately remembered by several generations of Manchester chemists who owe to her their early training in this branch and many of whom will recollect with pleasure the friendly hospitality of the Craven household.[23]

The Manchester University Chemical Society

The Chemical Society, formed in 1877, was originally one of the male bastions of the university. The first attempt to admit women came in 1906, as the Society Minutes recorded:

> It was proposed by Mr. E. W. Smith and seconded by Mr. Slade that "The Committee be asked to enquire into and report on the advisability of admitting women students into this society." On being put before the meeting the proposition was lost.[28]

There actually seemed to be two separate issues: the admission of women to the society and the admission of women to the student union building, which was under control of the male-only Union Committee. Sentiment seemed to shift, according to the Minutes of 5 February 1907, as the society invited Ida Smedley (see Chap. 2) as a guest speaker:

> It was then proposed by Dr. Hutton and seconded by Prof. Carpenter that "The Union Committee be asked to grant permission for Dr. Ida Smedley to attend the meetings in the Union." The proposal was carried. It was proposed by Dr. N. Smith, seconded by Mr. S. R. Best & carried that "The committee shall at some special meeting or at the general meeting present a report on the question of admitting women students to membership of this society."[29]

The reply to the request for Smedley to enter the union building was uncompromising: "The secretary read a letter from the Union secretaries stating that the Union Committee could not see their way to allow Dr. Ida Smedley to use the Union Rooms for any reason whatsoever."[30] In November 1907, the bar on women members was sustained: "… It was decided that women students be not admitted as members of the Society as it would spoil the social nature of the meetings."[31]

Something must have happened over the following 12-month period, as the society reversed itself in a motion of November 1908:

> An Extraordinary General Meeting was held on Friday, November 13th, to consider the following resolution: "That the words (in Rule VI) 'This is to be interpreted as applying to men students only' be rescinded." Professor Perkin was in the chair. The resolution was proposed by Dr. Meldrum seconded by Mr. Wood, and after a long discussion in which several members took part, the resolution was carried by 48 votes to 19. It was then proposed & passed unanimously that the alteration come into force at the beginning of the Session 1909–1910.[32]

At the 26 October 1909 meeting, 12 of the 35 new student members were women, and the Minutes noted that business began "After welcoming women students as new members of the Society ..."[33] It would seem that the barrier of location had been overcome by holding the Chemical Society meetings in the Women's Union instead; and on 7 December 1909, Smedley presented her paper to the Society.

Lucy Higginbotham

Among the women chemistry graduates from Manchester was Lucy Higginbotham.[34] Higginbotham was born on 6 November 1896, the daughter of a yarn merchant in Chorlton-cum-Hardy. She attended Penrhos College, Colwyn Bay, before completing her schooling at MHSG. Entering the University of Manchester in 1915, she graduated with First Class Honours in Chemistry in 1919 and was the recipient of the Leblanc Medal for that year.

The following year, Higginbotham worked on her M.Sc. with Henry Stephen, a research associate with Lapworth. From 1920 until 1922, she was Research Assistant to Lapworth, contributing to five of his research publications.

Then, she joined the staff of the British Cotton Industry Research Association at the Shirley Institute, Manchester, researching the minor constituents of cotton — particularly the complex mixture of substances present in cotton wax and their reaction products during the bleaching and finishing of cotton fibre. Her results were published in a series of papers in the *Textile Industry Journal* and in the *Memoirs of the Shirley Institute*.

In 1926, Higginbotham became ill and went for a medical operation to Berne, Switzerland, where she died on 22 November 1927. Her obituarist, Robert Fargher, observed:

> As a research worker she possessed unusual energy and a pronounced flair for the rapid exploration of a field of inquiry; outside her work, she was keenly interested in athletics, particularly tennis, golf, and motor cycling. She had many of the best characteristics of her native country, and her frank outspoken genial personality endeared her to her colleagues at the University and the Shirley Institute.[34(a)]

University College, Liverpool (University of Liverpool)

University College, Liverpool, came into existence in 1881, becoming one of the constituent colleges of Victoria University in 1884 and then going its own way in 1902 as the University of Liverpool (UL) following the dissolution of the Federation.[17] The three constituent colleges all shared a common philosophy, as David Jones described: "Middle class support and clientele, non-sectarianism, moderate fees, non-residence, the importance of evening classes and of part-time and non-degree students, and a strong desire to adapt to local needs were basic characteristics shared by all the civic colleges."[17] Thus, as with all the provincial universities, students in the early years largely came from the local district, which in Liverpool's case was the Merseyside area.

The University of Liverpool Chemical Society

The arrival of women chemistry students at UL had a considerable effect on the student chemistry culture. The Liverpool University Chemical Society (LUCS) had been founded in 1892,[35] and the social life of the society focused on the men-only Annual Dinner and the Annual Kneipe (Beer Party). The latter event was an evening spent in drinking beer, smoking, singing songs, and telling stories.

When the women petitioned to join the society in 1902, they requested a reduced subscription due to their exclusion from the male-only social functions. Their petition was rejected and they were barred from membership in the society. In response, the women promptly organised their own Women's Chemical Society. The admission of women to the LUCS was raised in a subsequent year (probably 1908), but again they were turned down. It was not until 1912 that they were finally admitted and a society dance was instituted. In 1914, they had their first woman speaker, Dorothy Baylis (see below), one of the graduating class. The same year, the men-only Kneipe was dropped and a Smoking Concert took its place. For those males who still abhorred the presence of women, there was the refuge of the Research Men's Club.[36]

Membership did not result in equality for women. As at Imperial College (see Chap. 3), the women chemistry students were expected to serve tea to their male colleagues. A cutting letter to the *LUCS Magazine* in 1923 commented: "Lady Chemists are overwhelmed by the extreme courtesy paid to them at Chem. Soc. teas. To the Victorian male mind, they still serve as Hewers of Bread and Drawers of Tea."[37]

In 1928, the *LUCS Magazine* carried an article on "Women and Chemistry".[38] In it, the anonymous author comments that:

> I often wonder why women take up chemistry. Can it be that they imagine they will become chemists? I shudder at the thought. … Women in the right setting are delightful creatures.

> A chemistry laboratory is not the right setting. A woman in a lab is as incongruous as a man at an afternoon tea party. … If it is impossible to have a special "female" lab, then let the flapper vote give England a women's University. …[38]

This article provoked an immediate response from a woman chemistry student, defending the presence of women in chemistry:

> Life at a University offers many attractions, not the least of which is, that should she find after many years that she is a superfluous woman she will always have a university training, and perhaps a degree, which are useful sort of things to have when one is thinking of earning one's living. …. Besides, Chemistry offers so many more possibilities than Arts. Engineering would, of course, be the ideal faculty for this attractive woman, but — it simply isn't done!![39]

In the closing remarks, she comments about men "… who would label their doors 'No Admittance to Women'."

Though the previous writer seemed to accept that a degree was a "back-up plan" in the event of failure to marry, the next issue carried a rebuttal with a more strongly feminist stance:

> The author [of the attack on women chemists] seems to forget that we are now living in the 20th century, when that which used to be a "man's job" is a man's job no longer. In almost every occupation women are equalling and have equalled men. … He evidently does not know that darning socks and rocking cradles went out with crinolines. …[40]

However, the author also realistically adds: "Women and men meet on equal terms and work on equal terms. At night, the man goes home to be waited on, while a woman goes home to do a 'woman's job.'"[40]

This third contribution seemed to end the correspondence, but the exchange clearly indicates the degree of hostility facing women students from some of their male chemistry colleagues.

Edith Morrison

Amongst the women pioneers at Liverpool was the first woman to obtain a Ph.D. and, as it happened, this Ph.D. was in chemistry.[41] The recipient, Edith Morrison, completed her B.Sc. at Liverpool in 1922 and her Ph.D. in 1924. The *LUCS Magazine* announced the event with pride:

> An event of a nature quite unprecedented in the annals of the Chemical Society took place in the large Lecture Theatre on January 16th. A degree of Ph.D. was introduced by Liverpool University in 1922 but not until December 1924 was the degree conferred on a woman. The Faculty of Science and especially the Department of Chemistry, is very proud that the first successful woman was a chemist — Dr Edith Morrison, who after two years of brilliant research work on Photosynthesis, under Prof. Baly, obtained a well-deserved degree.[42]

Morrison subsequently worked at Colman's Mustard Co. and married J. W. Corran.

Eileen Sadler (Mrs. Doran)

The career of Eileen Sybella Sadler[43] illustrated the difficulty of a woman chemist in finding an academic post except outside of chemistry or at a nonmainstream institution. Sadler was born in Wallasey on 10 November 1901, her father, Harold Sadler, being a chartered accountant. She was educated at Wallasey High School for Girls and then Malvern Girls' College. Entering UL in 1919, she graduated with a B.Sc. in chemistry in 1924, followed by an M.Sc. in 1926 and a Diploma in Education in 1928. That year, Sadler was appointed as Research Assistant in the UL Department of Pharmacology of the Faculty of Medicine. In 1929, she was promoted to Demonstrator, then teacher of Pharmacy in 1933, and Assistant Lecturer in 1934.

In 1935, Sadler resigned and accepted an appointment as Science Lecturer at the F. L. Calder College of Domestic Science, a college with ties to UL that later became a constituent college of the Liverpool Institute of Education, then Liverpool Polytechnic. Sadler married William Doran, who was a Lecturer in Organic Chemistry at UL, though, like Sadler, he only held an M.Sc. qualification.

Women Analytical Chemists

Analytical chemistry was an area favoured by some women chemists during the interwar period. In fact, a remarkable number of women chemists who graduated from UL chose this direction. We have chosen three individuals to exemplify this: Dorothy Baylis, Muriel Roberts, and Gertrude Andrew.

Dorothy Baylis[44] completed her B.Sc. at UL in 1914. After graduation, Baylis became an Analyst with Lever Bros., Port Sunlight., from 1914 until 1916. From 1916 to 1917, she worked as an Analyst to the West Riding of Yorkshire Rivers Board, Wakefield, followed by a period as Research Chemist to Brookes Chemicals, Ltd., Lightcliffe, from 1917 to 1919. During 1920 and 1921, Baylis was a Research Chemist at British Dyestuffs Corporation, Manchester; then, she changed paths and became a Lecturer in Pure and Applied Chemistry at Leicester College of Technology, simultaneously doing independent research towards her M.Sc. which she completed in 1926.

Whereas Baylis led a peripatetic life, Muriel Roberts[45] remained in one position for most of her career. Roberts was born in 1894 and completed her B.Sc. at UL in 1915. She became a member of the Society for Chemical Industry in 1924 and of the Analytical Society in 1931. Her name appears as Senior Analyst of the Liverpool City Corporation in 1932, and she still held this post at her retirement. She died on 29 June 1985, aged 91 years.

Gertrude Andrew,[46] daughter of William Andrew, clerk of New Brighton, was born on 9 October 1895 and educated at Wallasey High School, entering UL in 1913. After graduation in 1916, she worked as a Chemist for Cow & Gate, Winchester, Somerset, later becoming a Public Analyst. Andrew joined the Analytical Society in 1926. By 1948, she was a Senior Assistant Chemist with the County Laboratories, Staffordshire. On retirement, Andrew was rehired as an additional Public Analyst. She died on 28 February 1978, aged 82 years.

Yorkshire College (University of Leeds)

The Yorkshire College of Science was founded in 1875 to provide "instruction in such sciences and arts as are applicable or ancillary to the manufacturing, mining, engineering and agricultural industries of the County of York."[47] With the support of the Yorkshire Ladies Education Association, but against some opposition, arts courses were added in 1877 and the "of Science" was deleted from the name.[17] Leeds was admitted to the Federation in 1887. Like Manchester, Leeds initially had a separate Women's Department, but it seems to have become defunct early in the College's history.

Arthur Smithells

Of all the university chemistry professors, it was Arthur Smithells,[48] Chair of Chemistry at Leeds from 1885 until 1923,[49] who had been most active in promoting the science education of girls (see Chap. 1). The explanation for his interest dates back to his last days at Owens College, Manchester, before his appointment at Yorkshire College:

> Part of his last session (1882) at the Owens College was spent in conducting a course of lectures and practical work on chemistry at the neighbouring [Manchester] Girls' High School where he gained his first experience of teaching, an experience

> which may have directed his attention to the gaps in a girl's education, and implanted the germ of his present schemes for the scientific training of women.[50]

Smithells was obviously deeply influenced by his experience teaching high school girls, commenting: "... the first teaching I ever attempted was in a girls' high school, and I have at least a first-hand knowledge of a wrong way of doing it."[51]

Women Chemistry Students

The complaint was made in a letter of 1898 to the student magazine *The Gryphon* that the "industrious" and "enthusiastic" women students were showing up their laggardly male colleagues.[52] The presence of women in the chemistry laboratory was equally unpopular, as a correspondent in 1902 reported:

> ... Well might the new Hiawatha lament the depredations of the stool-snatcher, for the worst offenders are — tell it not to an Amazon — *snatcheresses*. And there is no redress — man must submit and stand. Look where you will, woman is in possession; aye, even of the very stink-cupboard.[53]

Interestingly, at Manchester, it had been the males who were the "stool-snatchers" in the labs, perhaps Mancunian women students being less assertive, as was expressed in the rhyme "A Lady's Lament" in the *Owens College Union Magazine* of 1902:

> And we, in weak appeal, our voices raise,
> If we want merely to retain a stool.
> The only way that we can hope to keep
> That "more primeval" beast our seats from filling,
> Is to inflict (small wonder that I weep!)
> A fine on each offender of a shilling.[54]

The Yorkshire College had a scientific society to which women were welcome, the first woman speaker noted being a Miss Findlay who spoke on "Recent Attempts in Colour Photography" on 16 December 1897.[55] The society changed its name to Leeds University Cavendish Society in 1910, with discussions solely on chemistry and physics topics. A report on a meeting of early 1916 indicates that women chemistry students had begun to play a more active role: "... The meeting was indeed a remarkable one, for the women students turned up in great numbers and joined in a little discussion."[56]

Though the Cavendish Society was the recognised body for discourses on the physical sciences, some male members decided that they wanted a women-free social venue, the Organic Lab "Club." This "Club" appears to have organised male-only dinners with songs, smoking, and musical solos, similar to the activities of the Manchester and Liverpool student chemistry societies. The "Club" was first reported in 1898[57] and was still active in 1905,[58] while in 1911 a "Chemists' Dinner" seems to have been organised in a similar male-only format.[59]

May Leslie (Mrs. Burr)

Leeds' most famous chemical alumna was May Sybil Leslie.[60] Leslie was born on 14 August 1887 in Yorkshire and studied chemistry at the University of Leeds. She graduated with first class honours in 1908, and was awarded an M.Sc. for research with Harry M. Dawson[61] the following year on the kinetics of the iodination of acetone, work that has since become a classic in its field.[62] In that same year, 1909, Leslie was awarded a scholarship, which she decided to use to work with Marie Curie[63] in Paris. Her letters from Paris to Smithells are among the few accounts of life in the early Curie laboratory.[64]

Leslie spent 1909 to 1911 with Curie, the only English woman in Curie's group.[65] Leslie's work involved the extraction of new elements from thorium. For a chemist used to working with grams of pure chemicals in beakers, the manipulation of kilogram quantities of minerals in huge jars and earthenware bowls must have been a completely new experience.

Returning to England, Leslie took a position with Ernest Rutherford at the Physical Laboratory of the Victoria University, Manchester. There she continued with her work on thorium, and extended her studies to actinium during the 1911–1912 period. After leaving Manchester, she spent 2 years as a science teacher at the Municipal High School for Girls in West Hartlepool. During this time, she managed to resume research with Dawson, this work being on ionisation in non-aqueous solvents. From 1914 to 1915, Leslie held a position as Assistant Lecturer and Demonstrator in Chemistry at the University College in Bangor, Wales.

In 1915, she entered the world of industrial chemistry, being hired to work at His Majesty's Factory in Litherland, Liverpool, a position that she obtained as a result of the call-up for military duty of the male research chemists. Her initial rank was that of Research Chemist, but in 1916 she was promoted to Chemist in Charge of Laboratory, a very high position for a woman at that time. Her research involved the elucidation of the pathway in the formation of nitric acid and the determination of the optimum industrial conditions for the process. This work was vital for the munitions industry, which required massive quantities of nitric acid for explosives production. In June 1917, the Litherland factory closed[66] and Leslie was transferred with the same rank to the H.M. Factory in Penrhyndeudraeth, North Wales. Leslie was awarded a D.Sc. degree in 1918 by the University of Leeds, mainly in recognition of her contribution to the war effort.

With the return of the surviving male chemists at the end of the First World War, Leslie lost her government position. She returned to the University of Leeds as Demonstrator in the Department of Chemistry in 1920, being promoted in the following year to Assistant Lecturer. Leslie then moved to the Department of Physical Chemistry in 1924 and was promoted to Lecturer in 1928. In 1923, Leslie had married Alfred Hamilton

Burr, a Lecturer in Chemistry at the Royal Technical College, Salford.[67] She had first met Burr at the H.M. Factory in Litherland, where he, too, had worked in 1916. She continued to be an active researcher at Leeds after marriage; in addition, the famous British chemist, J. Newton Friend, invited her to author one volume of the classic series *A Textbook of Inorganic Chemistry* and to co-author another.[68]

According to university records, Leslie resigned from her position at the university in 1929 due to health reasons. In 1931, when Burr was appointed head of the Chemistry Department at Coatbridge Technical College, Scotland, she moved with him. After Burr died in 1933, Leslie moved back to Leeds, resuming research work at the university. Her first project was the completion of Burr's unfinished research on wool dyes, then she returned to her own interest in the mechanisms of reactions. In addition to performing research work, she was employed as subwarden of a woman's hall of residence (Weetwood Hall) at the university from 1935 to 1937.

Leslie died at Bardsey, near Leeds, on 3 July 1937, having given up research only a month earlier. No cause of death was recorded, but it was quite possibly radiation-related considering her exposure to high levels of radioactivity during her research work in Paris. Her obituary in the *Yorkshire Post* noted that Leslie was "one of the University's most distinguished women graduates."[69] Her former supervisor, Dawson, commented that her research reputation was "deservedly high" and that as a teacher she was "exceptionally gifted."[70]

University College, Sheffield (University of Sheffield)

University College, Sheffield, was formed in 1897 by the merger of Firth College, Sheffield Technical School, and Sheffield School of Medicine.[71] University College, Sheffield, had applied to join Victoria University, but the request was rebuffed.[72] Upon the breakup of Victoria University, Sheffield had proposed a

Federal University of Yorkshire, having campuses at Leeds and Sheffield; but Leeds rejected the concept, opting instead to become an independent university. Sheffield then appealed for university status, which was granted in 1905.

The Chemistry Department in the First World War

In Chap. 12, we will see that the First World War resulted in a demand for women chemists — in particular, for the small-scale production of fine chemicals. Nowhere was this more evident than the chemistry department at the University of Sheffield under William Palmer Wynne.[73]

By 1915, Wynne was the sole remaining faculty member of the chemistry department, the others having departed for war work.[74] As a result, Wynne hired Emily Turner and Dorothy Bennett as Assistant Lecturers and Demonstrators (see below), and Annie M. Mathews as Demonstrator and Lecture Assistant.[75] All of the women held M.Sc. degrees in chemistry from Sheffield and, in addition to teaching duties, they synthesised large quantities of local anæsthetics for the war effort (see Chap. 12). With the war over, one of the returning Lecturers, James Kenner, married Mathews in 1918. Mathews resigned and was replaced by May Walsh for the 1918–1919 year. Turner and Bennett were to remain key figures in the chemistry department for ensuing decades. As R. L. Wain remarked: "The two of them were, to my mind, a well respected team, imposing strong discipline in classes. They were often referred to by students as 'the tartrate twins'."[76]

Emily Turner

Emily Gertrude Turner,[77] the second of two daughters, was born on 16 April 1888 to John Wesley Turner and Annie Hague of Manchester. She was educated at Wellgate Council School, then she completed a B.Sc. in chemistry at Sheffield in 1909, followed by an M.Sc. in 1911. During 1911–1912, Turner studied for an

Educational Diploma, and then taught at Newcastle-upon-Tyne.[78] It was while at Newcastle that Wynne invited her back as an Assistant Lecturer and Demonstrator at Sheffield.

Turner stayed at Sheffield from 1915 to 1953, and it was estimated that, over the years, she had given "… some 4,000 lectures and spent at least 20,000 hours in laboratory supervision" during her 38 years of service.[77(b)] Though her major responsibility was teaching, she was able to co-author six research papers between 1911 and 1941: three with Kenner, one with Wynne, one with G. M. Bennett, and one with Wynne and G. M. Bennett. Turner also acted as secretary to Wynne from about 1920 until his departure in 1931.

One of her former colleagues, Peter A. H. Wyatt, described working with her:

> She fitted into the old order in the chemistry departments of those days and I don't think you would find anybody with quite her work description in any British university these days. Thus she had no ambitions whatsoever to be a research scientist and seemed outwardly content with her role as a teacher of practical classes only at the elementary level, never giving any lectures apart from the talks at the beginning of the practical classes to describe the day's work (some which I also delivered). In consequence she was never promoted beyond the grade of Lecturer (though most people towards the end of her time rose to be at least Senior Lecturers) and it could be that she may have felt some pangs of disappointment about that. If so, she never said.[79]

However, one of her former students, H. J. V. Tyrrell, recalled that she did give chemistry lectures to the health science students:

> Until her retirement she was an institution in the Department, presiding at tea in the little staff room, and giving lectures, lavishly illustrated with experimental demonstrations, to the first-year medical and dental students, supplemented by practical classes.[80]

Turner's nephew, Jeffrey Turner, periodically visited her:

> ... as often as not [I] would meet my Aunt in her room adjoining the University Chem Labs, which she shared with Miss Dorothy Bennett, a life-long friend of E.G.T. and all the Turner families. EGT lived in a flat in Sheffield throughout her university career, returning to her parents and sister at weekends. Holidays would be taken, usually with her sister, to English countryside resorts. Equipped with a 1 inch Ordinance Survey map containing the resort they picked their way along many of the footpaths in their walks in the surrounding area.[81]

Turner retired in 1952. She then went to live with her sister Beatrice at Bawtry Road, Rotherham, where she died on 15 June 1956.

Dorothy Bennett (Mrs. Leighton)

Unfortunately, there is less information on the other "tartrate twin," Dorothy Marguerite Bennett.[82] Bennett was born in 1884 and completed her B.Sc. in 1909 (the same year as Turner), with her M.Sc. granted in 1910. Again, like Turner, she initially proceeded into school teaching until requested to join the chemistry department of Sheffield in 1915. Bennett, too, kept her position of Assistant Lecturer after the end of the war, rising to Senior Lecturer, and in 1926 she was also appointed Tutor for women students.

Bennett retired from her academic position in 1934, subsequently marrying Henry Birkett Leighton. However, she kept her post as Tutor for women students and added the role as warden of University (residence) Hall for Women (later called Halifax Hall) in 1936. She resigned from both these duties in 1947 and moved to Dartford. Bennett died on 11 May 1984 in Tunbridge Wells, 12 days after her 100th birthday.[83]

The Sheffield University Chemical Society

Women students had been admitted since 1886 by the antecedent institutions of University College, Sheffield.[72] Yet, as elsewhere, co-education did not necessarily imply that women students were accepted as equals. Attitudes towards the women chemistry students at Sheffield seem to have evolved from bemusement by their male colleagues in the early years through to hostility as the 20th century progressed.

In the Sheffield Student Newspaper, *Floreamus!*, of 1907, there is a semifictitious discussion between four male students about the effect on the university of the presence of women students. The science student described the women students as being some separate species:

> 'I have never found the ladies such a nuisance,' said the Science man, 'They never make all that noise in the labs., and they're always extremely grateful if you help them to set up their apparatus, or show them how to do an experiment.'[84]

Up until the First World War, the University of Sheffield Chemical Society had a significant proportion of women members, some of whom gave presentations to the Society. Nevertheless, according to the rules of the Society, women still had to perform their traditional roles: "Women Students Only: All are expected to help to prepare tea, but beware lest thou are late in commencing operations, and so the hungry ones are kept waiting."[85]

By the 1920s, the women students were being described in a very negative light. A letter to the student paper commented that, despite decades of co-education at Sheffield, the activities of the women students seemed to be "chiefly dancing and scandal-mongering," leading to the conclusion that: "Without formulating it as a rule, and conceding many exceptions, it is undeniable that the influence of the woman student is on the whole deterrent to serious thought."[86]

By 1949, the student Chemistry Society had become exclusively male. According to a male writer in *By-Product*, the Journal of the Sheffield University Chemical Society, the woman chemistry student is regarded as simply a husband-hunter:

> In these days of man-shortage a girl is often forced into a career other than that of hunting and training the male. Consequently advice seems desirable on occupations for that period between algebra and the altar. This week research chemistry is my subject — a pleasant career involving little work, few restrictions and close contact with eligible men. ... If these points are followed carefully, the young [female] research chemist should perfect the arts of entertainment and cultivate the graces of a hostess in sufficient time to make an excellent wife for her Professor![87]

Mason College (University of Birmingham)

Mason College of Science opened in 1880. At the inauguration of the new building in 1883, specific mention was made about the admission of women:

> Special interest was felt in the inauguration of the building on account of a feature which distinguishes it from any other Science College — namely, that it is open to women on exactly the same terms as men. All who have watched this experiment, if such it may be called, must be satisfied with the entirely successful way it has worked.[88]

In fact, according to the 1884–1885 *Mason Science College Calendar*, in the 1882–1883 session, the enrolment was 229 male and 137 female.

There was a sense of community among the women as they could meet in the Ladies' Common Room, as was described in 1899: "The liveliest time in the Common Room is between the hours of 1 and 2 p.m., and 4 and 5.30 p.m. At those hours students may be seen making tea, coffee, or cocoa, according to their tastes, and lounging in the arm-chairs …"[89]

Women Chemistry Students

The presence of women in the chemistry laboratories was first mentioned in 1883. A commentary describes how, if one was short of chemicals or equipment, one appropriated them from a neighbour's bench. The writer added: "The lady students, too, though they are a little longer in acquiring this beautiful virtue, soon learn how dependent all members of society are on one another."[90]

In another of the student magazines, *The Mermaid*, an account is given in 1909–1910 of the new woman chemistry student: "The modern Chemistry Girl in the smart, brand-new Lab is, you may be sure, quite up-to-date, and, if the truth be known, a source of perpetual worry to most of the rather dull men, who have uneasy visions of their names on the terminal lists preceded by those of a whole string of young girls."[91]

The Mason College Chemical Society was formed in 1884,[92] and among two of the early members were the first woman Associate of the Institute of Chemistry, Emily Lloyd (see Chap. 2), and the first woman Student Member, Rose Stern (see Chap. 2). However, according to the reports in the *Mason College Magazine*, two Birmingham women preceded them: Jessie Charles[93] and Constance Naden,[94] who were both very active in the Society.

Charles, born on 19 March 1865 to Andrew Charles, a hardware merchant, was educated privately before entering Mason College in 1882. She departed for Newnham in 1890 and, after completing Part 1 of the Science Tripos in 1893, she worked in Breslau and Leipzig. In her multi-faceted life, Charles (later Mrs. White) became a promoter of the Montessori School system, authoring a book on the subject.

Naden, who entered Mason College in 1882, led a colourful but short life. She had a long poem published in the *Mason College Magazine*, titled: "Free Thought in the Laboratory (Dedicated to the Demonstrator of Chemistry)." The rhyme began:

> My mind was calm, my heart was light,
> My doubts were few and fleeting,
> Till I attended yesternight
> An M. C. Chem. Soc. Meeting[95]

Naden continued in verse to describe the presentation, then reported how a member of the audience "in sad sepulchral tone"[95] had disputed the speaker's view of molecules, particularly, it can be inferred, the cyclic structure of benzene first proposed by Kekulé only 20 years earlier. Her Demonstrator, Thomas Turner, obviously contested the ring structure of benzene, as he replied in verse:

> A solemn man with tomb-like voice
> Would send Miss Naden greeting,
> And thank her for her pleasant rhyme
> On M.C. Chem. Soc. meeting
>
> He's no regrets of causing doubts
> of truth of benzene rings,
> for doubts should only lead to faith,
> in nobler, truer things.[96]

Naden left Mason College in 1887, but unfortunately died in 1889, as her obituarist described: "... after coming into the possession of a considerable fortune, she travelled throughout the Middle-East and South Asia. She contracted 'Indian demon-fever' & never completely recovered. During the last year, she lectured at Dartford on Women's Suffrage."[94(a)] A bust of Naden overlooks the archives room of the University of Birmingham.

Evelyn Hickmans

Evelyn Marion Hickmans,[97] born in Wolverhampton in 1883, obtained a B.Sc. at the University of Birmingham in 1905 and an M.Sc. in 1906. In 1909, she was a signatory of the 1909 letter to *Chemical News*, but she gave no affiliation. Curiously, there is no record of her from 1906 until 1919, apart from an unsuccessful application for a Demonstratorship at Bedford in 1906. From 1919 until 1922, she was a Lecturer in the Household Science Department of the University of Toronto, Canada, returning to England in 1923.

Hickman's cousin, Leonard Parsons, a Professor specialising in children's diseases at the Children's Hospital, Birmingham, asked her if she would help him with his investigations.[98] She started the following day, a week later being requested to undertake blood chemical analyses for the whole hospital. This she did, being appointed Head of Department. Between 1924 and 1956, Hickmans authored and co-authored 13 publications in a wide variety of studies, mostly on the relationship of abnormal blood chemistry to childhood diseases.

It was work in 1951 for which she became internationally renowned. At the time, Horst Bickel, a Research Fellow at the hospital, was visited by the mother of a 17-month old girl, Sheila, who was diagnosed as suffering from phenylketonuria (PKU). The mother refused to accept that nothing could be done for her daughter and, as a result of her persistance, Bickel, together with Hickmans, and John Gerrard devised a diet low in phenylanaline, which they had concluded was the culprit in the metabolic process of PKU sufferers. The girl recovered.

Their discovery was published in 1954, and the groundbreaking article has been considered so important that the journal *Acta Pædiatrica* reproduced the first page of the original paper in an issue of 2001 on the 47th anniversary of the original publication.[99] The three researchers received the John Scott Medal for contributions to "the comfort, welfare and happiness of mankind" from the City Trust, Philadelphia, in 1962 for their work on methods of controlling phenylketonuria.[100] Hickmans retired in 1953, visiting the hospital in 1987 to attend an anniversary meeting of the Clinical Chemistry Department to commemorate the 25th anniversary of the awarding of the John Scott Medal.[98]

University College, Bristol (University of Bristol)

In Chap. 1, we pointed out the crucial role played by science-active girls' schools in providing the grounding and enthusiasm to set their graduates on a path towards a chemistry degree. For Bristol, there were three schools in particular fitting this role: Red Maids' School[101] (founded in 1634), one of the oldest girls' schools in Britain; Clifton Girls' School (1878); and Redland High School[102] (1883).

To offer matriculants in the southwest of England the possibility of advanced academic courses, the Clifton Association for the Higher Education of Women was established about 1868.[103] When University College, Bristol, was founded in 1876, the Clifton Association transferred their students to the college, ensuring a significant female student population from the very beginning. Like the university colleges founded in other parts of England, University College, Bristol, offered degrees through the University of London until it obtained its own University Charter in 1909.

Both the male and female founders and supporters of University College, Bristol, had been adamant that women were to be admitted from the first as equals to men. However, equal did not mean co-educational: "The Lectures of the College will be open to the students of both sexes, a part of the lecture room being appropriated to women. Separate classes for women alone will also be held, in which the instruction will be of a more detailed and catechetical kind."[103] Women students occupied a separate sphere, as Marian F. Pease described in 1876: "Between lectures we sat in the small women's cloakroom. … It was furnished with three or four wooden chairs and a small deal table and with pegs for our heavy cloth waterproofs …"[104]

Women students were accepted at Bristol as long as they knew their place. One of the most shameful misogynist incidents in British university history occurred at University College, Bristol: the attack on the "Votes for Women" Office on Queen's Road, Bristol. After throwing the furniture through the broken windows, the male students then pulled bales of hay on sledges to the shop, ignited the hay, and watched the shop burn. The Editor of the Bristol student newspaper, *The Bristol Nonesuch*, justified the attack on the grounds that: "The arm of the law is just too short to reach these female fanatics, but we were able to supply the missing inches…"[105]

Women Chemistry Students

Marian Pease was one of the earliest women students to take chemistry courses. In her reminiscences, she commented on the chemistry laboratory of the time: "The chemical laboratory under Dr. Letts was housed in the attics. They seemed to be always making sulfuric acid gas and the house was perpetually full of the smell of rotten eggs…"[104]; while in 1898–1899, the student magazine, *The Magnet*, reported: "A research on the action of sodium in promoting the growth of plants is being carried on beneath one of the Laboratory windows under the direct supervision of the botanical gardener (i.e. Miss X. throws all her refuse metallic sodium out the window)."[106]

The Chemical Society admitted women, but there seemed to be a divide between the sexes as was noted in 1911: "An ordinary meeting of the Chemical or Physical Society is a somewhat dull affair. First of all, tea: two hungry parties, male and female, one on each side of a long table make the most of what is to be had, but ne'er exchange a word."[107]

Millicent Taylor

As we have seen, many of the provincial universities had a woman chemist who devoted her life to the chemistry department of that institution; and for Bristol, this was Millicent Taylor.[108] Taylor was born in October 1871 and attended Cheltenham Ladies' College (CLC) as a student between 1888 and 1893. It was from Cheltenham that she obtained an external B.Sc. (London) in 1893, the same year that she was appointed to the staff at the college. The following year, Taylor was made Head of the CLC Chemistry Department, and then Head of the Science Department in 1911.

Between 1898 and 1910, Taylor devoted most of her spare time to research work in organic and physical chemistry at University College, Bristol, producing a range of papers in those fields. On weekends, she would often cycle to and from the Bristol chemistry laboratories, an 80-mile round trip.[109] She signed the 1904 petition for admission of women to the Chemical Society and the 1909 letter to *Chemical News* (see Chap. 2). In addition, she was one of the first batch of women to gain admittance to the Chemical Society.

Taylor received an M.Sc. (Bristol) in 1910 and a D.Sc. (Bristol) in 1911. She then travelled to northern Ontario, Canada, a region where transportation was primitive and unaccompanied single-women visitors were unknown. Ever the academic, she wrote a monograph, *The Mining Camps of Cobalt and Porcupine*, describing her exploits.

During the First World War, Taylor was involved in the production of β-eucaine, and in 1917 she was appointed as a research chemist at H.M. Factory, Oldbury, returning to her post at Cheltenham at the end of the war in 1919. In 1921, she left CLC to accept an appointment as Demonstrator in Chemistry at the University of Bristol, being promoted to Lecturer in 1923. During the Spring and Summer terms of 1934–1935, Taylor was acting warden of Clifton Hill House, the Women's Hall of Residence. Retirement in 1937 was not the end for her, as she was given the use of a small laboratory in an army hut on the grounds of the Bristol Chemical buildings. Taylor continued research until an accident in November 1960, her death following in December of the same year at age 89. During her lifetime, she was the author and co-author of a total of 19 publications, the last being published when she was 80 years old.

Katherine Williams

Another woman chemist to be educated at King Edward VI High School for Girls (KEVI) was Katherine Isabella Williams.[110] Williams was born about 1848 and she became a student at University College, Bristol, in 1877 at the age of 29, passing the Cambridge Higher Local Examinations in 1885. In the 1880s, she collaborated with William Ramsay and then she embarked on her own research programme in food analysis. In his biography of Ramsay, William Tilden noted:

> At Bristol in the early days of the College there were but few advanced students capable of taking part in research. Among these Miss K. I. Williams, whose death took place in January 1917, deserves to be mentioned. Ramsay suggested to her an investigation into the composition of various food stuffs, cooked and uncooked, and this enquiry occupied her continued attention till the close of her life thirty-five years later.[111]

In Morris Travers' account of Ramsay's life and work, it was mentioned twice that Ramsay suggested Williams repeat the Cavendish experiment on air. It was the later repetition of Cavendish's experiment by others that resulted in the discovery of the noble gases:

> Miss Katherine Williams, who worked for many years in the department on the chemistry of cooked fish, came first under Ramsay. It is said that he suggested that she should repeat the Cavendish experiment on air, but she chose something easier, the determination of the oxygen dissolved in water (p. 69). Later, when he and Miss K. Williams, at Bristol, were investigating an alleged allotropic form of nitrogen (Proc. Chem. Soc., 1886), he says that he suggested that she should

> repeat the Cavendish experiment; but the matter went no further (p. 100).[112]

Williams signed the 1904 petition for admission of women to the Chemical Society and the 1909 letter to *Chemical News* (see Chap. 2). During her career, she authored 10 papers on the chemistry of food over 14 years, obtaining a B.Sc. (Bristol) by research in 1910 in her early 60s. She was in Switzerland when the First World War commenced and, according to her obituarist, the "anxiety and actual hardships she suffered before being able to return to England told severely on her constitution."[113] Williams died in January 1917 and unfortunately the full account of her research on food chemistry was never published, as the following comment in the obituary described:

> Sir William Ramsay and others had induced her to write a popular account of her work, and Sir William had promised to write a preface. This was completed about a year ago, but he had passed away before the promise could be carried out, and Miss Williams died before the work had gone to the publishers.[113]

Emily Fortey

Following after Williams, Emily Comber Fortey,[114] daughter of Henry Fortey, Inspector of Schools (India), was a student at Bristol from 1892, receiving a B.Sc. (London) in 1896. She was awarded a prestigious Science Research Scholarship of the Royal Commission for the Exhibition of 1851, which she used for research at Owens College, Manchester, over the period 1896–1898. Her research, together with that of the Russian chemist Vladimir Markovnikov, showed that the cyclohexane fractions from American, Galician, and Caucasian crude oil deposits were identical.

Upon her return to Bristol, Fortey commenced a five-year collaboration with Sydney Young,[115] and between 1899 and 1903 she co-authored seven papers with Young on fractional distillation. Much of the research formed the basis of Young's book, *Fractional Distillation*, published in 1903. Despite Fortey's name and work being mentioned extensively in the book, her name is not one of those whose contributions were acknowledged in the Preface.

During that period, she must have also undertaken research with Tilden, as he noted:

> Finally, in 1902, a careful study of mixtures of the lower alcohols with water was carried out by Miss E. C. Fortey and myself [Tilden], and our results, taken in conjunction with those of Konowalow, afford strong evidence that no hydrate of any alcohol is formed, at any rate at temperatures above 0°C.[116]

In her 8-year research career from 1896 to 1904, she authored 14 papers and several shorter communications. Fortey also signed the 1904 petition for admission of women to the Chemical Society. After 1904, she left science completely. She died in 1947.

Millicent King

The fourth Bristol woman chemist of note was Annie Millicent King.[117] King was born on 12 September 1900 in Dursley, Gloucestershire. Educated at Redland High School, she entered the University of Bristol intending to complete an arts degree. This was at the time of the First World War, and King changed to a chemistry degree to help meet the need for scientists. She graduated with a B.Sc. in chemistry and then completed a Ph.D. degree in 1922 under James William McBain.[118] Her work with McBain consisted of investigations into the detergent action of soaps and a study of the conductivity of glass surfaces in solutions of potassium chloride. When McBain left the University of Bristol in 1926, King became Secretary and Librarian to the

Chemistry Department at Bristol with the title of Research Assistant.

Until 1939, she continued her research on the specific heats and heats of crystallisation of a number of homologous series, including hydrocarbons, fatty acids, methyl and ethyl esters, and amides. The work was of importance in relation to the cause of the alternation in melting points of the homologous series. Her experimental work ceased at the beginning of the Second World War, when her duties as Librarian and Secretary took up all of her time. In addition, she was an ambulance driver during the air raids. She died on 17 December 1952.

Commentary

We have shown that at each of the provincial universities there were significant numbers of women students in chemistry. At most of the universities in the early years, there were one or more chemistry lecturers willing to be supportive of the women, as examples, Haworth at Newcastle and Birmingham; Lapworth at Manchester; Smithells at Leeds; and Wynne at Sheffield.

At the same time, many of the men students perceived chemistry as an exclusively male domain with its own social society, being hostile to the new arrivals (the riot at Bristol being the most extreme manifestation). Indeed, the male-only chemical-musical-and-social events resembled a men's club or fraternity rather than a chemical society at each of the northern universities. This point was actually stated by the Manchester University Chemical Society: "It was decided that women students be not admitted as members of the Society as it would spoil the social nature of the meetings."[31] Of note, at Manchester and Liverpool, it was the 1910 period when both Chemical Societies dramatically reversed themselves and admitted women chemists.

Some men obviously did not adapt to the new environment, as can be read from the quotes above in the various student magazines. There were also the male-only gatherings which each university chemistry department continued to possess: the Organic Lab "club" at Leeds (which had been the male sanctuary ever since the first arrival of women chemistry students) and the Research Men's Club at Liverpool; while at Manchester, the male chemists retreated to the safety of the Men's Student Union, which survived as a male-only environment until 1957.[119]

Notes

1. Robertson, C. G. (1939). The provincial universities. *Sociological Review* **31**: 248–259; Morse, E. J. (1992). English civic universities and the myth of decline. *History of Universities* **11**: 177–204.
2. Gibert, J. S. (1994). Women students and student life at England's civic universities before the First World War. *History of Education* **23**(4): 405–422.
3. Hird, M. (ed.) (1982). *Doves & Dons: A History of St. Mary's College Durham*. University of Durham.
4. Clemo, G. R. and Brown, N. S. (1956). Schools of chemistry in Great Britain and Ireland–XXII The University of Durham. *Journal of the Royal Institute of Chemistry* **80**: 14–21.
5. Gibby, C. W. (1986). Academic Durham in 1926. *Durham University Journal* **79**(new ser. 48): 1–6.
6. (a) University of Newcastle-upon-Tyne, Student Records; (b) Anon. (1974). Personal news: Deaths. *Chemistry in Britain* **10**: 271.
7. (a) Haworth, W. N. (1941). Grace Cumming Leitch. *Journal of the Chemical Society* 341; (b) "C.E.M." [Mallen, Catherine]. (1942). Dr. Grace Cumming Leitch. *The Northerner: The Magazine of Armstrong College* **42**(3): 24.
8. Hirst, E. L. (1951). Walter Norman Haworth, 1883–1950. *Obituary Notices of Fellows of the Royal Society* **7**: 372–404.
9. Lythgoe, B. and Swan, G. A. (1985). George Roger Clemo, 2 August 1889–2 March 1983. *Biographical Memoirs of Fellows of the Royal Society* **31**: 60–86.
10. (a) Buckley, S. (1985). Catherine Eleanor Elmes (née Mallen), 1896–1994. *Newnham College Roll Letter* 59; (b) White, A. B. (ed.)

(1979). *Newnham College Register, 1871–1971*, Vol. I, 1871–1923, 2nd ed. Newnham College, Cambridge, p. 265; (c) *Calendars*, University of Newcastle-upon-Tyne.
11. Note 3, Hird, unpaginated. We have been unable to ascertain the identity of the 'enterprising young lady.'
12. "A Female Girl." (1906). Debating society. *The Northerner: The Magazine of Armstrong College* **7**(2): 42–43.
13. "Umph." (1908). Woman — and man. *The Northerner: The Magazine of Armstrong College* **8**(4): 117–118.
14. Tylecote, M. (1941). *The Education of Women at Manchester University 1883 to 1933*. Manchester University Press, Manchester.
15. The Editor. (1887). Editorial notes. *Iris: The Magazine of the Department of Women, The Owens College* 5.
16. Anon. (1892). The Department for Women. *Owens College Magazine* **24**(3): 92–93.
17. Jones, D. R. (1988). *The Origins of the Civic Universities: Manchester, Leeds and Liverpool*. Routledge, London, pp. 50–63.
18. Cited in: Purvis, J. (1995). Student life. In Purvis, J. (ed.), *Women's History: Britain, 1850–1945*, St. Martin's Press, New York, p. 193; see also: LaPierre, J. (1990). The academic life of co-eds, 1880–1900. *Historical Studies in Education* **2**: 225–245.
19. Bailey, J. H. (1892). Types of college men — and women: VIII the lady student. *Owens College Magazine* **25**(1): 24–25. Reproduced by courtesy of the University Librarian and Director, the John Rylands University Library, The University of Manchester.
20. (a) Anon. (1947). Obituary: Mrs. Leonore Pearson. *Journal of the Royal Institute of Chemistry* **71**: 217; (b) University of Manchester, Student Records.
21. Robinson, R. (1947). Arthur Lapworth, 1872–1941. *Obituary Notices of Fellows of the Royal Society* **5**: 554–572.
22. Burkhardt, G. N. (1954). Schools of chemistry in Great Britain and Ireland–XIII The University of Manchester (Faculty of Science). *Journal of the Royal Institute of Chemistry* **78**: 448–460.
23. Anon. (1954). Obituary notes: May Badger Craven. *Journal of the Royal Institute of Chemistry* **78**: 105.
24. Wood, J. K. (1954). Schools of chemistry in Great Britain and Ireland–XXXII The Manchester College of Science and

Technology. *Journal of the Royal Institute of Chemistry* **78**: 755–762.
25. Gibson, C. S. (1941). Sir William Jackson Pope, 1870–1939. *Obituary Notices of Fellows of the Royal Society* **3**: 291–324.
26. Egerton, A. C. (1943). Frank Sturdy Sinnatt, 1880–1943. *Obituary Notices of Fellows of the Royal Society* **4**: 429–445.
27. Sinnatt, F. S. *et al.* (1922). *Coal and Allied Subjects*. H. F. & G. Witherby, London.
28. *Minutes 1905–1915*, Manchester University Chemical Society, meeting of October 25, 1906. Reproduced here and subsequently by courtesy of the University Librarian and Director, the John Rylands University Library, The University of Manchester.
29. *Minutes 1905–1915*, Manchester University Chemical Society, meeting of February 5, 1907.
30. *Minutes 1905–1915*, Manchester University Chemical Society, meeting of February 22, 1907.
31. Anon. (1907). Chemical Society notes. *Manchester University Magazine* 23.
32. *Minutes 1905–1915*, Manchester University Chemical Society, meeting of November 13, 1908. The 'Rule VI' itself was not given. The Society Rules changed periodically. Unfortunately, the Minute Book for 1892–1897 is missing from the Manchester University Archives and it is probably during this period that the Rule was proposed. There was no mention of 'men only' in the original Rules of 1878 and the addition might well have occurred when women were first admitted to the chemistry laboratories in 1892.
33. *Minutes 1905–1915*, Manchester University Chemical Society, meeting of October 26, 1909.
34. (a) Fargher, R. G. (1928). Obituary notices: Lucy Higginbotham. *Journal of the Chemical Society* 1056; (b) University of Manchester, Student Records.
35. Anon. (1929). The evolution of the Chemical Society. *Liverpool University Chemical Society Magazine*, New Series **9**(3): 7–8.
36. Anon. (1922). *Liverpool University Chemical Society Magazine, New Series* **3**(1): 30.
37. Anon. (1923). Rip-raps. *Liverpool University Chemical Society Magazine, New Series* **4**(1): 18.

38. Anon. (1928). Women and chemistry. *Liverpool University Chemical Society Magazine, New Series* **9**(1): 12–13.
39. "F.M.E." (1929). In defence of women. *Liverpool University Chemical Society Magazine, New Series* **9**(2): 15–16.
40. "A Woman Chemist." (1929). Women and chemistry — Part II. *Liverpool University Chemical Society Magazine, New Series* **9**(3): 9–11.
41. Edwards, L. P. (1999). *Women Students at the University of Liverpool: Their Academic Careers and Postgraduate Lives 1883–1937*. Ph.D. thesis, University of Liverpool (Ed. Lib Thesis 2153).
42. (Lenten Term, 1925). *Liverpool University Chemical Society Magazine* (2): 2; University of Liverpool, Student Records.
43. Personal communication, e-mail, A. Allen, Archivist, University of Liverpool, 20 September 2007.
44. Anon. (1926). Certificates of candidates for election at the ballot to be held at the ordinary scientific meeting on Thursday, 2nd December. *Proceedings of the Chemical Society* 114; University of Liverpool, Student Records; (19 Feb 1914). *Liverpool University Chemical Society Magazine*.
45. University of Liverpool, Student Records; Anon. (1985). Personal news; Deaths. *Chemistry in Britain* **21**: 800.
46. *Register*, Royal Institute of Chemistry, 1948; University of Liverpool, Student Records; *Liverpool University Chemical Society Magazine*, 1913–1916; Anon. (1978). News review. *Chemistry in Britain* **14**: 267.
47. Shimmin, A. N. (1954). *The University of Leeds — The First Half Century*. Cambridge University Press, Cambridge, p. 16.
48. Raper, H. S. (1940). Arthur Smithells, 1860–1939. *Obituary Notices of Fellows of the Royal Society* **3**: 96–107.
49. Challenger, F. (1953). Schools of chemistry in Great Britain and Ireland–IV The Chemistry Department of the University of Leeds. *Journal of the Royal Institute of Chemistry* **77**: 161–171.
50. "J.B.C." (1910). Our photograph. *The Gryphon: Journal of the University of Leeds* **13**(3): 37.
51. Smithells, A. (1912). Science in girls' schools. *School World* **14**: 460.
52. "F.C.G." (1898). Students: Lady students. *The Gryphon: Journal of the Yorkshire College* **1**(3): 44.

53. "Mel da Kahnt." (1902). The position of man in the laboratory. *The Gryphon: Journal of the Yorkshire College* **5**(5): 86.
54. "Miss Pickel." (1902). From the chemical side. *Owens College Union Magazine* **10**(81): 52. Reproduced by courtesy of the University Librarian and Director, the John Rylands University Library, The University of Manchester.
55. Anon. (1898). Scientific society. *The Gryphon: Journal of the Yorkshire College* **1**(2): 30.
56. "C.A.M." (1916). Cavendish Society. *The Gryphon: Journal of the University of Leeds* **19**(4): 64.
57. Anon. (1898). College news: The organic lab. *The Gryphon: Journal of the Yorkshire College* **1**(3): 44.
58. Anon. (1905). Convivial chemists. *The Gryphon: Journal of the University of Leeds* **8**(3): 46.
59. "H.E.W." (1911). Chemists' Dinner. *The Gryphon: Journal of the University of Leeds* **14**(5): 75.
60. (a) Rayner-Canham, M. (2004). Leslie [*married name* Burr], May Sybil. *Oxford Dictionary of National Biography*, Oxford University Press, http://www.oxforddnb.com/view/article/51604, accessed 27 September 2007; (b) Rayner-Canham, M. F. and Rayner-Canham, G. W. (1993). A chemist of some repute. *Chemistry in Britain* **29**: 206–208.
61. Whytlaw Gray, R. and Smith, G. F. (1940). Harry Medforth Dawson, 1876–1939. *Obituary Notices of Fellows of the Royal Society* **3**: 139–154.
62. Personal communication, e-mail, John Sichel, 31 August 1993; Dainton, F. S. (1993). Reputable memories. *Chemistry in Britain* **29**: 573.
63. (a) Quinn, S. (1995). *Marie Curie: A Life*. Simon and Schuster, New York; (b) Pflaum, R. (1989). *Grand Obsession: Marie Curie and Her World*. Doubleday.
64. Smithells Collection, University of Leeds.
65. Davis, J. L. (1995). The research school of Marie Curie in the Paris faculty, 1907–1914. *Annals of Science* **52**: 321–355.
66. Rogans, E. S. F. (1993). Reputable memories. *Chemistry in Britain* **29**: 573.
67. Anon. (1934). Obituary: Alfred Hamilton Burr. *Institute of Chemistry, Journal and Proceedings* 68.

68. (a) Burr, M. S. (née Leslie). (1925). The alkaline earth metals, Vol. 3. In Friend, J. N. (ed.), *A Textbook of Inorganic Chemistry; Part 1*, Charles Griffin, London; (b) Gregory, J. C. and Burr, M. S. (née Leslie). (1926). Beryllium and its congeners, Vol. 3. In Friend, J. N. (ed.), *A Textbook of Inorganic Chemistry; Part 2*, Charles Griffin, London.
69. (6 July 1937). *Yorkshire Post* 5.
70. Dawson, H. M. (1938). May Sybil Burr, 1887–1937. *Journal of the Chemical Society* 151–152.
71. Anon (1897). The foundation of University College, Sheffield. *Floreamus!* **1**: 3–5.
72. Chapman, A. W. (1955). *The Story of a Modern University: A History of the University of Sheffield*. Oxford University Press, Oxford.
73. Rodd, E. H. (1951). William Palmer Wynne, 1861–1950. *Obituary Notices of Fellows of the Royal Society* **7**: 519–531.
74. Haworth, R. D. and Stevens, T. S. (1956). Schools of chemistry in Great Britain and Ireland–XXIV The University of Sheffield. *Journal of the Royal Institute of Chemistry* **80**: 269–274.
75. Chapman, A. W. (1957–1958). The early days of the chemistry department. *By-Product: Journal of the University of Sheffield Chemical Society* **12**: 2–5.
76. Personal communication, letter, R. L. Wain, 27 February 1998. Turner and Bennett are referred to as the 'tartrate twins' in a variety of sources.
77. (a) Chapman, A. W. (1957). Emily Gertrude Turner 1888–1956. *Proceedings of the Chemical Society* 296; (b) Anon. (1956–1957). Obituary. *By-Product: Journal of the University of Sheffield Chemical Society* **10**: 4.
78. Turner Manuscripts, MS 280, University of Sheffield, letter from Turner to Dr. Rodd of biographical notes on W. P. Wynne that included asides on her own life.
79. Personal communication, letter, Peter A. H. Wyatt, Sheffield University, 24 March 1998.
80. Personal communication, letter, H. J. V. Tyrrell, 9 April 1998.
81. Turner Manuscripts, MS 280, Series 2, 1/1 family tree, University of Sheffield, biographical notes from Jeffrey Turner, nephew.

82. Anon. (1984). Obituary: Mrs. D. M. Leighton. *University of Sheffield Newsletter* **8**(13): 2.
83. (a) (4 May 1984). Our wonderful centenarians. *Kent and Sussex Courier*; (b) (25 May 1984). Obituary: Mrs. D. M. Leighton. *Kent and Sussex Courier.*
84. "Dizzy-Dotty" (1907). A dinner-table discussion. *Floreamus!* **3**: 129–132.
85. "M/M." (1912–1914). Rules for the guidance of first-year science students. *Floreamus!* **6**: 133.
86. "A.P.Q." (1920–1922). Correspondence. *Floreamus!* **10**: 56.
87. "Auntie Ethyl." (1949). Careers for women — I: Research chemistry. *By-Product: Journal of the University of Sheffield Chemical Society* **2**(2): 5–6.
88. Jordan, A. (1883). The Mason Science College: A sketch. *Mason College Magazine* **1**(2): 33–35.
89. Anon. (1899). The Ladies' Common Room. *Mason University College Magazine* **1**(2): 33–35.
90. "An Average Specimen." (1883). Another point of view. *Mason College Magazine* **1**(5): 112–116.
91. Gordon, A. (1909–1910). College encounters III: The chemistry girl. *The Mermaid: University of Birmingham* **6**: 151–153.
92. Anon. (1884). Mason College Chemical Society. *Mason College Magazine* **2**(1): 21–22.
93. White, A. B. (ed.) (1981). *Newnham College Register, 1871–1971*, Vol. II, 1924–1950, 2nd ed. Newnham College, Cambridge, p. 104.
94. (a) Anon. (1890). Constance C. W. Naden. *Mason College Magazine* **8**(3): 49–55; (b) Creese, M. R. S. (1998). *Ladies in the Laboratory? American and British Women in Science, 1800–1900*. Scarecrow Press, Lanham, Maryland, p. 362.
95. "C.C.W.N." [Naden, C.]. (1885). Free thought in the laboratory [Dedicated to the Demonstrator of Chemistry]. *Mason College Magazine* **3**(4): 83–84.
96. "T.T." [Turner, T.]. (1885). Reply. *Mason College Magazine* **3**(4): 84.
97. University of Birmingham, Student Records; Bedford College, Files of the Chemistry Department.
98. Green, A. *History of the Clinical Chemistry Department.* Birmingham Children's Hospital, unpublished manuscript. Birmingham Childrens Hospital is thanked for supplying a copy of the document.

99. (2001). Acta Paediatrica 47 years ago. *Acta Pædiatrica* **90**: 1356.
100. Fox, R. (1968). The John Scott Medal. *Proceedings of the American Philosophical Society* **112**: 416–430.
101. Vanes, J. (1984). *Apparelled in Red: The History of the Red Maids' School*. Governors of the Red Maids' School, Bristol.
102. Thomas, J. B. (1988). University College, Bristol: Pioneering teacher training for women. *History of Education* **17**: 55–70.
103. "E.S." (1876). University College, Bristol. *Journal of the Women's Educational Union* **4**: 124–126.
104. Pease, M. F. (23 February 1942). *Some Reminiscences of University College, Bristol*. Bristol University Archives.
105. The Editor (1913). Editorial. *The Bristol Nonesuch* **3**(8): 3.
106. Anon. (1898–1899). Chemistry. *The Magnet, University College, Bristol* **1**(1): 27–28.
107. Anon. (1911). The Chemical and Physical Societies' Social Evening. *The Bristol Nonesuch* **1**(2): 33.
108. Files in the Chemistry Library Archives, University of Bristol: (a) Millicent Taylor to the Senate of the University of London, December 1914, Application of Millicent Taylor, Head Lecturer in Chemistry at the Ladies' College, Cheltenham, Candidate for the Readership in Chemistry in the Home Science Department of the King's College for Women, University of London; (b) "D.T." (January 1961). Millicent Taylor, D.Sc., Incorporated Guild of the Cheltenham Ladies' College, leaflet 182; (c) Wallace E. S. (1961). An appreciation of Dr. Millicent Taylor, unpublished; (d) Letter, Dykes, D. W. (Administrative Assistant, University of Bristol) to a Mrs Reginald Temperley, Somerset, 6 January 1961, held in the University of Bristol Archives.
109. Baker, W. (1962). Millicent Taylor 1871–1960. *Proceedings of the Chemical Society* 94.
110. Note 94(b), Creese, pp. 266–267.
111. Tilden, W. A. (1918). *Sir William Ramsay: Memorials of His Life and Work*. Macmillan & Co., London, pp. 83–84.
112. Travers, M. W. (1956). *A Life of Sir William Ramsay*. Edward Arnold, London.
113. Anon. (1917). In memoriam. *The Bristol Nonesuch* **4**(18): 196–198.
114. (a) Anon. (1961). *Record of the Science Research Scholars of the Royal Commission for the Exhibition of 1851, 1891–1960*.

Published by Commissioners, London; (b) University of Bristol, Student Records; (c) Note 94(b), Creese, p. 271.
115. Atkins, W. R. G. (1938). Sydney Young, 1856–1937. *Obituary Notices of Fellows of the Royal Society* **2**: 370–379.
116. Note 111, Tilden, pp. 80–81.
117. Garner, W. E. (1954). Millicent King 1900–1952. *Journal of the Chemical Society* 2160.
118. Rideal, E. K. (1952–1953). James William McBain 1883–1953. *Obituary Notices of Fellows of the Royal Society* **8**: 529–547.
119. Pullan, B. and Abendstern, M. (2001). *A History of the University of Manchester, 1951–1973*. Manchester University Press, Manchester, p. 71.

Chapter 6

The Cambridge and Oxford Women's Colleges

Though the proportion of pioneering women students at the Oxbridge colleges was small, their influence on the ambitions of girls was immense. This is illustrated by a comment of Winifred Sturge and Theodora Clark, biographers of The Mount School, York:

> This brings us to Girton, opened in 1864. No other public event can have more influenced the Mount during our period. To Lydia Rouse [Headmistress at The Mount School] it was of greatest moment. She set constantly before her schoolgirls the prospect of college life as a practical ambition, and they strained to look through her window, standing not without effort on tiptoe.[1]

Admission to Cambridge

With the centuries-long reputation of Oxford and Cambridge, women educational activists saw associated women's colleges as providing instant credibility.[2] The first steps were initiated by Emily Davies, who was born in 1830, the daughter of an Anglican clergyman.[3] She became frustrated with the fact that her brothers had followed careers, while she was expected to remain home and undertake parish visiting with her father. In 1858, Davies met Barbara Leigh Smith Bodichon,[4] daughter of the Unitarian, Benjamin Leigh Smith. Davies and Bodichon formed an inspiring duo: Davies, the practical organiser; and Bodichon, the charismatic socialite.

In 1866, Davies authored a book expounding her views: *The Higher Education of Women.*[5] Following from this, with the help of a committee of well-wishers and fundraisers, she opened Benslow House, Hitchin, in 1869. Davies believed that the location of Hitchin between London and Cambridge would enable lecturers from both universities to visit and teach at her college. The impracticability of this plan soon became apparent as neither Cambridge nor London lecturers would travel the distances involved, so the decision was made to move the college nearer to Cambridge. A site was chosen at Girton, then two miles from Cambridge; and with fundraising by Bodichon, the building was completed and opened in 1873.[6]

Parallel with the development of Girton, a second college was being established much closer to Cambridge and with an alternative philosophy. This college owed its origin to Henry Sidgwick,[7] a Professor of Physics at Cambridge. In 1869, he had proposed that informal lectures be offered for women, and with their success it became apparent that the time had come for formal instruction for ladies. Sidgwick chose Anne Jemima Clough[8] to be the organiser of this endeavour.

Clough was a very different individual to Davies. Though her family was socially middle-class, a series of business disasters had left them in a precarious financial position. She had always been interested in the cause of education for women, one of her greatest triumphs being the organisation of lectures on astronomy to women and girls in Liverpool, Manchester, Sheffield, and Leeds, attracting a total of over 550 people. In 1872, Sidgwick rented and furnished a house at 74 Regent Street for Clough's college, but as the institution flourished it became necessary to find their own premises. As a result, in 1875, Newnham Hall was constructed at Newnham, a hamlet on the edge of the city of Cambridge.

The First Victory

Until 1879, women were admitted to the Tripos examinations only by permission of the individual examiners. At issue that year was the formal admission of women to the examinations and the issuance of a certificate noting their accomplishments (but no formal degree). The University Senate overwhelmingly supported the proposal. The news of the vote was awaited eagerly, as Eleanor Andrews, a student at Newnham, described:

> It was arranged as follows. Mrs. Sidgwick's [Eleanor Balfour] sister Lady Rayleigh[9] [Evelyn Balfour] was at the Senate House with her pony carriage & was to drive with the news at once. But some of the students had another plan. One was to get the news directly it was out, she then went to Clare Bridge, waved her handkerchief to another on King's Bridge, who signalled to another on horse back at the back of King's. She then galopped [sic] here at once with a white handkerchief tied on the end of her riding whip. Whereupon two others hoisted a flag on our roof, the gong was sounded and everyone clapped.[10]

Andrews, in this letter to her sisters at home, anticipated the next step: "When women get the Degrees (for this is only the thin end of the wedge) it will be nothing on this. We all feel it is the great crisis in the history of women's colleges."[10] She had no way of knowing another 69 years would elapse before the University of Cambridge allowed this.

The Newcastle Memorial

In 1880, a Girton student, Charlotte Angas Scott, although officially unplaced, was equal to the Eighth Wrangler (eighth in order of merit) in the first part of the Mathematical Tripos.[11] This achievement was seen as indicating that women could indeed emulate the highest attainments of the men undergraduates.

Yet the impetus for the formal recognition of women's qualifications came from outside the colleges, specifically from William Steadman Aldis and Mary Aldis. William Aldis had been an outstanding mathematics student at Trinity College; however, by university statute, his nonconformist religious beliefs precluded him from being granted a college fellowship. Instead, he was appointed as a Professor of Mathematics at the newly founded College of Physical Science at Newcastle.[12] The Aldis's were strong believers in women's rights, and they had fought for admission of women to the University of Durham.

Mary Aldis had spoken to sympathetic Tyneside audiences on the injustice meted out to women students at Cambridge, where their presence was essentially ignored and their academic successes unrecognised.[13] To attempt to remedy the injustice, the Aldis's circulated a petition, to become known as the Newcastle Memorial, exhorting the University of Cambridge to give women formal right of admission to university examinations and to confer formal degrees according to their results. The petition received a huge response, garnering over 8000 signatures.[14]

At the time, neither college had contemplated such a radical move; indeed, the two colleges learned of the document only after its circulation. If Newnham and Girton had been able to form a united front, events may have resulted in a much more satisfactory ending for them. However, Davies contended that women students had to go through the same hurdles and examinations as men, while Sidgwick and Clough were equally fervent in their belief that women's education should be tailored for their special needs and abilities. There was also the geographical separation of the two colleges that reinforced their mutual isolation: Newnham was nestled close to the men's colleges, while Girton was outside the city boundary. The memorial, therefore, only served to increase rancour between the administrators of the two women's colleges[14] and led to minimal action on the university's part. Shortly afterwards, William Aldis obtained an appointment at the University of Auckland, New Zealand, where he unsuccessfully proposed that the University of Auckland grant formal degrees to women graduates from Newnham and Girton.[13]

A second attempt to formalise degrees for women occurred in 1887 after another Girton student, Agnata Ramsay, had excelled in Classics.[2(b)] This venture, too, was unsuccessful. Ramsay was not the only woman to outperform the men, the most exceptional case being Philippa Garrett Fawcett, a Newnham mathematics student who placed above the Senior Wrangler in her year.[15]

The 1897 Battle

The first official attempt to give women degree status occurred in May 1897.[16] The modest proposal was to grant women "titular degrees"; that is, women could use the degree letters after their name and have unrestricted access to the university library, but were to receive none of the rights, such as membership of the university, which accompanied degree status for men. The opponents mobilised support, going to the lengths of chartering a special train on 21 May to carry nonresident MAs (mainly clergymen) who were eligible to vote from London King's Cross to Cambridge. In addition, free lunches were provided by the men's colleges for those opposed to the motion.

The Senate of the University held the official ballot, the results being 662 votes in favour and 1713 opposed. Sutherland described what ensued:

> Once the result of the ballot was announced, triumphant [male] undergraduates marched to Newnham [the closer woman's college], there to be faced down by the grim-faced women dons, assembled behind barred gates. They turned back to the construction of a huge celebratory bonfire in the Market Square, doing hundreds of pounds worth of damage.[2(a)]

Trinity College and the 'Steamboat Ladies'

Though the cause of women's rights had been thwarted at Cambridge, for a brief period, namely 1904–1907, women Oxbridge graduates were able to obtain a degree from Trinity College, Dublin. And it was predominantly graduates of Girton and Newnham Colleges, Cambridge, and Somerville College, Oxford, who took advantage of the opportunity.

The experiences of women in Dublin had largely paralleled those in England. Alexandra College had been founded in 1869, modelled on Queen's College, Harley Street (see Chap. 4). Then, in 1878, the Royal University of Ireland had opened its doors to women at the same time as the University of London.[17] The direct access to a full university undermined Alexandra College and, at the same time, put pressure on Trinity College to allow women to register for degrees. Following a lengthy battle, the first women students entered the doors of Trinity College in January 1904.

For many years, Trinity had enjoyed close relationships with Oxford and Cambridge, in particular, the reciprocal recognition of examinations. Thus, it seemed unfair that Irish women who had successfully completed the degree requirements at Oxford or Cambridge should not have a formal title as the forthcoming women graduates of Trinity would have. To right this wrong, the new Provost of Trinity College, Anthony Traill, a strong supporter of women's higher education, proposed that Oxbridge women graduates be granted Trinity degrees. In June 1904, the Senate approved such a move with a termination of December 1907, by which time Trinity would produce its first class of women graduates.[18]

Traill and the Senate had assumed that, among the women Oxbridge graduates, only those of Irish ancestry would take advantage of the opportunity; but events proved otherwise. Word spread rapidly among British women. All of the women's colleges at both Oxford and Cambridge encouraged their graduates to take advantage of the situation. The Clothmakers' Company, which had provided scholarships for students at Girton, Newnham, and Somerville, announced that it would pay the Commencement (Graduation ceremony) fees of any women who wished to obtain a degree from Trinity.

The whole expedition could be accomplished in as little as 34 hours using the fast mailboats. Women graduates often arranged to travel as a group of friends from the same graduation year in a college, or from the same school in which they were currently teaching. A former student of Manchester High School for Girls reported:

> On Saturday, April 29th, [1905] a good number (87) of women from Oxford and Cambridge — mostly from the latter University — assembled in "No. 6" to don their gowns and hoods. It was a merry gathering; old friends who had not met since College days greeted each other.[19]

With this flood of women arriving at the gates of Trinity, attempts were made to abrogate the agreement, but Traill insisted that the Senate could not renege on its commitment. He mollified the Senate by promising that all fees collected would be committed to the construction of a women's residence at Trinity.

During the near-four-year period this dispensation was in effect, approximately 700 Oxbridge women availed themselves of the opportunity to purchase a formal degree. Following the expiration of the agreement, Girton and Newnham students jointly petitioned the Trinity Senate to renew the privilege, but this was denied. Subsequent petitions, even one by two Indian princesses, were also dismissed.[20] The "window of opportunity" had closed.

The 1920 Battle

In 1919, a Syndicate was formed that in 1920 brought forth two options: Report A, which proposed full membership for women; and Report B, which proposed a separate university for women. This time, the women of Cambridge were ready and organised, forming a joint Girton and Newnham Committee.[21] A total of 3413 former women students at Cambridge were sent a Memorial requesting support for Report A. From those mailed, 2631 signatures were obtained, and the document with a list of signatories was forwarded to the 7000 members of the Senate. Agnes Conway, a former Newnhamite, divided the country into regions and organised women lobbyists to encourage Senators in their area to support Report A. Leaflets were prepared explaining why the women's colleges deserved equal status and why a separate women's university was unacceptable.

The debate became public in the Correspondence column of *The Times*. Letters in support of women's rights came from many sources including Eleanor Balfour (Mrs. Sidgwick),[22] Sophia Jex-Blake (see Chap. 8), F. Gowland Hopkins (see Chap. 9), and, of particular importance, a stirring statement signed jointly by two Cambridge Professors, Ernest Rutherford in Physics and William Pope in Chemistry:

> For our part, we welcome the presence of women in our laboratories on the ground that residence in this University is intended to fit the rising generation to take its proper place in the outside world, where, to an ever increasing extent, men and women are being called upon to work harmoniously side by side in every department of human affairs.[23]

The official day of voting was 8 December 1920. The Council of Newnham College offered hospitality for all those supporting the cause of Report A. Despite the lobbying and proselytising by Newnhamites and Girtonians, when the Senate met to give its verdict, Report A was defeated by 904 votes to 712, while Report B was defeated by 146 to 50.

Embarrassed by Cambridge having become the only university to bar women from degrees (Oxford having succumbed that same year, see below), the Council of the Senate put forward a pair of linked motions, the first having provisos to appease some of the less rigid opponents. The first motion was that women be admitted to formal membership, provided that the number of women students was limited to 500; that they were members of women's colleges; and that women would not be members of Senate. In the event this motion failed, then women would be admitted to the title of degree, but they would have no privileges. When the vote was taken on 20 October 1921, the first motion was defeated by 908 votes to 694, while the second passed by 1011 to 369.

Following the vote, the Reverend Pussy Hart aroused the assembled 1300 to 1500 male undergraduates outside Senate to head for Newnham. Though the authorities expected some sort of demonstration, the size and speed overwhelmed them. The Memorial Gates of Newnham College were shut and barred, but this time the mob, using a hand cart, smashed the lower part of the gates. In addition, some of the rioters turned their attention to Clough Hall, forcing open the Hall gates and breaking a few windows.[24] One of the women students of the time, Mary Roberts (Mrs. Henn), recalled that: "... [Newnham] College temporarily resembled a beleaguered fortress which we were forbidden to leave."[25]

The following day was very different, as Roberts reported:

> Next day the University was steeped in gloom and guilt, to an extent I imagine seen neither before nor since. Apart from the official and collective apology of the University, Miss Clough was inundated with contrite statements from every conceivable group of [male] undergraduates, athletic and otherwise. Certainly the captain of the University rugger team rose to unprecedented heights of oratory in his plea for contributions to the fund to make good the damage and wipe out the stain.[25]

And so another attempt at equal rights for women students had failed.

Success in 1947

Twenty-seven years passed before the issue arose again. A Second World War had come and gone, while women at Newnham and Girton were behaving in every other way as if they were full members of the university. The issue had become a considerable embarrassment to the university authorities. Another Syndicate reported in 1947 and, to the relief of all, its recommendations giving women equal rights passed without dissent.

With the arrival of 1948 and after 67 years of controversy, women finally had formal degree status with all of the rights and privileges thereto attached. Dorothy Dale (Mrs. Thacker) described the new reality:

> For examinations we sat with the men students and our names were in the same alphabetical list as theirs. Degrees were conferred on us in the Senate House by the Vice-Chancellor and we became eligible for all University posts and prizes; we were now full members of the University on an equal footing with men.[26]

Girton and Newnham Colleges

Life at the women's colleges, particularly in the early years, was very circumscribed.[27] A former student, "M.L.," described the regularity of life at Girton in 1888: "The daily routine is — Prayers at eight, breakfast from eight to nine, study hours from nine to one, lunch when you like between twelve and three, noise hours from one to three, [the students then attended lectures in the afternoon,] dinner at six, study hours again from half-past seven to nine, and then noise hours again from nine until half-past ten."[28]

In those early years, the women were often taught separately. To instruct Girton students, the lecturers had a considerable distance to travel with such lectures being given in the afternoon, as their classes at the University in Cambridge were held in the morning. The Girton women therefore spent their mornings reading. Some lecturers already allowed mixed classes, another former student, "J.W.B.," reporting: "Notably among these are the Natural Science students, who have the reputation of never being in their rooms except during the evening, the afternoon generally being spent on practical work, either in Cambridge or the laboratory here."[29]

The publication of the Tripos results was the highlight of the year. In 1882, the success of Natural Science students in the examinations provided especial excitement: "The scene last year, when news was brought out that we had obtained *three* first classes in Natural Science, was a never-to-be-forgotten one; the wildest excitement ensued, bells were set ringing, and many other frivolous things were done in the heat of our enthusiasm."[29]

Even in the early 20th century, the chaperone rules were enforced. The issue of modifying these rules arose in 1917, provoking a dispute between the traditionalists and moderate reformers:

> We were discussing whether a Newnham student could be allowed to take an afternoon walk with an [male] undergraduate, and a senior member said, 'Well ... yes ... if she is engaged to him.' To which a friend of mine on the committee replied that it seemed a heavy price to pay for a walk.[30]

Women Chemistry Students

At Girton and Newnham, women students could learn of the principles of science, but the laboratory experience was another question altogether. There was already a shortage of laboratory space at Cambridge, and the intrusion of women only exacerbated the situation. During the 1870s, Philip Main generously organised early morning practical chemistry sessions for women students at his laboratory in St. John's College.[31]

His classes ceased in 1879 with the completion of chemical laboratories at Newnham and Girton. With their own facilities within the walls of the women's colleges, women chemistry students no longer needed chaperones to attend the laboratory sessions. A chemistry student, Marguerite Ball, recalled that the standard attire of Newnham students was also worn in the laboratory:

> We wore almost a uniform of white blouse, and tweed coat and skirt. The blouse should be of 'nun's veiling' and the collar fixed with a plain rolled gold pin; but such blouses were expensive! When I wore my first, my clumsy spilling of acid in the lab. made large holes in the front which no darning would conceal... But it was still all right with the coat on; and not on the warmest spring day might that coat be removed, either at lectures or in the street, so all was well.[32]

However, after 1913, upon the retirement of Ida Freund (see below), women chemistry students from Newnham had to undertake the experimental component of their courses in the University Chemistry Laboratories. The women at Girton were able to keep their own laboratory until Beatrice Thomas (see below) retired in 1935. In the University Chemistry Laboratories, an entirely different atmosphere prevailed, as the arrival of women students was against the wishes of the male laboratory staff. They made life difficult for the women pioneers, as Ball remembered: "... the lab boys took a delight in leaving some essential bit of apparatus out of our lists so that we had to walk the whole length of the lab to the store to ask for it. An ordeal for some of us, especially as they appeared to be too busy to attend to us for several minutes while we waited at the door."[32]

Ida Freund

Though the smaller one, Newnham was regarded as the more science-oriented of the two Cambridge women's colleges. It was

Newnham's own Chemistry Laboratory, in the middle of the College grounds, where Ida Freund[33] reigned supreme. Born on 5 April 1863 in Austria, she was raised by her maternal grandparents following the death of her mother. She studied at the State School and the State Training College for Teachers in Vienna.[34] Following her grandmother's death in 1881, Freund moved to Britain to join her uncle and guardian, the violinist Ludwig Strauss.

It was arranged that Freund would be sent to Girton to complete her education, a decision that she bitterly opposed at the time. Completing her undergraduate studies in 1886, Freund accepted a one-year Lectureship at the Cambridge Training College for Women (later Hughes Hall). The next year, she was offered a position as Demonstrator at Newnham, and she was promoted to Lecturer in Chemistry in 1890.

In a letter to her mother, Catherine Holt, one of the students of 1889, mentioned Freund:

> I attended my first lecture today; it was Chemistry; there were about 8 students from this college and three from Girton; ... Afterwards we adjourned for a couple of hours to the laboratory here; Miss Freund is the presiding genius, a jolly, stout German, whose clothes are falling in rags off her back. We made lots of horrible smells and got back here for lunch at a quarter past one.[35]

Hilda Wilson, who entered Newnham in 1905, recalled how every year, just before the examinations, Freund would summon her chemistry students for a study session.[33] For the 1907 study session on the lives of important chemists, each student was provided with a large box of chocolates containing a written biography of a famous chemist. The following year when the periodic table was the focus, a large periodic table was provided with each element location consisting of an iced cake showing its name and atomic weight in icing, while the group numbers were made of chocolate and the dividing lines were rows of candy sticks.

During her youth, a cycling accident resulted in Freund having a leg amputated, but this did not affect her mobility, as Ball recalled:

> Miss Freund was a terror to the first-year student, with her sharp rebuke for thoughtless mistakes. One grew to love her as time went on, though we laughed at her emphatic and odd use of English. Yet how brave she was trundling her crippled and, I am sure, often painful body about in her invalid chair smiling, urging, scolding us along to "zat goal to which we are all travelling which is ze Tripos."[32]

Not only did Freund move freely about campus, but she became a fervent traveller, "wheelchairing" her way around England, Scotland, Germany, Austria, Switzerland, and Italy.

Apart from her teaching activities, Freund performed research on the theory of solutions that culminated in a substantial paper, and in 1904 she addressed the Cambridge University Chemistry Club on the topic of double salts.[36] Her most renowned work, however, was a chemistry text, *The Study of Chemical Composition*,[37] which remained popular for many years.[38] The historian of chemistry, M. M. Pattison Muir, commented that her text "is to be classed among the really great works of chemical literature,"[39] and the book itself was reprinted in 1968 as a classic in the history of chemistry.[40] Then in 1904, she wrote a manual of laboratory procedures, *The Experimental Basis of Chemistry*, which could be used to illustrate chemical concepts.[41]

Freund was devoted to Cambridge, as the following story shows. For an essay on the early history of the atomic theory, she was awarded the University's Gamble Prize in 1903. She donated the money to Girton College and the Balfour Laboratory for the purchase of books and scientific apparatus, provided that it was spent on "luxuries and not necessities."[34(a)]

Freund was active in feminist causes, being a strong supporter of women's suffrage, and she was one of the three organisers of the petition for admission to the Chemical Society (see Chap. 2). She died on 15 May 1914 and, following her passing, the Ida Freund Memorial Fund was instituted at the University of Cambridge to further train women teachers in the physical sciences.

Dorothy Marshall

During the 1897–1906 period, Dorothy Blanche Louisa Marshall[42] was the Demonstrator in the Girton Chemistry Laboratory. Marshall, born on 12 December 1868 in London, was the daughter of Julian Marshall, connoisseur and collector, and Florence Ashton Thomas, musician and author. Marshall was educated at King Edward VI High School for Girls, Birmingham (KEVI), and then went to Bedford College in 1886. From there, she transferred to University College, London (UCL), studying chemistry, physics, and electrical technology and graduating with a B.Sc. in 1891.

Staying at UCL as a postgraduate researcher until 1894, Marshall studied heats of vaporisation of liquids, her work resulting in three lengthy publications, one of which was co-authored with William Ramsay. Following a one-year Demonstratorship at Newnham, she was appointed as Demonstrator in Chemistry at Girton in 1896, and then promoted to Resident Lecturer in Chemistry in 1897. During her time at Girton, she signed the 1904 petition for the admission of women to the Chemical Society (see Chap. 2).

Marshall left Girton in 1906 to become Senior Science Lecturer at Avery Hill Training College when it was founded in 1906. Avery Hill, which would become the most prestigious of the women's teacher training colleges, did not, as it happened, teach science until 1930.[43] Soon after being appointed as Acting Principal in February 1907, Marshall became ill (as both her predecessor and successor as Principal resigned due to illness, it is possible that "illness" was a way out of a thankless position). Following her resignation, Marshall became Senior Science

Mistress of Huddersfield Municipal High School in 1908. She moved south in 1913, taking a position as Chemistry Mistress at Clapham High School (a GPDSC school). Like so many other women chemists (see Chap. 12), Marshall started war work in 1916, in her case, with the National Physical Laboratory (NPL) as a Scientific Research Assistant, where she stayed for the remainder of her working life.

Beatrice Thomas

Marshall was followed at Girton by another former KEVI student, Mary Beatrice Thomas.[44] Just as Freund had reigned supreme at Newnham, so her close friend Thomas was a dominant figure in Chemistry at Girton from 1902 until 1935. Thomas was born on 15 October 1873 in Birmingham, one of two daughters of 'Wild William' Thomas, a surgeon and later Professor of Anatomy at Mason College, Birmingham.

With her father's support, but opposition from her mother, Thomas followed the well-trod trail from KEVI to Newnham (see Chap. 1) in 1894, where she studied chemistry and physiology, completing the degree requirements in 1898. After graduation, she was a Demonstrator in Chemistry for two years at the Royal Holloway College; the subsequent year, she held a Priestley Scholarship at the University of Birmingham. From 1902 to 1906, she was a Demonstrator in Chemistry at Girton College, under Marshall. Thomas was appointed as Lecturer and Director of Studies in Natural Science at Girton in 1906, following Marshall's departure, a post that Thomas held until 1935.

Initially, Thomas was able to undertake some research; but in later years, with the pressure of teaching and a hostility to women in the University Chemistry Research Laboratory, she focused her life on her students. A student, "D.M.P.," described how Thomas was a scientist and a feminist:

> ... an eager devotee of the gospel of dedication to that search for truth and for scientific knowledge which, at the end of the

> Victorian age, had for many ardent spirits taken the place of orthodox religion. We found her unyielding in her demands that we should put academic tasks before all other interests, that we should maintain the tradition of the earnest pioneers, who never for one moment allowed themselves to forget their goal of proving and demonstrating that no distinction can be drawn between the intellectual powers of men and women, once they are offered similar educational opportunities.[44(a)]

Thomas, who had been a chemistry student under Freund, became close friends with her former Demonstrator. Upon Freund's death, it was Thomas who organised the Ida Freund Memorial Fund and co-edited Freund's unfinished *Experimental Basis of Chemistry*.[41] Though she taught at larger, classics-oriented Girton, her sympathies lay more with the smaller, science-oriented Newnham. There must have been some sharing between the two women's colleges, for one of her obituarists, a former Newnham student, "D.M.A.," commented: "Fortunately for Newnham science students, Miss Thomas was able to undertake our supervision in Chemistry so that we were able to appreciate her qualities as a teacher."[44(b)]

Known as "Tommy" by her students, Thomas demanded the highest standards. She was never perturbed by any laboratory incident. In a touching posthumous letter to Thomas, Dorothy Russell remembered:

> I recall how, on one occasion, a moth elected to commit suicide in my flask of standardised acid. This naturally did not escape your eye! You regarded the corpse with considerable disfavour and merely said: "Of course that acid will have to be re-standardised."[44(c)]

Thomas was an ardent supporter of women's rights, signing both the petition for admission to the Chemical Society in 1904 and the letter to *Chemical News* in 1909 (see Chap. 2). However, women at Cambridge had to be careful not to let their sympathies become too obvious, as "D.M.A." noted in her obituary of Thomas:

> The position of women in the University was uncertainly poised, and it was still very necessary to avoid adverse criticism, so no member of the women's colleges could join the W.S.P.U. [Women's Social and Political Union] — the "Militant Suffragettes" led by Christabel Pankhurst — but most of us belonged to the N.U.W.S.S. [National Union of Women's Suffrage Societies] which pursued the same aims by non-violent and more legitimate methods.[44(b)]

When Thomas purchased a house, she found an avenue to make a personal protest. The passive resisters, including Thomas, refused to pay the education rates levied on householders on the basis of nonrepresentation. As "D.M.P." described: "Time after time, ... she submitted to the arrival of the bailiffs and ensuing sale of her goods, without offering obstruction, but with signal success in making the authorities appear ridiculous."[44(a)]

Outside of the laboratory, physically frail Thomas was an avid hiker through all weathers. Nothing fazed her. During one night in the Second World War, a bomb dropped near her house, causing the roof to collapse. A policeman forced his way up to her bedroom. He ordered her to get up and leave as the roof was no more, to which she replied: "I am quite aware of that, but it *has* fallen, so why should I get up *now*?"[44(a)]

Cambridge had been Thomas's life, and even in retirement she welcomed visits from her former students and from the daughter of Ida Smedley (see Chap. 2), who called her "Aunt Bea." Also, she attended meetings of the Faculty of Physics and Chemistry regularly until her health declined further. Her rebelliousness continued in altercations with nationalised gas services, doctors, and any other figure of authority who crossed her path, until her death on 14 June 1954.

Mary Johnson (Mrs. Clark)

Mary Johnson,[45] a graduate of Newnham, became one of the first women researchers in the Cambridge Chemistry Laboratories. Born on 14 January 1895, Johnson was educated at Bede College School, Sunderland. She had actually wanted to read mathematics at Newnham when she entered in 1913, but inadequate preparation at school resulted in a change of plans and she took chemistry instead. When she completed the Part II of the Natural Science Tripos in 1917, she placed above all of the men in her year.

Johnson was awarded a Newnham Bathurst Research Scholarship in 1917–1918 for research in organic chemistry, but she could not be registered for a Ph.D. as, at the time, women were ineligible for higher degrees. From 1920 to 1925, she was a researcher in the Cambridge Chemistry Laboratories holding various research fellowships. Her work on the dyestuffs of the pyrazolone series, undertaken with Charles Stanley Gibson,[46] was published in 1921.

According to family lore, being a senior in the laboratory, Johnson was asked who she wanted to work across the bench from her. She responded that she didn't care "so long as he is a good tidy worker"[45(b)] and, in due course, Leslie Marshall Clark was assigned to the bench. In 1925, they announced their engagement and Johnson submitted her resignation. They were married on 26 February 1926 and had two daughters, both of whom followed in her footsteps to Newnham.

Again according to family history, Clark, who was running a research laboratory for Imperial Chemical Industries at the beginning of the Second World War, was quoted as saying that: "since she was a better chemist than he was, he would have her in his laboratory like a shot, if it wasn't for the children."[45(b)] Later in the war, she briefly taught chemistry at Howell's School in North Wales. Unfortunately, Johnson had a massive stroke in 1945 that severely incapacitated her until her death on 26 April 1969.

Delia Simpson (Mrs. Agar)

During our time frame, the most prominent woman researcher in the Cambridge Chemistry Laboratories was the spectroscopist, Delia Margaret Simpson.[47] Born on 5 February 1912 in London, the daughter of Robert Simpson, a sanitary inspector, and Delia Maud Pope, a teacher, she was educated at the Haberdashers' Aske's Hatcham Girls' School. Simpson entered Newnham in 1930, completing her studies in 1934.

Simpson was a Bathurst Research Student from 1934 to 1936, following which, with a Travelling Scholarship, she spent 1937 to 1939 undertaking research at the University of Vienna. Her initial research had been in the field of optical rotary dispersion and circular dichroism, but she discovered that she found spectroscopy much more interesting. Before the Second World War, she studied the spectra of small molecules in the vacuum ultraviolet; but with the onset of war, she shifted to infrared spectroscopy. This technique was used to analyse samples of enemy fuels for the Air Ministry, in particular to identify the use of synthetic oils in place of natural oil, which, because of the naval blockades, was in very short supply in Germany.

This war work was undertaken with the women chemists of Bedford. As discussed in Chap. 4, Bedford personnel and students had been evacuated to Cambridge, and for the duration of their stay Simpson held the position of Demonstrator and Assistant Lecturer in Chemistry for Bedford College. Following the departure of the Bedford students in 1944, she was appointed as College Lecturer in Chemistry at Newnham and was elected to a Teaching Fellowship.

Simpson continued research after the war, working for a time with biologists on fluorescence spectroscopy. Her major focus continued to be infrared and Raman spectroscopy, resulting in a total of 20 papers over her career, including a two-part review of the spectra of hydrocarbons in *Quarterly Reviews of the Chemical Society*. However, in 1954, she was appointed as the Director of Studies in Natural Sciences, which required her to spend most of her time teaching at Newnham and Downing Colleges. In 1952, Simpson married the electrochemist John Newton Agar, a Fellow of Sydney Sussex College, Cambridge. After retirement in 1977, she continued to attend spectroscopy seminars until close to her death on 29 March 1998, aged 86 years.

Sosheila Ram

Of all the early chemistry graduates of the Cambridge women's colleges, one of the most remarkable was Sosheila Ram.[48] Ram, born on 29 November 1899 in Edinburgh, was the elder daughter of Labbhu (Murgai) Ram, a doctor, and Sukhda Ram. She was educated at Mexborough Secondary School, Yorkshire, and entered Newnham in 1918. Completing the first part of the Natural Science Tripos in 1921, she spent 1921–1922 at St. Mary's Training College, London, from which she received a Certificate in Education from the University of London. Then, she travelled for her first time to India, where she was appointed as a Lecturer replacing Alice Bain (see Chap. 7) in the Department of Science at the Lady Hardinge Medical College for Women, a college and associated hospital that was set up in 1916 and was attached to the University of Delhi.[49] She returned on two occasions to Cambridge, where she continued her studies, being granted an M.A. in 1925.

In 1938, Ram left Lady Hardinge Medical College to become a Consulting and Analytical Chemist with Gindlay & Co., the first woman of Indian heritage to establish and run her own laboratory. Her research in analytical chemistry, some of which was published in *The Analyst*, resulted in her election to Fellowship of the Royal Institute of Chemistry in 1945. In addition to her new career, she resumed teaching at the Lady Hardinge Medical College as a Lecturer in Biochemistry for the 1942–1944 period.

Throughout her time in India, Ram had endeavoured to further the education of women and to spread knowledge of modern dietetics. Thus, in 1953, she decided to devote the rest of her life to social work at remote foothill communities in the Himalayas, using Dagshai as a base. This outreach work was to lead to her death on 28 March 1957, as the following account describes:

> Her social work won for her the love and gratitude of many, both men and women, but her interest in prohibition brought her the animosity of a few individuals. One moonlight night, on her return from a visit of mercy to a village home, amidst the glorious scenery of the Simla Hills, she was ambushed and most cruelly molested. Within a few weeks she was dead.[48(b)]

Oxford Women's Colleges

Though the authorities at Oxford were almost as recalcitrant as those at Cambridge, the progress of women's rights was not accompanied by the sort of demonstrations of overt hostility that occurred twice at the gates of Newnham. The first event in the Oxford saga was the offering of classes for women in 1865 by Mark Pattison, Rector of Lincoln College, together with some other men and women.[2(b)] These classes were modelled on those which had been offered for women by the Midland Institute, Birmingham, and the ultimate aim was the establishment of regular post-secondary classes (or a college) for women in every large town in Britain. The courses at Oxford languished, but were then revived by a committee of five women in 1873.

That same year, Balliol and Worcester Colleges of the University of Oxford were offering scholarships to school students based on the results of the Senior Local examinations. A student by the name of A. M. A. H. Rogers had placed in the First Class of Junior candidates in 1871, and then headed the Senior candidate list in 1873.[50] Henry Daniel of Worcester College wrote to Rogers' father offering a scholarship to Worcester College for the successful student. Informed that the student in question was Annie Mary Anne Henley Rogers, the offer was hastily withdrawn. As Vera Brittain commented: "Thus, for the first time in history, the question of admitting women to Oxford University was raised. It was, of course, dropped as hurriedly as a red-hot coal ..."[50]

Signing herself always, "A.M.A.H.R.," Rogers was to subsequently triumph and to continue her role as a thorn in the flesh of the powers-that-were at Oxford, becoming an expert in the University Statutes, particularly as they pertained to women's rights.[51] Brittain observed: "If the women at Oxford could be said to owe their triumph to any one individual, the credit is hers. She was their forerunner, their expert, their champion, and the symbol of their struggle."[50]

Perhaps prodded by the Rogers incident, in 1875 separate examinations for women over 18 were established, corresponding to the Oxford examinations: Responsions, Moderations, and Final Schools. Rogers was one of the first to take advantage of the opportunity, obtaining honours in Latin and Greek in 1877, then ancient history in 1879. In 1878, an Association for the Education of Women (the AEW) was formed to further the advanced education of women in Oxford, and Rogers was elected to the Committee.

There was an interesting parallel between the cause of women's education at Cambridge and at Oxford. At Cambridge, as we have seen, Henry Sidgwick and his wife, Eleanor Balfour, had been key "agents of change." At Oxford, it was Henry's younger brother, Arthur Sidgwick,[52] and his wife, Charlotte Wilson, who were at the centre of the battle. Arthur held the position of Secretary of the AEW from 1882 to 1907, and the Presidency from 1907 to 1915.[53]

The Founding of the Colleges

The first of the Oxford women's colleges, Lady Margaret Hall (LMH), opened its doors in 1879. It was founded by Elizabeth Wordsworth, greatniece of the poet William Wordsworth, and it was named after Lady Margaret Beaufort, a learned medieval noblewoman and mother of King Henry VII. LMH was the first venture into higher education by Anglican high churchmen, and all students admitted were required to be Anglicans.[54] Ethel Harvey, a student in 1903, described how every morning there were mandatory prayers at 8.05 a.m. in the chapel and then prayers again at 8:30 p.m. She added: "Everybody is expected to attend Divine Service at least once on Sundays, and Miss Wordsworth's Bible Class after dinner."[55]

The religious requirement of LMH initiated the formation of Somerville College in 1879, named after the pioneering woman scholar and scientist, Mary Somerville.[56] The founders of Somerville, which included William Sidgwick (the eldest of the Sidgwick brothers) and the chemist A. G. Vernon Harcourt, insisted that there may be no religious tests or obligations for Somerville women students. In fact, it transpired that over one third of the students at Somerville were nonconformists.[57] This made Somerville unique among the Oxford women's colleges. In some ways, Somerville seemed closer philosophically to Newnham and Girton than its sister Oxford colleges. For example, Newnham, Girton, and Somerville were the three colleges from which nearly all of the "Steamboat Ladies" came. Somerville, with its links to university liberalism, was the first to offer opportunities for research to Oxford women students.

St. Hugh's Hall was founded in 1886 by Elizabeth Wordsworth as a private initiative, and named after Hugh of Avalon, Bishop of Lincoln in the 12th century. Wordsworth was concerned that daughters of impecunious families, such as the clergy, could not afford the fees of Lady Margaret Hall, and therefore had to attend a nonreligious-based college: Somerville, Newnham, or Girton. She hoped that the college would provide a steady supply of church workers and teachers. Students at St. Hugh's attended the same classes and programs as those at Lady Margaret Hall, and underwent equivalent religious instruction.

Wordsworth was one of several influential anti-suffrage women at Oxford. In fact, as Julia Bush has concluded, feminism did not provide the driving force at Oxford for women's advancement.[58] Many women spouses of Oxford academics had been signatories of the 1889 anti-suffrage Appeal; while Mary Ward, wife of a Fellow of Brasenose College, was the organiser of the Women's National Anti-Suffrage League in 1908. Bush explains the philosophy of these women as follows: "… improvement in women's education were regarded as a means of reinforcing women's ordained role in society, and enhancing its influence without any dangerous claims to gender equality."[58]

St. Hilda's was opened in 1893, also as an Anglican College. The founder was Dorothea Beale of Cheltenham Ladies' College (CLC) fame (see Chap. 1), who chose the name in recognition of the 7th-century St. Hilda of Whitby, head of the most important place of learning of its time. There were close ties between the institutions, with CLC having representatives on the St. Hilda's Council until 1955, and, up to 1962, about 10% of all students ever at St. Hilda's were Cheltonians.[59]

At the time, many Anglicans had deep reservations about the higher education of women. This was reflected in an early emphasis on training to women for teaching, social, or missionary work. All three of the Anglican Colleges suffered financially, as Janet Howarth commented: "Anglican ambivalence towards women's higher education had various consequences for the Church of England halls at Oxford. One was that they were much poorer than Girton, Newnham or Somerville: church people, although they had given generously to Keble, were slow to respond to appeals for money to bring women students to Oxford."[60]

When LMH and Somerville were founded in 1879, there were 35 women students living in Oxford with parents or friends who wanted to continue their education. Initially called the "unattached," in 1891 they were formally named the Society of Oxford Home-Students (SOHS). A band of strong supporters (including Arthur Sidgwick) argued that, by living in lodgings or at home, Oxford would be able to attract talented women whose families could not afford the fees of the residential colleges. Over time, the SOHS enabled women graduates from universities around the world to study for graduate degrees at Oxford. In 1942, the SOHS became the St. Anne's Society, and then St. Anne's College in 1952.

Admission to the University

On Trinity Sunday, 1884, Dean Burgon preached a sermon in New College Chapel that closed with a message for women: "Inferior to us God made you, and inferior to the end of time you will remain. But you are not the worse off for that."[61] Nevertheless, there were no dramatic incidents on the long pathway towards the acceptance of women students at Oxford as there was at Cambridge. Brittain contended that the Oxford method of publishing graduation lists in alphabetical order rather than the Cambridge method of descending marks meant that women who excelled at Oxford were not so obvious.[62]

By the end of 1894, University examinations had been opened to women. However, the offering of a formal B.A. degree to women was far more contentious. When the matter came to a vote on 3 March 1896, the women's cause was defeated in Congregation, the governing body, by 215 votes to 140.[63] Though the vote took place in a civil manner, Rogers observed: "... we have more actual enemies to the firm establishment of women's education in Oxford than we thought."[64] A second unsuccessful attempt was made to entitle women to degree status in 1908. Rogers, by then Tutor in Classics at St. Hugh's, was once more one of the proponents.[2(b)] However, a small victory occurred on 1 November 1910 when the women's colleges were granted the title of Recognized Societies, and the previously invisible women students were accepted as Registered Women.[65]

Oxford: A Men's University

It was not until May 1920 that, without opposition, the final statute passed to permit women to graduate with full membership of the University. The fact that Parliament had passed the

Sex Disqualification (Removal) Act in 1919 was certainly no coincidence, just as it had been 1920 when the Chemical Society finally admitted women chemists (see Chap. 2).

But this was not the end of the story. On 14 June 1927, fearful of the rising proportions of women, Congregation debated and passed a motion by 229 votes to 164 that: "The University has a right to remain predominantly a men's university."[66] In addition, this motion limited the number of women students in residence to 840 and prohibited the foundation of any new women's college if it would make the proportion of women greater than 20% (the quota for women was raised to 970 in 1948, but not abolished until 1957).

Howarth has commented on this decision:

> Why did Congregation make a point of denying them [women's colleges] this parity? Not, it seems because of any general backlash of feeling against women at Oxford. The [student] Union had, it is true, debated and passed earlier in the year a motion that 'the women's colleges should be razed to the ground', but Dame Lucy [Sutherland], who was guest-speaker in that debate, described it as a light-hearted affair, full of bad jokes, after which she struck up a lasting friendship with the proposer of the motion.[66]

One of the major reasons for the vote, according to Howarth, was "... pressure on library and lecture space and on lodgings in crowded post-war Oxford." The animosity towards women during the debate at Congregation cannot be ascribed to such a rational argument, and, in any event, a lack of lodging space had no relation to the women's residential colleges. Instead, the vote seems more a specific example of the anti-feminist wave of the late 1920s and early 1930s that we will examine in Chap. 13. The secondary reason provided by Howarth is, in our view, the central factor, namely: "But in 1927 the 'fundamental question', as one friendly male don had put it, was 'feminine influence and the *monstrous regiment of women*'."[66] In other words, many Oxford men saw the increase in numbers of women students as destroying the essential masculine identity of the University.

Women Chemistry Students

The Oxford Colleges did not serve the same role as those at Cambridge, as A. Mary Baylay reported back to students at North London Collegiate School: "At Oxford, unlike Cambridge, most of the lectures attended are in the men's Colleges, we are coached by tutors of the University, we read in the Bodleian and Taylorian; in the final dread week we assemble in the lobby of the Schools with pallid men in black coats and white ties..."[67] Balliol, Exeter, and Corpus Christi were the first colleges to admit women; and by 1897, only Magdalen still forbade its fellows from allowing women to attend their lectures.

The necessity to travel to the men's colleges for lectures put a particular burden on students at Lady Margaret Hall, which, like Girton at Cambridge, was sited about a half-hour outside of the city. As a result, the "new women" with their bicycles had a distinct advantage, as Isobel Sargeant recalled: "One's morning energy in cycling grimly to all the lectures recommended by an optimistic tutor is only equalled by one's evening energy in cycling even more grimly to all the 'right' societies."[68]

There was a significantly lower proportion of women students in the early years studying the Natural Sciences at the four residential Oxford colleges than at the two Cambridge colleges, as can be seen from Table 6.1. In the late 19th and early

Table 6.1. A comparison of the percentage women students taking Natural Sciences at Cambridge and Oxford.

	1881–1883	1891–1893	1901–1903	1911–1913
Cambridge	12.4	12.4	14.1	12.3
Oxford	1.7	3.3	6.8	4.3

20th centuries, there was a common perception of "Oxford for arts and Cambridge for sciences." Howarth has shown that the reasons for the weakness of science at Oxford are complex, though she contended that chemistry at Oxford was one of the few stronger sciences.[69]

In the early years, Somerville was the home of most of the women chemists. This may have been due to students with nontraditional goals being more likely to have come from nonconformist backgrounds. In addition, the influence of co-founder Vernon Harcourt and the presence on the Somerville Council of a series of science professors' wives[70] would have encouraged a scientific focus. Somerville was unique at Oxford in having a women's Scientific and Philosophical Society, which flourished from 1892 until 1897.[71] The Somerville students set up a small science museum and held meetings with a speaker, followed by appropriate experiments or practical demonstrations. The Society even published an annual record of its learned proceedings.

The earliest woman chemistry student at Oxford to gain prominence was Margaret Seward[56] (see Chap. 3), a Somervillean. In 1885, Seward had been the first woman to obtain a first class in the final Honour School of Natural Science. Seward had a classmate, Mary Watson,[71] who also studied chemistry. Watson, born in 1856, the daughter of John Watson, was educated at home and at St. John's Wood High School. She entered Somerville in 1879 on a Clothworkers' Scholarship, completing honours geology for women in 1882 and honours chemistry for women in 1883. She was then appointed as science mistress at CLC. In 1886, she married John Styles and resigned her position. She died in 1933.

A couple of years behind Seward and Watson was Mary Florence Rich. Rich[72] was born on 18 October 1865, the daughter of Thomas Rich, and she was educated at Haberdashers' Aske's Hatcham School for Girls. In 1884, she too entered Somerville as a Clothworkers' Scholar. Seward coached Rich in physics and acted as a chaperone when necessary. Pauline Adams described Rich's experiences:

> In her second year, Miss Rich was the only woman in a class of eight studying quantitative analysis in Mr. Harcourt's [A. G. Vernon Harcourt] laboratory at Christ Church. Mr. Harcourt, whose refusal to lecture separately to women students had hastened their admission to university lectures, had equally strong objections to allowing chaperones in his laboratory; an arrangement was therefore arrived at, by which, at times when Miss Rich was in the laboratory, Miss Seward carried out research with Mr. W. H. Pendlebury on the rate of chemical change.[56]

Rich completed an Honours in Natural Science (Chemistry) in 1887, becoming a "steamboat lady" in 1904 to collect a degree from the University of Dublin. In the meantime, she had held posts as Science Mistress at Howell's School, Llandaff (1887–1888); Ladies' College, Grantham (1889–1892); and Roedean School (1892–1905). Then, in 1906, she was appointed as a Principal at Granville School, Leicester. In 1922, she moved back into academia as Research Assistant to the Professor of Botany, Queen Mary College, London, a position that she held until her death in 1939.

The Alembic Club

As we have seen in earlier chapters, most universities had a chemistry club or society. Near one end of the scale was the Manchester University Chemical Society, which fought the admission of women until 1909; at the other end, the Chemical Club of Cambridge University, which seems to have accepted women members without comment. In fact, two of the early members, Ida Freund and M. Beatrice Thomas, presented research papers to a meeting of the Club in 1904.[36]

The attitude at the Oxford chemical society, known as the Alembic Club, could not have been more different from that at

Cambridge. The male-only Alembic Club was divided into a Senior Club for graduates and faculty, and a Junior Club for undergraduates. Both Clubs held occasional open meetings, but, in addition, each week there were members-only seminars. These seminars were a focus of the life of the chemistry department. In 1932, during the fourth year of her undergraduate degree at Oxford, Dorothy Crowfoot (see Chap. 9) discovered the existence of these meetings and that, being a woman, she was barred from them.[73] This exclusion particularly rankled with her when her supervisor presented her own research to a meeting from which she was prevented from attending.

The situation was no better when Crowfoot later returned from Cambridge to Oxford as a fellow and tutor. The Senior Alembic Club ignored her existence.[73] On one occasion, she arrived early for an open session of the Club and entered the room while the closed session was still in progress. One of the members lifted her off the ground and ejected her from the room. It was not until 1950 that the Club voted to admit women as members, following an acrimonious debate in which some of the older members threatened to resign, though only Dalziel Hammick actually did resign, later to rejoin.[74] A decade later, the Club had a woman President, Muriel Tomlinson of St. Hilda's College (see below).

Women Chemists and the Dyson–Perrins Laboratory

Despite the misogyny of the Alembic Club, a significant number of women chemists worked in the Dyson–Perrins Laboratory (DP) in the 1930s. Oxford had hoped that the arrival of Frederick Soddy[75] from Aberdeen (subsequently to be rejoined by his research assistant, Ada Hitchins; see Chap. 7) would usher in a school of radiochemistry, but Soddy's attentions were more on economics. It was to be organic chemistry that mainly provided Oxford with its chemical claim to fame. The impetus was the construction of the DP, "for long the envy of all other university chemistry departments."[76] And it was the DP which attracted brilliant young organic chemists to Oxford, particularly, as Blebis Bleaney described, when Robert Robinson[77] arrived from Manchester: "... his arrival with a young team produced a great upsurge in research on the chemistry and synthesis of natural products."[78]

Robinson had arrived at Oxford in 1930 with his chemist-spouse, Gertrude Walsh (Mrs. Robinson, see Chap. 11), who worked with him in the laboratory. However, she was not the first woman to undertake research in the DP, as Elinor Katherine Ewbank[79] had arrived in 1920. Ewbank was born in Ryde and educated at Highfield School, Hendon, entering Lady Margaret Hall in 1899. She obtained a first class in chemistry in 1903 and then undertook research with Edward Charles Cyril Baly at the Spectroscopy Laboratory, University College, London, co-authoring two papers with him in 1905. During the First World War, Ewbank was a nurse in Russia and Italy. From 1919 to 1920, she worked in the organic chemistry research division of the Department of Scientific and Industrial Research. Then, from 1921 until at least 1930, she undertook research at the DP, the majority of her seven publications being co-authored with Nevil Sidgwick.[80] She died in 1958.

Dalziel Hammick,[81] despite his opposition to women in the Alembic Club, took on several women researchers, including Dorothy Marion Langrish (LMH, 1937), Beryl Rose Hamilton (St. Hugh's, 1938), Ruth Bessie Maltby Yule (St. Hugh's, 1940), and Dorothea Grove (Somerville, 1948), prior to 1950. Each of the women only co-authored one publication with him and, apart from Grove (see Chap. 3), we have no detailed information on them. In 1938, we do know that Hamilton was an Analyst at Imperial Chemical Industries (General Chemicals) Ltd. in Runcorn, and the following year married a Mr. Davis.[82]

Sydney Plant[83] was another DP researcher to take on women students, one of whom was Kathleen Margaret Rogers.[84] Rogers was born on 30 December 1911 and educated at St. Swithun's School, before entering St. Hilda's in 1932. She completed her B.Sc. in 1935, her research at the DP being published in 1935 and 1936. Her marriage to D. Penfold in 1938 terminated her academic life. However, after her two sons had grown up, she took a position as a teacher at The Croft House School in Shillingstone.

Two of the women chemists of the DP found subsequent employment at the Royal Holloway College (RHC): Katherine Isobel Ross[85] and Elisabeth Mary Wykeham Lavington.[86] Ross was educated at Birkhamstead School for Girls and completed her B.Sc. at St. Hilda's College in 1930. Though her publications from the DP are dated 1931, the actual research must have been accomplished the year earlier, for she was appointed Assistant Lecturer in Chemistry at the RHC in 1930, replacing Marjorie Wilson-Smith (see Chap. 11). In 1934, Ross resigned from RHC and joined Imperial Chemical Industries at Billingham-on-Tees, where she worked until 1936 when she married Mr. C. Wilson. Lavington was a graduate of St. Hugh's College, and her DP research was published in 1934. She was a Demonstrator and Assistant Lecturer in Chemistry at RHC in 1936, as a replacement for Ross.

The first woman chemist we could identify as holding an academic position at Oxford was Elizabeth Monica Openshaw Farrow.[87] Farrow was born on 23 August 1899, the daughter of Rev. J. W. Farrow, and was educated at Oxford High School (a GPDSC school). She entered St. Hugh's College in 1918 and graduated in 1922. In 1923, she was appointed as tutor in Natural Science at St. Hugh's and as Lecturer and Demonstrator in the Department of Chemistry, being made a Senior Demonstrator two years later. Then, in 1929, Farrow was promoted to University Demonstrator in Chemistry and elected fellow of St. Hugh's College. She resigned her positions in 1930 upon her marriage to E. F. Cutliffe, following which she had a son and a daughter, the daughter following in her mother's footsteps, entering St. Hugh's in 1949 to read chemistry. Farrow died on 12 December 1954.

The first woman chemist appointed as a Fellow at Oxford was Margaret Augusta Leishman.[88] Leishman was born on 3 June

1903, the daughter of General Sir William Leishman, and she was educated at Croham Hurst School, Croydon. She graduated with a B.Sc. in chemistry from St. Hilda's College in 1926, with her research at the DP being published in 1932. In 1927, she was appointed Demonstrator in Inorganic and Physical Chemistry at Bedford College under James Spencer (see Chap. 4). Her sojourn at Bedford was brief, as in 1930 she was invited to return to Oxford as a Demonstrator in Chemistry and Lecturer at St. Hugh's College, followed by an appointment as fellow and tutor at St. Hugh's in 1932. In 1937, she accepted a position as Assistant Mistress at St. Swithun's School, Winchester. She remained in this position until retirement, dying in 1976.

The DP even attracted postgraduate women chemists from elsewhere, the most notable being the Japanese chemist, Chika Kuroda.[89] Kuroda, born in Japan on 24 March 1884, arrived at Oxford in 1921 as a Japan Minister of Education overseas student. She spent two years working with W. H. Perkin, Jr.,[90] before returning to her homeland to become Japan's most well-known woman chemistry researcher.

Muriel Tomlinson

The one women chemist who, more than any other, gained recognition in the Oxford chemistry laboratories was Muriel Louise Tomlinson.[91] She was born on 17 July 1909 in Birmingham, daughter of William Seckerson Tomlinson and Annie Wall. Her initial interest in chemistry began when she was at King's High School, Warwick, and it had a very unusual cause, as Tomlinson recounted:

> When I was at school I was first attracted to chemistry because of the delightful blue colour the word conjured up for me. To me all words have colour: the colours are perhaps less vivid than they were when I was a child and I remember being amazed when I discovered that to most people words are not

> coloured at all and that this 'colour-thinking' is experienced by relatively few people.[92]

This "colour-thinking," as Tomlinson called it, is a rare condition known as synesthesia, in which stimulation of one sense causes an unusual response in another.[93]

Entering St. Hugh's College in 1928, she was advised by her tutor to take Part I of her examinations at the end of her second year. She accepted the advice, then discovered that none of the men were taking the examination until their third year. Tomlinson recalled: "I was very frightened but it was too late to change. & I carried on."[92] She obtained a first class. Despite the hard work, she still found time to master the art of punting, and as her friend Margaret Christie observed: "... forty years later she could still put young colleagues to shame on the river."[91(a)]

Tomlinson graduated in 1931, staying at Oxford as Senior Scholar and Gilchrist Student at Lady Margaret Hall, and undertaking research at the DP with Plant on indoles and carbazoles. After obtaining a D.Phil. in 1933, she spent two years collaborating with Robinson, the last year holding a Mary Somerville Research Fellowship at Somerville.

In 1935, Tomlinson obtained an appointment at Girton College, Cambridge, first as Lecturer, then as fellow, teaching the main organic chemistry general course and organising the university practical classes in organic chemistry. She continued to do this through the years of the Second World War, an especially difficult task with the arrival of the students from Queen Mary College, Bedford College, and Guy's Hospital, as these London Colleges were evacuated to Cambridge.

Tomlinson was invited back to Oxford as fellow at St. Hilda's in 1946, and then became university Lecturer in 1948. One of her students in the 1950s, Margaret Goodgame, recalled:

> She was keen on practical work, and expected us to work in the labs most of the hours they were open — 11 am to 7 pm Monday to Friday plus Saturday morning. Lectures were usually 9–11 am.

> Essays and other tutorial work were to be done at other times. She was an assiduous demonstrator and a stickler for good practice in the lab, coming down hard on anything she considered sloppy. Consequently when she was demonstrating many of the male students used to slip off to the library (or wherever!). But for the girls, who all worked together at one end of the large laboratory, it was better to be there, even if doing something badly, than not to be there.[94]

At the same time, Tomlinson resumed organic chemistry research with Robinson and Plant at the DP. Goodgame recalled that, during her time at Oxford, Tomlinson was the only women in the DP, in "what was very much a man's world."[94] Tomlinson's name appears on a total of 29 research papers: four on her own; five with Robinson; and eight with her life-long friend Plant, with Tomlinson also writing Plant's obituary for the *Journal of the Chemical Society*.

A great traveller, she was usually accompanied by either of her friends, Mary Cartwright or Helen Gardner. Christie recalled: "She was a good, if aggressive driver with a liking for large powerful Volvos."[91(a)] After retirement in 1975, she returned to Warwick, where she had spent her childhood, and she became Chairman of the Board of Governors of the Junior Division of King's High School. She died on 17 May 1977.

Rosemary Murray

Murray's span covers the latter part of our time frame, and her story shows how far women chemists had come since the opening of Oxford to the first women students. Alice Rosemary Murray[95] was born on 28 July 1913 at Havant, daughter of Admiral Arthur John Layard Murray and Ellen Maxwell Spooner, and granddaughter of William Spooner, Warden of New College, Oxford. She was educated at Downe House, Newbury, and entered Lady Margaret Hall to read chemistry, completing a B.Sc. in 1936 and a D.Phil. in 1938.

Murray was appointed Assistant Lecturer at Royal Holloway College in 1939, but early in 1940 she was granted leave of absence to join the Admiralty Research Department at Portsmouth.[96] During that summer, she obtained a post as Lecturer in the Chemistry Department at the University of Sheffield, where she also carried out research in organic chemistry for the Ministry of Supply. In 1942, finding life at Sheffield frustrating, Murray joined the Women's Royal Naval Service, undertaking a variety of roles and rising to the rank of Chief Officer.[97]

In 1946, Murray received a telegram from the Mistress of Girton College asking if Murray would like to be interviewed for a position as chemist. Murray recalled:

> Although I had not considered what I would do after the war, this seemed too good an opportunity to miss, so off I went in wren uniform to Cambridge and had various interviews and was shown the chemistry laboratories by the professor. I gather later that I caused quite a stir among the technicians — a wren officer in uniform going round the chemistry lab.[97]

Initially appointed Lecturer in Chemistry, Murray was promoted in 1949 to a Fellowship, and Tutor and Demonstrator in Organic Chemistry. She continued teaching and research until she was appointed Tutor in Charge of the newly established New Hall. In 1964, her post was renamed President, and in 1974 New Hall became a College of the University of Cambridge. She remained President until 1981, also holding the post of Vice-Chancellor of the University of Cambridge from 1975 until 1977. That year, she was made Dame Commander of the Order of the British Empire. Murray died on 7 October 2004.

Commentary

Though we traditionally consider Oxford and Cambridge as similar institutions, for pioneer women chemists, they offered very different experiences. At Cambridge, Newnham and Girton provided a self-contained "haven" in which the women had their own classes and, more importantly, their own chemistry laboratories with women staff, particularly Freund, Marshall, and Thomas. At Oxford, undergraduate women students experienced the male environment and culture in both lectures and laboratories. Women were not even admitted to the Balliol-Trinity Chemistry Laboratories until 1904[98] and, in addition, exclusion from the Alembic Club must have added to the negative atmosphere for women chemistry students.

The situation seemed to be reversed for the research laboratories. The Oxford Dyson–Perrins Research Laboratory had women researchers throughout the period of our study, while in the Cambridge Chemistry Research Laboratories there did not appear to be any mentors for women students; for example, Mary Johnson published her research as sole author, while Delia Simpson undertook her research in the Physics Laboratory. For Cambridge, it was biochemistry which provided the supportive research atmosphere, as we show in Chap. 8.

Notes

1. Sturge, H. W. and Clark, T. (1931). *The Mount School-York: 1785 to 1814, 1831–1931*. J. M. Dent, London, pp. 108–109.
2. (a) Sutherland, G. (1994). Emily Davies, the Sidgwicks and the education of women in Cambridge. In Mason, R. (ed.), *Cambridge Minds*, Cambridge University Press, Cambridge, pp. 34–47; (b) Kamm, J. (1965). *Hope Deferred: Girls' Education in English History*. Methuen & Co., London, pp. 250–270; (c) Steinbach, S. (2004). *Women in England 1760–1914: A Social History*. Palgrave Macmillan, New York, pp. 176–183; (d) Brooke, C. N. L. (1993). *A History of the University of Cambridge, Volume IV: 1870–1990*. Cambridge University Press, Cambridge, pp. 301–330.
3. Bennett, D. (1990). *Emily Davies and the Liberation of Women 1830–1921*. Andre Deutsch, London.
4. Hirsch, P. (2000). Barbara Leigh Smith Bodichon: Feminist leader and founder of the first University College for Women. In Hilton, M.

and Hirsch, P. (eds.), *Practical Visionaries: Women, Education and Social Progress 1790–1930*, Longman/Pearson Education, Harlow, pp. 84–100.

5. Davies, E. (1866). *The Higher Education of Women*. Alexander Strahan, London and New York.
6. Bradbrook, M. C. (1969). *'That Infidel Place': A Short History of Girton College 1869–1969*. Chatto and Windus, London.
7. Collini, S. (2004). Sidgwick, Henry (1838–1900). *Oxford Dictionary of National Biography*, Oxford University Press, http://www.oxforddnb.com/view/article/25517, accessed 11 Dec 2007.
8. (a) Gallant, M. P. (1997). Against the odds: Anne Jemima Clough and women's education in England. *History of Education* **26**: 145–164; (b) Sutherland, G. (2000). Anne Jemima Clough and Blanche Athena Clough: Creating educational institutions for women. In Hilton, M. and Hirsch, P. (eds.), *Practical Visionaries: Women, Education and Social Progress 1790–1930*, Longman/Pearson Education, Harlow, pp. 101–114.
9. Howard, N. (1964). Eleanor Mildred Sidgwick and the Rayleighs. *Applied Optics* **3**: 1120–1124.
10. Andrews, E. A. (1988). Hurrah! We have won! In Phillips, A. (ed.), *A Newnham Anthology*, 2nd ed., Newnham College, Cambridge, pp. 17–19.
11. Kenschaft, P. (1987). Charlotte Angas Scott, 1858–1931. *The College Mathematics Journal* **18**: 98–110; Kenschaft, P. (1987). Charlotte Angas Scott. In Grinstein, L. S. and Campbell, P. J. (eds.), *Women of Mathematics*, Greenwood Press, Westport, Connecticut, pp. 193–203.
12. Nield, D. A. (1983). Professors of Mathematics at Auckland University College: The missionary, the businessman, the story-teller and the salesman. *New Zealand Mathematical Society Newsletter* 27, http://www.massey.ac.nz/~wwifs/mathnews/centrefolds/27/Apr1983.shtml, accessed 14 October 2007.
13. Sinclair, K. (1983). *A History of the University of Auckland 1883–1983*. Auckland University Press, Auckland, New Zealand.
14. McWilliams-Tullberg, R. (1975). *Women at Cambridge: A Men's University — Though of a Mixed Type*. Gollanz, London, pp. 73–78.
15. Series, C. (1997–1998). And what became of the women? *Mathematical Spectrum* **30**: 49–52.

16. Shils, E. and Blacker, C. (1996). Preface. In Shils, E. and Blacker, C. (eds.), *Cambridge Women: Twelve Portraits*, Cambridge University Press, Cambridge, pp. xii–xiv.
17. McDowell, R. B. and Webb, D. A. (1982). *Trinity College Dublin 1592–1952: An Academic History*. Cambridge University Press, Cambridge, pp. 344–353.
18. Parkes, S. M. (1996). Trinity College, Dublin and the 'Steamboat Ladies' 1904–1907. In Masson, M. R. and Simonton, D. (eds.), *Women and Higher Education: Past, Present and Future*, Aberdeen University Press, Aberdeen, 1996.
19. "C.C." [Coignou, C.] (1905). Commencement at Trinity College, Dublin. *The Magazine of the Manchester High School* 50–51.
20. Bailey, K. C. (1947). *A History of Trinity College Dublin 1892–1945*. The University Press, Dublin, p. 12.
21. Strachey, J. P. (1921). Admission to membership. *Newnham College Roll Letter* 33–36.
22. Fowler, H. (2004). Sidgwick [née Balfour], Eleanor Mildred (1845–1936). *Oxford Dictionary of National Biography*, Oxford University Press, http://www.oxforddnb.com/view/article/36086, accessed 11 Dec 2007.
23. Rutherford, E. and Pope, W. (8 December 1920). Letter to The Editor. *The Times* (London), p. 8.
24. Clough, B. A. (1922). October 20th. *Newnham College Roll Letter* 48–53.
25. Henn, M. E. (1988). The breaking of the gates. In Phillips, A. (ed.), *A Newnham Anthology*, 2nd ed., Newnham College, Cambridge, pp. 150–151.
26. Thacker, D. (1988). Women in laboratories. In Phillips, A. (ed.), *A Newnham Anthology*, 2nd ed., Newnham College, Cambridge, p. 80.
27. Platt, H. (July 1952). Open letters and articles: Three old girls. *Magazine of the Manchester High School* 17.
28. "M.L." (March 1888). Life at Girton. *Magazine of the Manchester High School* 217–220. For a full account of daily life at Newnham at that period, see: Field, E. E. (1889). Newnham College, Cambridge. *The College, Dundee* **1**(5): 133–138.
29. "J.W.B." (1883). Girton College. *Magazine of the Manchester High School* 70–73.

30. Wallas, M. G. (1988). A restless generation. In Phillips, A. (ed.), *A Newnham Anthology*, 2nd ed., Newnham College, Cambridge, pp. 116–119.
31. Gardner, A. (1921). *A Short History of Newnham College*. Cambridge University Press, Cambridge, p. 37.
32. Ball, M. D. (1988). Newnham scientists. In Phillips, A. (ed.), *A Newnham Anthology*, 2nd ed., Newnham College, Cambridge, pp. 76–78.
33. Wilson, H. (1988). Miss Freund. In Phillips, A. (ed.), *A Newnham Anthology*, 2nd ed., Newnham College, Cambridge, pp. 71–72.
34. (a) "E.W." [Walsh, E.] (1914). Ida Freund. *Girton Review* 9–13; (b) *Ida Freund Memorial: Vacation Course for Women Teachers of Physics*. Address given by Mrs. Sidgwick, Litt.D., LL.D., at Newnham College on Saturday, 7 August 1915, to the students attending the course.
35. Letter, Holt, C. D. to Holt, L., 12 October 1889. In Cockburn, E. O. (ed.) (1987), *Letters from Newnham College 1889–1892*, Newnham College, Cambridge, p. 11.
36. Berry, A. J. and Moelwyn-Hughes, E. A. (1963). Chemistry at Cambridge from 1901 to 1910. *Proceedings of the Chemical Society* 357–363.
37. Freund, I. (1904). *The Study of Chemical Composition. An Account of Its Method and Historical Development, with Illustrative Quotations*. The University Press, Cambridge.
38. Hill, M. and Dronsfield, A. (2004). Ida Freund — Pioneer of women's education in chemistry. *Education in Chemistry* **41**(5): 136–137.
39. Quoted in Gardiner, M. I. (1914). In Memoriam — Ida Freund. *Newnham College Roll Letter* 34–38.
40. Note 37, Freund, reprint edition, Dover Publications, New York, 1968.
41. Freund, I. (1904). Hutchinson, A. and Thomas, M. B. (eds.), *The Experimental Basis of Chemistry; Suggestions for a Series of Experiments Illustrative of the Fundamental Principles of Chemistry*, The University Press, Cambridge.
42. (a) Butler, K. T. and McMorran, H. I. (eds.) (1948). *Girton College Register, 1869–1946*. Girton College, Cambridge, p. 638; (b) Creese, M. R. S. (1998). *Ladies in the Laboratory: American and British*

Women in Science 1800–1900. Scarecrow Press, Lanham, Maryland, p. 279.

43. Shorney, D. (1989). *Teachers in Training, 1906–1985: A History of Avery Hill College*. Thames Polytechnic, London, p. 53.
44. (a) "D.M.P." (1954). In memoriam: Mary Beatrice Thomas 1873–1954. *Girton Review* (Michaelmas Term): 14–24; (b) "D.M.A." [Adcock, D. M.] (1955). Mary Beatrice Thomas, 1873–1954. *Newnham College Roll Letter* 37–38; (c) Russell, D. S. (1954). Posthumous letter to Miss M. B. Thomas. *Girton Review* (Michaelmas Term): 24–25.
45. (a) White, A. B. (ed.) (1979). *Newnham College Register, 1871–1971*, Vol. I, 1871–1923, 2nd ed. Newnham College, Cambridge, p. 37; (b) Newnham College Archives, Curriculum vitae and family history supplied by Margaret Spufford (daughter), 1998.
46. Simonsen. J. L. (1950). Charles Stanley Gibson, 1884–1950. *Obituary Notices of Fellows of the Royal Society* **7**: 114–137.
47. (a) White, A. B. (ed.) (1981). *Newnham College Register, 1871–1971*, Vol. II, 1924–1950, 2nd ed. Newnham College, Cambridge, p. 83; (b) Newton, A. A. (1999). Delia Agar, 1912–1998. *Newnham College Roll Letter* 101–103; (c) Lynden-Bell, R. M. and Sheppard, N. (1999). The scientific work of Delia Agar. *Newnham College Roll Letter* 103–105; (d) Mason, J. (1999). *Newnham College Roll Letter* 105.
48. (a) White, A. B. (ed.) (1979). *Newnham College Register, 1871–1971*, Vol. I, 1871–1923, 2nd ed. Newnham College, Cambridge, p. 292; (b) Anon. (1958). Sosheila Ram, 1899–1957. *Newnham College Roll Letter* 45–46.
49. Mathur, N. N. (1998). Indian medical colleges: Lady Hardinge College, New Delhi. *National Medical Journal of India* **11**(2): 97–100.
50. Brittain, V. (1960). *The Women at Oxford: A Fragment of History*. George G. Harrap, London, pp. 18–19.
51. Rogers, A. M. A. H. (1938). *Degrees by Degrees: The Story of the Admission of Oxford Women Students to Membership of the University*. Oxford University Press, Oxford.
52. Howarth, J. (2004). Sidgwick, Arthur (1840–1920). *Oxford Dictionary of National Biography*, Oxford University Press,

http://www.oxforddnb.com/view/article/48597, accessed 11 Dec 2007.
53. Howarth, J. (1984). 'In Oxford but ... not of Oxford': The women's colleges. In Brock, M. G. and Curthoys, M. C. (eds.), *The History of the University of Oxford, Vol. VII, Nineteenth-Century Oxford, Part 2*, Clarendon Press, Oxford, p. 249.
54. Howarth, J. (1994). Women. In Harrison, B. (ed.), *The History of the University of Oxford, Vol. VIII, The Twentieth Century*, Clarendon Press, Oxford, p. 346.
55. Harvey, E. (July 1903). College letter. *Magazine of the Manchester High School* 70–73.
56. Adams, P. (1996). *Somerville for Women: An Oxford College 1879–1993*. Oxford University Press, Oxford, pp. 38–40. By permission of Oxford University Press.
57. Note 53, Howarth, p. 246 and p. 350.
58. Bush, J. (2005). 'Special strengths for their own special duties': Women, higher education and gender conservatism in Late Victorian Britain. *History of Education* **34**(4): 387–405.
59. Rayner, M. E. (1993). *The Centenary History of St. Hilda's College, Oxford*. St. Hilda's College, Oxford, p. 110.
60. Howarth, J. (1986). Anglican perspectives on gender: Some reflections on the centenary of St. Hugh's College, Oxford. *Oxford Review of Education* **12**(3): 299–304.
61. Note 50, Brittain, p. 69.
62. Note 50, Brittain, p. 70.
63. Note 53, Howarth, p. 266.
64. Cited in Note 60, Howarth, p. 267.
65. Rogers, A. M. A. H. (1911). The present position of women at the University of Oxford. *Journal of Education* **33** (new series): 677.
66. Note 54, Howarth, pp. 356–358. By permission of Oxford University Press.
67. Baylay, A. M. (1901). Some aspects of life in the women's colleges at Oxford. *Our Magazine: North London Collegiate School for Girls* **26**: 40–45.
68. Sargeant, I. (1952). Open letters and articles: Three old girls. *Magazine of the Manchester High School* 17–18.
69. Howarth, J. (1987). Science education in Late-Victorian Oxford: A curious case of failure? *English Historical Review* **102**: 334–371.

70. Note 53, Howarth, p. 283.
71. Note 56, Adams, pp. 128–129.
72. P. Adams, Librarian and Archivist, Somerville College, is thanked for providing biographical notes on Watson and Rich.
73. Ferry, G. (1998). *Dorothy Hodgkin: A Life*. Granta Books, London, pp. 132–134.
74. Smith, J. C. (n.d.). *The Development of Organic Chemistry at Oxford. Part II: The Robinson Era 1930–1955*. Unpublished, p. 54; H. Anderson, Memorial University, is thanked for a copy of this manuscript.
75. Fleck, A. (1957). Frederick Soddy. Born Eastbourne 2 September 1877 died Brighton 26 September 1956. *Biographical Memoirs of Fellows of the Royal Society* **3**: 203–216.
76. Hartley, H. (1955). Schools of chemistry in Great Britain and Ireland–XVI The University of Oxford, Part II. *Journal of the Royal Institute of Chemistry* **79**: 176–184. See also: Knowles, J. (2003). The Dyson Perrins Laboratory at Oxford. *Organic and Biomolecular Chemistry* **1**: 3625–3627.
77. Todd, Lord, and Cornforth, J. W. (1976). Robert Robinson. 13 September 1886–8 February 1975. *Biographical Memoirs of Fellows of the Royal Society* **22**: 414–527.
78. Bleaney, B. (1994). The physical sciences in Oxford, 1918–1939 and earlier. *Notes and Records of the Royal Society, London* **48**(2): 247–261.
79. R. Staples, Archivist, Lady Margaret Hall College, is thanked for providing brief biographical notes on Ewbank.
80. Tizard, H. T. (1954). Nevil Vincent Sidgwick, 1873–1952. *Obituary Notices of Fellows of the Royal Society* **9**(1): 236–258.
81. Bowen, E. J. (1967). Dalziel Llewellyn Hammick, 1887–1966. *Biographical Memoirs of Fellows of the Royal Society* **13**: 107–124.
82. *Calendars*, University of Oxford; (1938). *Register*, Royal Institute of Chemistry.
83. Tomlinson, M. (1956). Obituary notices: Sydney Glenn Preston Plant, 1896–1955. *Journal of the Chemical Society* 1920–1922.
84. Verity, G., Brown, B. and Rayner, M. E. (1930). *St Hilda's College, Oxford Centenary Register 1893–1993*.
85. (1931). Forms of recommendation for Fellowship. The ballot will be held at the ordinary scientific meeting on Thursday, 7 May.

Proceedings of the Chemical Society 33; Verity, G., Brown, B. and Rayner, M. E. (1930). *St Hilda's College, Oxford Centenary Register 1893–1993*; (November 1930). *College Letter*, Royal Holloway College Association, p. 16.

86. (a) *Calendars*, University of Oxford; (b) (1936). Forms of recommendation for Fellowship. The ballot will be held at the ordinary scientific meeting on Thursday, 3 December. *Proceedings of the Chemical Society* 78.
87. (a) Anon. (1954–1955). Elizabeth Monica Openshaw Cutliffe (née Farrow). *St. Hugh's College Chronicle* (27): 19–20; (b) (1925). Certificates of candidates for election at the ballot to be held at the ordinary scientific meeting on Thursday, 3rd December. *Proceedings of the Chemical Society* 130; (c) Note 84.
88. Note 84; Bedford College, Staff Records.
89. (a) Haines, M. C. (2001). Kuroda, Chika. *International Women in Science: A Biographical Dictionary to 1950*, ABC-CLIO, Santa Barbara, California, p. 164; (b) Tsugawa, A. and Konami, Y. (2002). Women scientists in Japan — History and today. *Proceedings of 12th International Conference of Women Engineers and Scientists*, Ottawa, Ontario, July 27–31.
90. "J.F.T." (1931). Obituary notices: William Henry Perkin — 1860–1929. *Proceedings of the Royal Society of London, Series A* **130**: i–xii.
91. (a) Christie, M. (1990–1991). Obituaries: Muriel Louise Tomlinson. *St. Hilda's College Report and Chronicle* 30–32; (b) (1991). *Girton College Newsletter* 54–55. (c) Butler, K. T. and McMorran, H. I. (eds.) (1948). *Girton College Register, 1869–1946*. Girton College, Cambridge, p. 661.
92. Hand-written draft of a speech given by Muriel Tomlinson at the St. Hilda's College Gaudy, 25 June 1977, copy provided by E. Boardman, Archivist, St. Hilda's College.
93. Hubbard, E. M. and Ramachandran, V. S. (2005). Neurocognitive mechanisms of synthesia. *Neuron* **48**(3): 509–520.
94. Personal communication, e-mail, Margaret Goodgame, 7 December 2006.
95. Haines, C. M. C. (2001). Murray, Alice Rosemary. *International Women in Science: A Biographical Dictionary to 1950*, ABC-CLIO, Santa Barbara, California, pp. 217–218.

96. (1940). *College Letter, Royal Holloway College Association*, 12.
97. Walton, K. D. (1996). In retrospect: Dame Rosemary Murray. In Walton, K. D. (ed.), *Against the Tide: Career Paths of Women Leaders in American and British Higher Education*, Phi Delta Kappa Educational Foundation, Bloomington, Indiana.
98. Bowen, E. J. (1970). The Balliol-Trinity Laboratories, Oxford 1853–1940. *Notes and Records of the Royal Society of London* **25**(2): 228–229.

Chapter 7

Universities in Scotland and Wales

Up to this point, we have reviewed the lives and work of women chemists in England. We showed how certain girls' schools played a major role in providing the teaching environment that encouraged girls to choose chemistry degree programmes at university. In this chapter, we shall see that the cause of higher education of girls in Scotland and Wales followed different paths from that in England.

Within Scotland, there was a diversity of approaches, from the autonomous women's Queen Margaret College of the University of Glasgow to the tougher fight for acceptance at the University of Edinburgh. In Wales, though girls' secondary education was encouraged, one school of the Girls' Public Day School Company (GPDSC) did not survive, and it was predominantly English women chemists who entered the University Colleges of Wales.

Girls' Education in Scotland

From the early 19th century, education for girls, either at day or boarding school, was centred around the three cities of Glasgow, Edinburgh, and Aberdeen. The offering of specific subjects by individual teachers at their homes was a particular feature of the age, as Mrs. Furlong commented in 1855:

> In every street might, at the period to which I refer [1815–1835], be seen some delicate girl, hurrying from class to class (according

> as each teacher found their partizans), which were held at their respective houses. These young creatures were generally encumbered with a load of books, and the time lost in their endeavour to make the respective master's classes hours meet, being found at length an evil which called for some better plan ….[1]

Scotland did not have the equivalent of England's Miss Buss or Miss Beale (see Chap. 1); instead, the young ladies' institutions were administered by men, and men made up the majority of the teaching staff. The Scottish Institution for the Education of Young Ladies (SIEYL) in Edinburgh particularly emphasised the teaching of science. For example, in 1835, the school report noted:

> The explanations were accompanied by a great variety of experiments and practical illustrations, and in some of the classes, an examination followed the lecture. In the chemical classes, the ladies answered many questions, and practiced, under Dr. Reid's superintendence, many useful experiments.[1]

The rival of SIEYL, the Edinburgh Ladies' Institution for the Southern Districts, was equally boastful of its science offerings, including chemistry. On the other side of the country, the Glasgow Institution for the Education of Young Ladies offered physical science from 1847 and even chemistry in 1855, though by the mid-1860s science had disappeared from the basic curriculum.[1]

Despite the lack of charismatic women headmistresses, some girls' schools did promote the teaching of chemistry. For example, Park Street Girls School, Glasgow, constructed a chemistry laboratory in 1884. Gillian Avery noted: "It cost £60 to equip, and being on the top floor, there were disastrous results when enthusiastic chemists left the taps running. 'Such episodes added greatly to the gaiety of the girls'."[2] Avery added that a new laboratory was subsequently constructed that was "the envy of some of the boys schools."[2]

Entry of Women to Scottish Universities

In Scotland, by the late 1850s and early 1860s, there was considerable discussion in the pages of newspapers about the social advantages and the personal justice of admitting middle-class women into paid employment. There was a particularly wide support for the training of women in medicine, as it was felt that it was more suitable for women and adolescent girls to be intimately examined by women.[3] In addition, it was felt that middle-class women medical practitioners could bring hygienic and moral guidance to working-class women.

The Scottish universities became dragged into the issue when Elizabeth Garrett applied for admission to St. Andrews University in 1861,[4] and again when Sophia Jex-Blake[5] and others applied for admission to Edinburgh University in 1869. In both cases, the universities initially approved the women's admission, but then withdrew their consent when some faculty members at each institution voiced their antagonism to the presence of women on campus. In 1872, the Edinburgh case went to the Scottish High Court, which ruled that women had the right to attend university classes at the University of Edinburgh. The University appealed and, in 1874, a bare majority of the full Court of Session ruled against women's admission.

A Private Member's Bill was then introduced at Westminster that would allow the universities in Scotland to admit women as students and permit them to grant degrees to women. The Universities of Edinburgh and Glasgow were the focus of much of the opposition to the Bill, though many faculty members at both institutions supported the women's cause. The bill was defeated in March 1875 by 194 votes to 151. The case for the opposition was based on three issues: the inadvisability of permitting women into practical medical classes; the impropriety of mixed classes; and, following from this, the impossibility of professors providing parallel classes for men and women students.

The ladies' educational associations in each of the Scottish cities campaigned for a higher qualification for women. It was

Table 7.1. The increase in the proportion of women students at Scottish universities between the 1890s and the 1900s (as percentages).

University	1895–1896	1907–1908
Aberdeen	5	31
Edinburgh	9	18
Glasgow	9	24
St. Andrews	18	40

St. Andrews who led the way by introducing the Lady Licentiate in Arts (LLA).[6] This qualification proved hugely popular, attracting women students from across Scotland, much to the annoyance of the other Scottish universities. In 1889, the Universities (Scotland) Bill finally permitted the universities to admit and graduate women, and it was from Edinburgh that the first eight women graduated in 1890.[7] The struggle was over, and by the beginning of the 20th century a substantial proportion of students were women (Table 7.1).[3]

Life of Scottish Women Students

On the basis of their father's occupation, women students came predominantly from the middle class, like their male colleagues.[8] Some of the upper-class women used university as more of a finishing school, and this was particularly true of some of those attending Queen Margaret College, Glasgow.[9] Through oral interviews, Sheila Hamilton has shown how life as a middle-class Scottish woman student in the 1920s was totally different from that of a working-class woman student.[10] The middle-class student, Mrs. M., led an active social life:

> Dancing was very much to the fore in the 1920s and this was reflected within the social life of Mrs. M. She attended sixty-three dances in one winter, although these were not all university functions. Even by the inter-war period, chaperonage

> still occurred at ... faculty dances, faculty professors' wives taking on the role. ... Chaperoning was a means of maintaining an image of a social way of life which was becoming less the normal pattern of everyday life at the time. Similarly, one could say that a place like the Masson Hall of residence [at Edinburgh University] also, to a certain extent, maintained the image of gentility and wealth by the duplication of the home situation, that is, in having a maid service in the halls and, significantly by the holding of "At Home" days [for gentleman callers].[11]

Whereas Mrs. M's father was a graduate of Edinburgh and her three siblings (all girls) all also went to university, the working-class interviewee came from a very different background. Mrs. G., one of six children, was the first and only member of her family to go to university. She recalled that the only money she had was the half-crown her eldest brother gave her every week. By contrast, half a crown was the entrance fee each time Mrs. M. went to the "Palais de Dance."

Mrs. M. had been active in the Women's Union, while Mrs. G. saw it as an institution for the elite, preferring to eat her lunchtime sandwiches in a cloakroom for women students. To Mrs. G., Mrs. M. represented an alien species:

> A great deal of the superiority that they seemed to us to have was due to their poise and sophistication and speech .... Our speech would be pretty common .... Many of these women were at ... (private schools) [Mrs. M. had attended St. Leonard's School in St. Andrews] ... they came you see with all the confidence and if you went to meetings they would stand up and speak ... and you met people like these students I am talking about who made you feel as if you were a complete ignoramus. I don't mean that they meant to think that, but that's how within yourself you felt.[12]

There was also a difference in the family duties of the daughter at university. Mrs. M. was required to pay social calls with her mother, which was time-consuming; but the cook, the housemaid, the table maid, the cleaning woman, the nanny, and the gardener took care of all the chores. By contrast, Mrs. G. had to continue with her share of "women's work" in her household:

> Mrs. G. had to share a bedroom with her [two] sisters and was expected to help out with the household chores at the evenings and weekends. One weekend in particular, she recalled that she was nearly in tears because she had all the men's shirts to iron for four brothers and father, and she should have been studying for an examination.[12]

University of St. Andrews

The University of St. Andrews was founded in the Kingdom of Fife in 1411, making it the oldest university in Scotland and the third oldest in the United Kingdom. The College of St. Salvator was created in 1450, and the College of St. Leonard in 1512. The two Colleges were combined in 1747 to form the United College of St. Salvator and St. Leonard in the University of St. Andrews.[13]

Life for Women Students

St. Andrews was the only Scottish University to strongly champion the admission of women. As S. Sloan commented in the pages of *The Girl's Realm*:

> Not only did St. Andrews University early open its doors to women, it has treated them royally ever since; and today [1911] it holds the unique position of being the one college in which men and women students stand upon an absolute equality. ... There are those who have shaken their heads over a Senate which thus decrees that Jill shall enter into the full college life enjoyed by Jack; but time has proved it wise in its generation, and the University of St. Andrews is justified of

> all her children. Sharing alike the work and play of the men gives the women a broader, fuller outlook upon life; while the gentler, more refining influence of the girl students is by no means lost upon even the wilder spirits among the male undergraduates.[14]

Nevertheless, it did not mean that women students were necessarily welcomed by the male contingent, according to a complaint in the *St. Andrews University Magazine* of 1890:

> It seems to me that I scarcely know a young lady now-a-days who has not, to some undefined extent, a craving in the direction of higher education and deep thought, which renders the pleasure of her companionship somewhat dubious. One can pardon a lady who desires to read Plantus in the original so long as one has not the privilege of her acquaintance... It is one of the most trying ordeals I can think of, to be in the company of a *femme savant* for more than a few minutes at a time...[15]

Women Chemistry Students

Thomas Purdie,[16] who had been elected to the Chair of Chemistry in 1884, saw the future as being organic chemistry and, in particular, stereochemistry. This proved a brilliant move, producing research on carbohydrates that became renowned throughout Europe.[13] With Purdie's resignation in 1909, his successor was his former student, James Colquhoun Irvine.[17] Irvine continued to forge ahead with the studies on sugar structures, and among his research students were several women chemists: Agnes M. Moodie (see below), M. E. Dobson, B. M. Paterson, Ettie S. Steele (see below), Helen S. Gilchrist (see below), and J. K. Rutherford. It is not surprising that Irvine had so many women researchers, as by the 1910–1911 academic year, the first-year cohort in chemistry consisted of 36% women (27 male, 15 female).[18] Irvine was appointed

Principal of the University in 1921; nevertheless, researchers worked with him until 1935.

Agnes Moodie

One of the early women chemistry graduates was Agnes Marion Moodie,[19] who completed her M.A. in 1902 and a B.Sc. in 1903. Moodie, a petitioner for admission of women to the Chemical Society (see Chap. 2), stayed on at St. Andrews to undertake research with Irvine, co-authoring five publications with him on alkylated sugars. She was awarded a Carnegie Scholarship in 1907, but we know nothing of the intervening years until she retired from the Ministry of Education in 1946, moving to Hove, Sussex.

Ettie Steele

In the history of the Chemistry Department of St. Andrews, there was one woman who made an indelible imprint — Ettie Stewart Steele.[20] Steele obtained an M.A. followed by a B.Sc. in 1914, and then commenced research with Irvine on the structure of mannitol.[21] Appointed as University Assistant in 1919, she received a Ph.D. in 1920 for her research work, following which she was promoted to Lecturer and Assistant in 1921, and to Lecturer in 1924. Steele contributed to five research papers on carbohydrates, including one under her name alone.

When Irvine became Principal of the University in 1921 and Robert Robinson arrived to assume the Professorship in Chemistry, Robinson noted: "I worked in the private laboratory also occupied partly by Dr. Catherine [sic] Steele, a colleague of Irvine's who kept alive some of his work on sugars, but this petered out on account of the onerous duties [of Irvine]."[22] She continued as Lecturer until her retirement in 1956.

In 1930, Steele was appointed Warden of the women's students' residence, then called Chatten House (later McIntosh

Hall), a post she held until 1959. Her duties filled much of her time, as her obituarist noted:

> Each alumnus who knew Ettie Steele will have a particular and personal recollection — of her penetrating blue eyes, mirrored but not eclipsed by the blue of her Ph.D. gown, which she wore with such pride and distinction; of her firmness in dealing with young male intruders in McIntosh Hall; of her infinite kindness and concern for students, a caring interest maintained to the end of her life as her Christmas correspondence bore witness.[20]

Steele filled a third role, that of secretary to Irvine, during all of his years as Principal. The University of St. Andrews was her life:

> No day passed, even in retirement, when the institution was not central to her thoughts, and from her flat in St. Andrews she jealously watched over the affairs of the University and was firm in her comments upon current issues — comments which were trenchant and apposite. ... To Ettie Steele life was full of fortunate and happy events. In her own eyes her association with the University was one of supreme privilege and she considered herself unworthy of the responsibilities she carried with such competence and humour.[20]

Steele died in July 1983.

Ishbel Campbell

Another pioneer, Ishbel Grace MacNaughton Campbell,[23] was to fill a similar role to Steele, but at University College, Southampton. Campbell was born in 1906, and graduated from St. Andrews with a B.Sc. in 1927 and a Ph.D. in 1931. She then spent a year at Cornell University, New York, where she held one of the first Commonwealth Fellowships awarded to a woman.

On her return, Campbell accepted a Lectureship in Chemistry at Swanley Horticultural College. In 1936, she joined Bedford College as a Demonstrator and Teacher, before making her final move in 1938 to University College, Southampton, first as Lecturer and later as Reader of the University of Southampton.[24]

One of her former students, Martin Hocking, recalled:

> "Ish" was experimentally well known for her ability to coax more-or-less pure crystals of a new substance from tiny amounts of solution of an unlikely looking, gluey reaction product. It was rumoured that her success was the beneficiary of traces of her cigarette ash that provided nuclei in the crystallization test tube to help initiate the crystallization process aided by temperature changes and by scratching the side of the tube with a glass rod.[25]

Hocking noted also that Campbell provided students with a sense of community:

> "Ish" as she was colloquially known, was a slight, but physically tough, proud Scottish spinster who regularly played tennis until well into her 80s. She also loved long walks in the hills around Southampton, once a year or so inviting a small group of students to join her. These walks were followed by a most enjoyable picnic at some high point in the walk, where the view could be enjoyed along with the conversation.[25]

During the Second World War, despite heavy air attacks on the town, teaching continued, with evening classes for part-time students being transferred to weekends. Spare time was spent on research into chemical warfare.

While at Southampton, Campbell undertook a tremendous quantity of research, primarily on organometallic compounds of groups V and VI. A total of 19 publications resulted from her work at Southampton between 1945 and 1966: 15 with her as senior author and 4 with her as sole author. According to Hocking, Campbell never really retired: "Long after her official retirement Ish enthusiastically gave us a tour of the new medical faculty at Southampton and where she had volunteered to teach courses in chemistry to new medical students. 'It keeps me young' she said, and it certainly worked!"[25] Campbell died on 10 October 1997 at the age of 91.

Helen Gilchrist

A later graduate was Helen Simpson Gilchrist.[26] Gilchrist completed a B.Sc. degree in Botany and Chemistry in 1917, and that same year became a Berry Scholar in Chemistry to pursue chemical research. The following year, Gilchrist was awarded a Carnegie Scholarship; but soon after moved to the Ministry of Munitions, where she undertook research connected with the war, which included the preparation of novocaine and "mustard gas." In 1919, she resumed her Carnegie Scholarship and then, in October 1919, became a Research Chemist under the Food and Investigation Board of the Department of Scientific and Industrial Research (DSIR), a government research agency which had been instituted in 1917. It was for her research with DSIR that Gilchrist was awarded a Ph.D. During this time, Gilchrist was also a University Assistant in Chemistry at the University of St. Andrews.

In 1927, Gilchrist applied for a position at the Royal Institution. W. H. Bragg, Professor at the Royal Institution, wrote to Irvine:

> ... I think I could promise that she would not be merely a hewer of wood and drawer of water. The work she would be doing would give her a chance of carrying out investigations of her own though, of course, she would be here for a definite purpose. I always like the people here to publish under their own names as much as possible....[27]

Gilchrist accepted the position, working on the separation of long-chain hydrocarbons with the Austrian scientist Berta Karlik,[28] and then on the synthesis of fats containing a carbohydrate unit.

University College, Dundee

University College, Dundee (the University of Dundee since 1967), was founded in 1881 as a satellite campus of the University of St. Andrews with the mission of "promoting the education of persons of both sexes."[29]

It was the organic chemist Alexander McKenzie,[30] Professor of Chemistry, who attracted a significant number of women chemistry researchers. During his "reign" from 1914 to 1938, they included Isobel Smith (see below), Nellie Walker (see below), Mary Lesslie (see Chap. 4), Agnes Grant Mitchell, Ethel Luis (see below), and E. R. L. Gow.

Life for Women Students

Though University College, Dundee, attracted a high proportion of women students, gender relations were not always smooth, particularly in the 1920s as "An Undergraduette" commented:

> Before I came to College I had no idea any community of men could be so collectively dull. ... The rock-bottom trouble is that you University men continue to treat us University women as survivals of the Victorian age. You continue your pre-historic drawing-room attitudes in class-room, dance and corridor, simpering like a bunch of be-whiskered Lord Dundrearys before us, and forgetting that you are addressing the new, jazz-infected, bobbed and madcap daughters of this generation. Oh, I know there are some Victorian survivals, sitting primly acidulate in the Women's Union, but the rest of us don't pay attention to them, and when they do cut in with their reproofs and reminders, you can

> take it from me, we moderns just ignore them. They always have the Christian Union to occupy most of their spare time.[31]

It is interesting to note this division among the women students of the "moderns" and the "traditionalists." Another correspondent, "Alphabeta," sounded even more bitter about her male colleagues:

> Time was when I thought I should get me a husband and be happy. Husbands are not nearly so difficult to get as most men fondly imagine, and I feel sure that, had I been so inclined, I could have been married once legally and twice bigamously within the past two years. But I have changed, and no longer do I care for men. Rather would I be an old maid and live my life alone, uncared for and uncuddled, but free from the baneful influence of men and their stupid brutish ways. That is what a University education has done for me. ... As for Varsity men, I should no more think of accepting one of their illustrious number in marriage than I would of knitting a jumper of jute.[32]

Even in the 1940s, women's presence at University was being questioned, as this male correspondent wrote:

> To our science women I have a few words to say. "Beauty is truth, truth beauty, that is all you need to know," but you seem to prefer truth to beauty, at least that pale semblance of truth which drips in honeyed words from the mouths of bored lecturers. A little application of the rouge of Hu-fuan-chang t...d ex aq. p.c. will obtain you a better home, a larger family, and a much more assured income than a vain tickling of the receptive areas of the cerebrum with so-called wisdom of Lavoisier, Dalton, and Marat. You have come to the university to study Science, you say? The proper study of mankind — and womankind too — is man *s*cientius.[33]

Nellie Walker (Mrs. Wishart)

The first woman staff member was Nellie Walker.[34] The daughter of William Walker, she was educated at Dundee High School and entered University College, Dundee, in 1908 at age 17. She obtained an M.A. (St. Andrews) in 1911 and a B.Sc. (St. Andrews) in 1913. For the next 2 years she was a research scholar, and then briefly in 1915 undertook war work with the Royal Society War Committee. Later that year, she obtained a position as Demonstrator in Chemistry at Bedford College. In 1918, she returned to Dundee, where she completed a Ph.D. (St. Andrews) in 1920.

Walker was then appointed as University Assistant. The editor of *The College: Magazine of University College, Dundee* commented:

> For two years a Carnegie research scholar, she has now been appointed assistant in our own Chemistry Department, and we rejoice at her promotion. Successful as a student, she will be as successful in the capacity of teacher. In the social life of the College, too, Miss Walker has always taken an active part. No society for which she is eligible has lacked her support... Quiet, unassuming, deliberate in all her words and actions, she has a very real influence on all around her. If she be no brilliant conversationalist, to her at least belongs that gift so priceless, yet so rare, the genius for listening.[35]

In 1921, Walker was promoted to Lecturer and Assistant, a post she held until 1927 when she resigned and married W. G. Wishart.

Isobel Agnes Smith

Isobel Agnes Smith[36] was the stalwart of Dundee, just as Ettie Steele was the "backbone" of St. Andrews Chemistry Department. Born in 1897, daughter of Robert Graham Smith, she entered Dundee in 1914 at the age of 17. Completing an M.A. (St. Andrews) at Dundee in 1918, a B.Sc. in 1920, and a Ph.D. in 1922, she was awarded Carnegie Fellowships for 1922–1923 and 1923–1924.

Smith was appointed Assistant Lecturer in 1928 and was promoted to Lecturer in 1937. She co-authored 13 research papers over the period from 1922 to 1943, all on optical activity of organic compounds. She held the rank of Senior Lecturer at the time of her retirement in 1961.

Ethel Luis

For the duration of the Second World War, a second woman was appointed to the chemistry teaching staff at Dundee: Ethel Margaret Luis.[37] Luis was born on 28 August 1898, daughter of Theo. G. Lewis, Spinner and Manufacturer of Bloomfield, Dundee. She entered the Royal Holloway College in 1918, obtaining an honours B.Sc. (London) in 1921; while for the 1923–1925 period, Luis is listed as attending the Royal College of Science, Imperial College.

Luis returned to Scotland to commence a Ph.D. (St. Andrews) with McKenzie, completing it in 1931. She continued research at Dundee, authoring and co-authoring nine papers between 1929 and 1941, and was appointed Demonstrator in 1938. In 1939, at the start of the Second World War, Luis was promoted to Assistant Lecturer to replace a male faculty who had departed on war duties, her employment being terminated with the end of the War in 1945. Luis died on 30 May 1998 at Broughty Ferry, aged 98 years.

Women Chemistry Students

During the early years of the College, *The College: Magazine of University College, Dundee* contained periodic remarks on women chemists. In the 1890s, *The College* describes a ladies-only room: “Coming, as the greater number of us do, from the unsavoury and often poisonous atmosphere of heated laboratories, to enjoy an hour of chat and gossip (with its accompaniment of afternoon tea), we long for a complete change of environment."[38]

The novelty of women in chemistry prompted the comment: "The junior chemistry class is a study of blouses."[39] and also: "The fame of Marie Curie seems to have stimulated the female sex to pursue the study of chemistry. The advanced class is composed of four women students, and one solitary male. Changed days indeed."[40]

The pages of the magazine were used for cutting comments about the women chemistry students: "Rumour has it that thc female section is experiencing difficulty in sucking up liquids into pipettes. They say they don't know when to stop. That's the worst of practicing with straws in the summer."[41] Objections to the behaviour of women students during glass-blowing was reported: "We would remind the female elementary chemists that bulb-blowing can be done quite well without screeching and howling, or any sort of accompaniment."[42]

The absence of women from an alcoholic social was also noted:

> This [chemistry] society, ever mindful of the comforts of its members, prefaced the month of December with an instruction by Mr Young on the merits and demerits of alcohol in its many and varied forms. It was notable that none of the women were present. Perhaps, after the manner of their sex, they wished to convey the suggestion that they needed no instruction in these matters.[43]

Heather Beveridge

In Chap. 2, we discussed the signatories of the 1909 letter to *Chemical News* concerning the request for women to be admitted to the Chemical Society. Listed in alphabetical order, the first signatory was Heather Henderson Beveridge.[44] Beveridge, daughter of Rev. John Beveridge, entered University College, Dundee, in 1902 at the age of 16. After graduation, she became a Carnegie Research Scholar undertaking research with James Walker. When he moved to the University of Edinburgh, Beveridge moved with him continuing research there.

University of Aberdeen

The first university in Aberdeen was founded in 1495, later to become King's College. During Reformation, King's College remained "true to the ancient faith"; thus, in 1593, the reformist Marischal College was established.[45] These competing degree-granting institutions were not merged until 1860.

Aberdeen was, if anything, less female-friendly than Edinburgh. Individual professors had allowed women to attend courses; in fact, it was said that one young woman had attended her father's regular chemistry classes in the 1820s.[46] However, in 1873, Aberdeen University students voted against university degrees being open to women, and the (male) students periodically voted the same way for the next two decades.

Despite the efforts of the Aberdeen Ladies' Educational Association and others, women's access to university in the northeast lagged behind that in other parts of Scotland, as Lindy Moore had concluded: "... Aberdeen's efforts to promote their [women's] higher education were too late, too tentative, and intrinsically unsuited to the requirements of most girls in north-east Scotland."[46] Thus, in 1895, Aberdeen had the lowest percentage of female students in Scotland, though by 1907–1909 it had overtaken Glasgow and Edinburgh. The situation, specifically in science, also changed in the first decade of the 20th century: science had low popularity among women in 1901, but top popularity by 1911.[46]

Life for Women Students

Mary Paton Ramsay commented on the life of first-year women students, and in particular on the segregated seating for women students:

> The Session opens about the third week of October, and the eager Bajanella (the feminised form of Bajan, or first year's student) begins her College life. The gown and trencher are not compulsory, but most girls prefer to wear them. The scarlet gown, and the trencher with scarlet tassel (the men wear the tassel black) is becoming to most girls .... As one's curriculum draws to a close, one looks back with mingled amusement and melancholy to the excitement of these early days of College life; the shyness of the first encounters in the crowded women students' room, and the dread of being late for a lecture and of having to pass the crowded benches of men students in order to reach the ladies' seats beyond at the back.[47]

Ramsey noted that the greater proportion came from "the country" and resided in lodgings. As to gender relations at Aberdeen, she proclaimed them to be far superior than elsewhere:

> A strong spirit of unity, a keen "esprit de corps," exists among the women students, ... All the great Scholarships, Fellowships, Bursaries, and Prizes are open equally to men and women, and though competition is keen, the courtesy shown by the men students to the women students has always been perfect. The same courtesy has not always been shown to women students at other Scottish Universities, ...[47]

Women Chemistry Students

As we mentioned earlier, a significant number of women studied chemistry at Aberdeen in the earlier part of the 20th century, and this is quantified in Table 7.2.[48] The drop in male

Table 7.2. Chemistry enrolments at the University of Aberdeen, 1911–1912 to 1920–1921 (Note 48).

Session	Males	Females
1911–1912	25	5
1912–1913	26	15
1913–1914	23	9
1914–1915	17	16
1915–1916	18	17
1916–1917	4	17
1917–1918	4	8
1918–1919	17	18
1919–1920	29	16
1920–1921	16	7
1921–1922	21	7

enrolment for 1916–1918 resulted from the conscription for the First World War. More surprising is the drop in female enrolment of 1917–1918, which presumably resulted from some women students also relinquishing their studies for war-related work, though there is no indication of what this was. In addition to the negative effect of the War on students, it did open up positions for the first women staff — but on a temporary basis, as Kenneth Page noted:

> In order to meet staff shortages in 1917, history was made by the appointment of the first women ever to hold official posts in the department. Miss Jeannie Ross and Miss Beatrice Simpson both joined the department as assistants during the winter term of 1917. ... In 1918 Dr. Marion Richards took up the post of senior assistant in chemistry.[48]

Ada Hitchins (Mrs. Stephens)

Ada Florence Remfry Hitchins[49] was the principal research student of Frederick Soddy. It was her toil, isolating lead samples from uranium ores, which contributed to the discovery of isotopes and Soddy's subsequent Nobel Prize.[50] Hitchins was born on 26 June 1891 in Devon, and obtained a B.Sc. from the University of Glasgow in 1913.[51] She commenced working with Soddy in 1913 and moved with him to Aberdeen in 1914. It was anticipated that Soddy, when appointed to the Chair of Chemistry at Aberdeen, would build up a strong school of radiochemistry[49]; but this did not happen, as Soddy's other student, John A. Cranston,[52] remarked: "During his sojourn in Aberdeen, 1914–19, I am not aware of any other work done by Soddy in the field of radioactivity — apart from supervising the completion of some work by his research student, Miss Hitchens [*sic*], on the growth of radium from uranium."[53]

At the time, Soddy was seeking evidence that lead from thorium ores had different atomic weights from "normal" lead. When Soddy announced the discovery of a sample of lead of atomic mass 207.74, he acknowledged the contribution of Hitchins for the separation and analysis work. Thus, Hitchins' precise and accurate measurements on the atomic masses of lead from different sources were among the first evidence for the existence of isotopes.[51] In addition, Hitchins took over the research on protactinium from Cranston when the latter was drafted for the First World War.

In 1916, Hitchins herself was drafted to work in the Admiralty Steel Analysis Laboratories.[54] When the former male occupants of the analytical laboratories returned upon the end of hostilities, Hitchins lost her position. However, the wartime analytical experience enabled her to find employment as a chemist with a Sheffield steel works until Soddy, then at Oxford, obtained funding to rehire her.[55]

Despite a Nobel Prize in Chemistry, Soddy had great difficulty in attracting graduate students to work with him at Oxford[56]; thus, Hitchins played a crucial role in Soddy's research programme. Initially appointed as technical assistant, she was promoted to private research assistant in 1922. Soddy noted: "... she has also charge of my radioactive materials ... and has worked up considerable quantities of radioactive residues and other materials for general use."[57]

Hitchins finally left Soddy's employ in 1927, emigrating to Kenya and becoming Government Assayer and Chemist for the Mining and Geological Department of the Colonial Government in Nairobi. After her retirement in 1946, she married John Ross Stephens of Amboni Bend Farm, Nyeri Station, Kenya. Subsequently returning to England, she died on 4 January 1972 in Bristol.[58]

Beatrice Simpson

Beatrice Weir Simpson[59] was born on 15 February 1893 in Inveravon, Banffshire, the daughter of James S. Simpson, an excise officer. Simpson completed her M.A. in 1913 and her B.Sc. in 1917. Following graduation, she was appointed temporary Assistant in the Chemistry Department upon the recommendation of Soddy.[60]

From 1923 until her retirement in 1956, Simpson was on the staff of the Rowett Research Institute in Aberdeen. The Rowett Institute had been proposed in 1913 when a Joint Committee of the University of Aberdeen and the North Scotland College of Agriculture decided upon the need for an Institute for Research in Animal Nutrition; however, research did not commence until 1919. Much of Simpson's work was concentrated on the role of iodine in nutrition.[61] She died on 14 February 1981, aged 87 years.

Marion Richards

The Rowett Institute attracted a significant number of women researchers in the field of biochemistry, another of the early notables being Marion B. Richards.[59] Richards, daughter of Robert Richards, a commercial traveller, obtained her M.A. in 1907 and a B.Sc. in 1909. She was the first woman chemistry graduate of Aberdeen to undertake research, completing a D.Sc. with Professor Freundlich at the University of Leipzig in 1916.

From 1916 to 1918, the University records state that she "taught and lectured at Leeds,"[59] while Page stated: "Since the beginning of the war she had held teaching appointments in Wales and England."[48] Unfortunately there are no further details to be found. In 1918, she was appointed Senior Assistant in Chemistry to Soddy at Aberdeen, taking over the responsibility for the organic chemistry course. She joined the staff of the Rowett Institute as a biochemist in 1920, a move forced upon her when Soddy left Aberdeen for Oxford. In the 1930s, the Biochemistry Department of the Rowett consisted solely of William Godden (the Head), Richards, and Simpson.[62] Richards remained at the Rowett until her retirement in 1951.[63]

Alice Bain

The first woman to complete a chemistry degree at Aberdeen was Alice Mary Bain.[59] Bain, daughter of John Bain, was born on 31 August 1875 in Greenock. Entering University in 1898, she completed an M.A. in 1899 and a B.Sc. in 1901. Upon graduation, she was appointed Junior Science Mistress at Croydon High School (a GPDSC school). In 1903, Bain went to the Northern Polytechnic, London, as Assistant Lecturer and Demonstrator in Chemistry, her research on optically active compounds being with William Mills[64] and resulting in four publications. In 1916, Bain travelled to India, where she was appointed Professor of Chemistry at Lady Hardinge Medical College, Delhi. She transferred in 1923 to the University of Delhi, where she held the position of Reader in Chemistry.[65] Bain was still teaching there in 1928, but no later records of her could be traced.

University of Edinburgh

The University of Edinburgh is actually the youngest of the "old" Scottish universities, founded in 1583 as the College of Edinburgh. It then became the College of James VI, and finally the University of Edinburgh.[66]

Life for Women Students

It was at Edinburgh that the only serious opposition to the admission of women occurred, and the event came to be known as the "Surgeons' Hall Riot."[67] Five women, including Jex-Blake (see Chap. 4) and Edith Pechey (see below), were admitted to the Medical School in 1869. They were the first female medical undergraduates at any British university — though they had to fund their own segregated lectures.

The Riot of 18 November 1870 occurred when the women took their anatomy examination and, like the Bristol (see Chap. 5) and Newnham (see Chap. 6) riots, showed the misogynist attitude of many of the male students. The rioters had been called together by a rabble-rousing anti-women pamphlet from the "Chemistry Class of the University" with the encouragement of one of their Professors.[68] As the five women approached the Surgeons' Hall, they were mobbed by drunken students. Then, at the end of the examination, the women refused an offer to stealthily exit through a back door. A group of sympathetic Irish male students then offered to escort them out through the front entrance. The burly escorts ensured that the only harm to the women was that of being splattered by mud and rotten eggs and having their dresses torn.

The most vociferous supporter of the women was Robert Wilson[68] of the Royal Medical Society of Edinburgh. He sent a letter to Pechey following the riot:

> I wish to warn you that you are to be mobbed again on Monday. A regular conspiracy has been, I fear, set on foot for that purpose. I have made what I hope to be efficient arrangements for your protection. I had a meeting with Micky O'Halloran who is leader of a formidable band, known in College as "The Irish Brigade" and he has consented to tell off a detachment of his set for duty on Monday. Micky was the formidable hero with the big red moustache who stood by us on Friday and whose presence with us rather disappointed the

> rioters who, I think, calculated on the aid both of himself and his set. May I venture to hint my belief that the real cause of the riots is the way some of the professors run you down in their lectures. However you and your friends need not fear, as far as Monday is concerned. You will be taken good care of.[69]

Though Jex-Blake and Pechey passed, the University authorities cited University regulations that only allowed medical degrees to be awarded to men. As a result, the British Medical Association refused to register the women as doctors.

It was not just in medicine that women students were treated badly at Edinburgh. In 1896, a plea was made to treat women students in a more civilised fashion:

> It is unfortunately a matter of not infrequent observance that the treatment of lady students at the hands of their *confrères* has been utterly out of keeping with the cherished canons of gentlemanly conduct. It is a painful fact that ever since the portals of our University were opened to the lady students, as the result between sweet reason and dogmatic rigidity, they have been the victims of gratuitous annoyance. Their entry into the class-rooms opens the floodgates of British chivalry. The tapers and tadpoles of the back benches begin to howl and screech with all the lustiness of rural louts, and seem to be as much amused as if they saw a picked company of the far-famed Dahomeyan Amazons march in all the glory of their military attire.[70]

Women Chemistry Students

Thomas Charles Hope was appointed as the sole Lecturer in Chemistry in 1797. In addition to teaching the regular students, in 1826 he introduced "a Short Course of Lectures for Ladies and Gentlemen."[71] The presence of women on campus was not appreciated by many academics; for example, Lord Cockburn wrote to T. F. Kennedy:

> The fashionable place here now is the College; where Dr Thomas Charles Hope lectures to ladies on Chemistry. He receives 300 of them by a back window, which he has converted into a door. Each of them brings a beau, and the ladies declare that there was never anything so delightful as these chemical flirtations.[72]

Edith Pechey (Mrs. Pechey-Phipson)

Before Mary Edith Pechey[73] entered medical school, she had been an outstanding chemistry student in the 1869–1870 session.[74] Pechey, born on 7 October 1845, was the daughter of William Pechey, a Baptist minister at Langham, near Colchester, and Sarah Rotton, a well-educated Nonconformist lawyer's daughter. Pechey worked as a governess and teacher before joining Jex-Blake's efforts to enter medical school at Edinburgh.

Chemistry was required for admission to medical school, and the Professor of Chemistry, Alexander Crum Brown,[75] gave separate lectures to the women students, insisting they were identical to those which he was concurrently giving to the men. Pechey attained third place in the chemistry examinations. Forty years earlier, Hope had instituted awards known as the Hope Scholarships, and these were presented to the top four students in the first-year chemistry examinations. The recipients were entitled to free use of the facilities of the University chemistry laboratory for the next term. The two students above Pechey in the list were repeating the course and were therefore ineligible.

Though Crum Brown had been initially sympathetic towards the women, his friends and colleagues resented the appearance of women at the venerable institution. As Jex-Blake's biographer, Shirley Roberts, added: "Many of the male [chemistry] students also turned hostile when they saw that the women were capable of outstripping them in competitive examinations."[76] Crum Brown then proclaimed that Pechey was ineligible as she had been taught in a separate class, contradicting his earlier statements. Having said this, Crum Brown felt unable to issue the women the usual certificates of attendance required for admission to medical school, instead giving them his own certificates, which Jex-Blake referred to as the Professor's "strawberry jam labels."[74] The issue rapidly escalated, gaining national and even international attention. Finally, by a one-vote margin, the University Senate approved the issuing of attendance certificates to the women; while also by a margin of one vote, the Senate denied Pechey the Hope Scholarship. To make up for the loss of the Scholarship, Pechey and Jex-Blake set up their own chemical laboratory at 15 Buccleuch Place in which the women could undertake their practical work.

Pechey then went to the University of Berne, completing an M.D. in 1877. For the next 6 years, she practiced medicine in Leeds; then, Elizabeth Garrett Anderson suggested that Pechey should go to Bombay (now Mumbai) to become Senior Medical Officer at the Cana Hospital for Women and Children. Pechey arrived in Mumbai in 1883, quickly becoming fluent in Hindi. She married Herbert Musgrave Phipson, a reformer as well as wine merchant and naturalist, in 1889. They returned to Britain in 1905, Pechey becoming active in the Leeds suffrage movement. She died in 1908 at Folkestone, Kent.

Christina Miller

There was one star student who became a key member of the staff of the Chemistry Department — Christina Cruickshank Miller.[77] Christina Miller was born on 29 August 1899 in Coatbridge, the elder of two sisters. In grade school, Miller developed an interest and aptitude towards mathematics. At the time, she was told that, as a woman, she could only use her mathematical aptitude towards a career as a school teacher. Unfortunately, she had become deaf as a result of childhood measles and rubella, and this was considered a major impediment to a school-teaching career. She read an article in a magazine that mentioned the employment potential for women as analytical chemists, and this possibility determined her future.

Miller won an entrance scholarship to the University of Edinburgh, but to give her a better training for an industrial position, Miller combined the three-year degree programme from Edinburgh with a four-year industrial chemistry diploma programme from Heriot-Watt College. Heriot-Watt College, which dated from 1885, had been founded in 1821 as the "School of Arts of Edinburgh for the Education of Mechanics in Such Branches of Physical Science as are of Practical Application in their several trades," the first institute in Britain for the education of manual workers. Women were welcomed at Heriot-Watt as early as 1869, 20 years ahead of other institutions. Chemistry was offered as an evening course, accounting for Miller's ability to take the diploma simultaneously with the degree from Edinburgh.

At the beginning of her graduating year, Miller discussed her plans for the future with chemist James Walker.[78] Walker advised her to learn German, then a prerequisite for research due to the importance of the German chemical literature. Thus, in her final year, in addition to taking classes, acting as laboratory demonstrator, and performing a research project on organic compounds of arsenic and mercury, she also taught herself German. Despite her hearing difficulty (and her workload), she won the class medal in the university advanced chemistry course, graduated with special distinction in chemistry, and gained a scholarship which allowed her to pursue a higher degree.

Walker accepted Miller as a graduate student and, from 1921 until the completion of her Ph.D. in 1924, she worked with him on the process of diffusion in solution. She was then awarded a two-year Carnegie Research Fellowship to undertake independent research. The position enabled her to study a problem that had long fascinated chemists: the glow produced when tetraphosphorus hexaoxide is oxidized. During this time, she applied for a lectureship at Bedford, but was rejected on the grounds of her deafness. Walker suggested that Miller accept a post as an assistant in the chemistry department at the University of Edinburgh. The position involved the supervision of undergraduate students, but she was allowed to continue with research in her spare time.

By 1929, Miller's thorough research on the oxides of phosphorus showed that traces of elemental phosphorus caused the glow, not the oxides themselves. As a result of her five publications, Miller was awarded a D.Sc. degree, the prestigious Keith Prize by the Royal Society of Edinburgh, and a Lectureship with tenure. Shortly afterwards, during her research, a glass bulb exploded, the fragments blinding her in one eye. Deciding that avenue of research was too dangerous, Miller developed an interest in micro- and semimicro-analysis, and in 1933 she was appointed Director of the inorganic laboratory. During the next 28 years, she became renowned for her innovative undergraduate microscale analytical techniques. She was always looking to refine and improve analytical methods, and her many research papers were devoted to this topic, such as the definitive work on 8-hydroxyquinoline as a reagent for magnesium.

Sadly, ill health in the form of otosclerosis, together with family commitments (care of a relative), caused Miller to take early retirement in 1961. Never one to remain idle, she pursued interests in genealogy and in the history of Edinburgh. Her biographer, Robert Chalmers, commented:

> It is arguable that had she been a man ... she might have become one of the UK's first professors of analytical chemistry. At a time when analytical chemistry was practically non-existent in UK universities she provided courses that would stand comparison with the best available today. ... Her work was highly esteemed by many internationally renowned analysts.[77(a)]

She died on 16 July 2001 at the age of 101.

Elizabeth Kempson (Mrs. Percival, Mrs. McDowell)

The only woman to be mentioned in the history of the Chemistry Department at the University of Edinburgh was Elizabeth E. Kempson.[79] Born on 3 January 1906 in Coventry, she was educated at Wolverhampton High School and obtained an honours degree in chemistry at the University of Birmingham in 1928. She stayed in Birmingham, undertaking research with Norman Haworth.[80] Haworth's senior research assistant was Edmund George Vincent Percival, and in 1934 Kempson and Percival married.

That same year, Percival was appointed to a lectureship at the University of Edinburgh, where Kempson started research with him on the synthesis and reactions of carbohydrates, for which she received a Ph.D. in 1941. With the founding of the Scottish Seaweed Research Association, their interest turned to marine polysaccharides. Percival died in 1951 and Kempson took over the research, being appointed Lecturer, while raising two children. For her contributions to polysaccharide chemistry, she was elected a Fellow of the Royal Society of Edinburgh.

In 1962, she married Richard McDowell of Alginate Industries Ltd., and together they wrote the monograph, *Chemistry and Enzymology of Seaweed Polysaccharides*.[81] Kempson then moved to Royal Holloway College as an Honorary Lecturer, where she built up a new research group. Her collaborators included 26 Ph.D. students, and resulted in more than 100 publications.

During her lifetime, an issue of the journal *Carbohydrate Research* was dedicated to her, as a tribute to her work. In the introduction, one of her former students, Helmut Weigel, commented:

> Dr. Percival is an ardent traveller. In her journeyings, which have taken her to all continents, she always combines a taxing lecture programme with the collection of seaweeds — sometimes a very hazardous occupation — and with visits to former

> students and friends. Her zest for travel has made her an enthusiastic camper, and not many of us can claim, as she can, to have cooked a three-course meal whilst stranded by a blizzard on top of the Atlas mountains. ... She plays tennis, enjoys her piano and, in the Royal Holloway College, Dr. Percival has been President of the Women's Club. However, her chief interest remains the pursuit of chemistry, and she takes pleasure in communicating her own enthusiasm to younger students of chemistry.[79(a)]

Kempson died on 16 April 1997, aged 91 years.

University of Glasgow and the Royal Technical College, Glasgow

The University of Glasgow (UG), the second-oldest university in Scotland, was founded in 1451.[82] In 1796, UG was joined in Glasgow by a new institution: the Royal Technical College (RTC), now the University of Strathclyde. The RTC was the oldest technical college in the world and the first to open its doors to women.[83] Students at the RTC were granted degrees from the University of Glasgow.

Queen Margaret College

The University of Glasgow was unique in having a separate women's college for many years. It was in 1868 that Mrs. Campbell of Tullichewan organised a series of "lectures for ladies"[84] that were taught by professors of UG. In 1877, the Glasgow Association for the Higher Education of Women was founded, and they organised courses for women, some of the lectures being given at the University while others were given in rented rooms. With the success of this endeavour, the association was incorporated in 1883 as Queen Margaret College (QMC-UG) for the education exclusively of women. The name was chosen to commemorate the 11th-century Queen Margaret, who was reputed to have brought education and culture to Scotland.

As seems so often to have been the case, exceptional women appeared to take the reins of the institution. The first was Janet Anne Galloway, who became the first Secretary of the College, as the administrative head was then called. A fervent anti-suffragist, she contended that women needed to be educated for their traditional roles and she opposed the hiring of women lecturers. Her successor in 1909, Frances Helen Melville, Mistress of the College (as the post had been retitled) and former Lecturer at Cheltenham Ladies' College (CLC), held largely contrary views: an active suffragist, Melville argued that women had to be educated so that they could follow professional lives equal to men.

The College steadily expanded, adding science laboratories in 1888 and a medical school in 1890. In 1892, the College was merged into the University of Glasgow. A student reported in 1902: "As a general rule the women attend the same classes as the men, occupying the front benches."[85]

By the 1920s and 1930s, one would have expected that the presence of women at Glasgow was an established fact, but this was not the case. In 1924, Miss Sheavyn, Director of Women Students at the Victoria University of Manchester, sent a letter to Melville enquiring whether QMC-UG would like to participate in a survey on women students at university and particularly on any health issues. Melville's reply was very negative:

> I fear that the women in this University are not yet sufficiently regarded as a matter of course to make it safe to begin an investigation upon them. In any case, I should have doubts about such an investigation being closely associated with medical men, whom I have, so far, seldom found unbiased in their judgements as to the causes of any ill-health in women students. They invariably put it down to the "effects of study", whereas a little enquiry would show a plurality of causes at work, including the effects of much dancing with late hours, in conjunction with study.[86]

The success of QMC-UG was its own downfall. The number of women students increased from 857 in 1917–1918 to 1708 in 1929–1930. With limited space and facilities at QMC-UG, more and more of the courses had to be offered in the large lecture halls on the UG campus, a 15-minute walk away. As a result, by the 1920s and 1930s women were taking more courses on the UG campus than at QMC-UG. With women students spending much of their time trudging back and forth between QMC-UG and UG for their classes, the University authorities decided to absorb the women students into the University itself, and the College ceased to have a separate existence in 1934. The only Scottish experiment of an autonomous women's college had come to an end.[87]

Margaret Sutherland

Margaret Millen Jeffs Sutherland[88] was the first woman Lecturer in chemistry at the Royal Technical College. Daughter of Andrew Sutherland, a manufacturer, she was born in 1881 and attended the Lenzie Academy, Lenzie, and the Westbourne Terrace School, Glasgow. She obtained a B.Sc. (Glasgow) in 1908 and then became a researcher at UG from 1908 to 1912.

In 1913, she was appointed as Assistant to the Professor of Chemistry at the RTC, then in 1920 was promoted to Lecturer, and in 1942 to Senior Lecturer. During her career, she co-authored a total of 12 publications on organic reactions, the first five with George Gerald Henderson[89] on the chemistry of terpenes. Henderson had been appointed in 1889 as the Lecturer in Chemistry at Queen Margaret College, and held the position for 3 years before being promoted to Chair in Chemistry at the RTC.[83]

Sutherland then had six publications with Forsyth James Wilson,[90] Henderson's successor, on semicarbazones and acridines. Sutherland closed her obituary of Wilson with the statement: "It has been a privilege of the writer to collaborate with Professor Wilson in much of his later research and she therefore takes this opportunity to pay tribute to a kind and inspiring teacher and friend."[90] She retired in 1947 and died in 1972 in her hometown of Lenzie in nearby Stirlingshire.

Women Chemistry Students

The University of Glasgow had a student chemistry society, the Alchemists' Club, founded in 1916 by Catherine F. Davidson (later Mrs. Jones).[91] The Club was formalized in 1918, at which time one of the two Vice-Presidents was a woman (Davidson), as were four of the six Ordinary Members of Council.[92]

The Club had regular social events, particularly dances. For example, in 1925, it was reported that:

> A new reaction was tried by several women students in the vicinity of Lab. III on 19th. November. It is called the French Tango Reaction. Several complicated movements of the Ions took place, usually they moved in pairs. The Results of the experiment were demonstrated at the Dance that evening.[93]

The dances were not always well received by parents. A Mrs. V. A. Grant complained to *The Alchemist* that the goal of "social intercourse among its members" was not being met:

> The experience of my dear daughter, Audrey, simply testifies to that. ... Are the Dance Committee promoting social intercourse when they allow my daughter to dance with the same partner for the whole evening? Whilst discussing this dance, I may say that a dance is rather prolonged if carried on till 4 a.m. My daughter did not arrive home until 4.15 a.m.[94]

By 1929, the welcome for women seemed to be fading when *The Alchemist* carried an "imaginary" conversation between two Glasgow University male chemistry students:

> Coeducation would not be so bad if the women stayed at Q.M. and only emerged perhaps for tennis, golf or dances .... What

> woman, think you, is intelligent enough to grasp the principles of chemistry — you've only got to look at the exam. lists. No, woman's brain is too feeble, her temperament too volatile, her nature too erratic, for her to be of any use to Chemistry or Industry, even if, by some lucky fluke, she does get her degree. I repeat, women ought to be abolished from the labs., yea, even from the University.[95]

No rebuttal could be found in later issues from women chemists, and it is curious that the author claims women students were at the bottom of the exam lists when the usual complaint was that women chemists dominated the upper ranks.

Ruth Pirret

Ruth Pirret[96] was the first woman graduate in science from the University of Glasgow, obtaining a B.Sc. honours degree in Pure Science (mainly chemistry) in 1898. She was an outstanding student throughout her undergraduate work, and after graduation she became a school teacher in Arbroath.[97] Starting in October 1909, Pirret worked for at least 6 months with Soddy. Lord Fleck, another researcher with Soddy at the time, commented: "T. D. MacKenzie was the research student who was the most important worker in those early years. Miss Ruth Pirret was also a contributor, but on a less prominent scale."[98]

Pirret's studies were on the ratio of uranium to radium in minerals, an extension of the work of Ellen Gleditsch,[99] the famous Swedish researcher in radiochemistry. It had been argued at the time that the older the mineral, the higher the proportion of radium, and Pirret and Soddy confirmed some of Gleditsch's findings. However, during their studies, they made a much more important discovery: that uranium existed in nature as a mixture of two isotopes. As they stated in the paper: "We are therefore faced with the possibility that uranium may be a mixture of two elements of atomic weights 238.5 and

234.5, which, like ionium, thorium, and radio-thorium [thorium isotopes 230, 232, and 228], are chemically so alike that they cannot be separated."[100]

Mary Andross

Among the early graduates from UG was Mary Ann Andross.[101] Born in 1893 in Ayrshire, she was the daughter of Henry Andross, a cashier. Andross graduated from UG in 1916, before joining Henderson's Chemistry Department at the RTC and then returning with him to UG. It is not clear whether she taught or was a researcher during this period.

Subsequently, Andross became an outstanding teacher of domestic science as Head of the Science Department of the Glasgow and West of Scotland College of Domestic Science, where she remained for 40 years until retirement in 1965. In addition to pioneering courses for the training of dieticians, she was an active researcher on the chemical composition and nutritional value of foodstuffs, publishing five papers solely under her name. Andross was elected Fellow of the Institute of Food Science Technology. During the Second World War, she promoted the use of rose hips as a source of vitamin C.

Andross was a colourful character, as her obituarist, David Cuthbertson, noted.[101(a)] In addition, she was very active in social events: "The revival of the Ramsay Dinner in Glasgow was largely due to her enthusiasm and energy and she acted as convenor on many an occasion. … Many would have liked her to be Chairman of the Glasgow Section of the Society for Chemical Industry, but she emphatically refused the honour."[101(a)]

In her youth, Andross had been extremely athletic. While still quite young, she was seriously injured by a car and lost much of the use of her legs. This did not diminish her interest in sports, in particular, fishing on the island of Harris. Her independence is illustrated by the story of how a tyre on her car had a puncture one dark night in the country. She changed the tyre without any light source, storing the grimy wheel nuts in her mouth for fear of losing them.

Andross died on 22 February 1968.

Education in Wales

There were three women in particular who championed the case for the education of girls in Wales: Margaret Williams, Dilys Davies, and Elizabeth Hughes.[102] Margaret Hay Williams (later Lady Verney) became the President of the Association for Promoting the Education of Girls in Wales during the late 1880s. Dilys Davies (later Mrs. Glynne Jones), one of the Secretaries of the Organization until 1898, had taught at the North London Collegiate School for Girls in 1887 under Buss. Elizabeth Phillips Hughes was educated and then briefly taught at CLC under Beale. Thus, the education of girls in Wales, too, can be linked back to Buss and Beale (see Chap. 1).

To provide an academic schooling for girls, one of the Girls' Public Day Schools (see Chap. 1) had been opened in Swansea in 1888, but it had closed in 1895, during which time they had unsuccessfully tried to spread the ideas of the GPDSC in Wales.[103] Lacking an equivalent to the GPDSC, the only Welsh school to have nurtured women chemists seems to have been Howell's School, Llandaff (which joined the GPDSC in 1980). It was in 1537 that Thomas Howell had left a substantial sum of money to the Drapers' Company to provide dowries "every yere for maydens for ever."[104] From this charitable bequest, the Drapers' Company founded Howell's School in 1860, a second school being subsequently opened in Denbigh, North Wales.

For higher education, four university colleges were founded in the late 19th century: the University College of Wales, Aberystwyth; the University College of North Wales, Bangor; the University College of South Wales and Monmouthshire, Cardiff; and University College, Swansea. These Colleges were combined in 1894 under the name of the University of Wales. The Charter was clearly progressive on equal opportunities for women: "Women shall be eligible equally with men for admittance to any degree which the University is by this our charter authorised to confer; every office hereby created in the University and membership of every authority hereby constituted shall be open to women equally with men."[105]

Nevertheless, in the early years, perhaps due to the lack of academic schools for Welsh girls, the women who populated the Welsh university colleges came primarily from such English schools as Cheltenham Ladies' College; North London Collegiate School; Wyggeston School, Leicester; Orme's School, Newcastle-under-Lyme; Haberdashers' Aske's Hatcham School; and the Girls' High Schools in northwest England such as those at Blackburn, Bury, and Southport.[102]

University College of North Wales, Bangor

The location for a North Wales college was settled in 1884, with Bangor being chosen as the location.[106] Though co-educational, life for the women students was strictly controlled. J. Gwynne Williams elaborated:

> Bangor was by no means unusual in drawing up careful rules to govern relations between men and women students. If they talked to one another for more than a few moments even between lectures, they were in danger of being reported for "skylarking" and in lectures they sat quite separately..... There was a rigid system of chaperonage, the women's warden attending all mixed gatherings.[107]

From 1903 until his death in 1930, Kennedy Joseph Previté Orton[108] held the Chair of Chemistry, and the women chemists who came to Bangor all undertook research with him (including Phyllis McKie, see Chap. 12). Orton was a strong believer in the historical context of chemistry, and a quote by one of his obituarists is of note as it mentions the book written by Ida Freund (see Chap. 6):

> Historical chemical characters were made to live. Who will ever forget the lecture on Mendeléeffs prediction of the properties

of the missing elements and their fulfilment by later discoveries, a discourse largely based on the account in Ida Freund's "The Study of Chemical Composition" — a unique treatise for which Orton had the highest regard.[108(a)]

Alice Smith

One of Orton's students was Alice Emily Smith. Smith was born on 18 June 1871, daughter of Thomas Smith, Commission Agent of Warrenpoint, County Down, Northern Ireland; and she was educated at Crescent House School, Bedford.[109] She entered University College of North Wales, Bangor, in 1897 and completed a B.Sc. (London) in chemistry in 1901. Smith was awarded an 1851 Scholarship, which she chose to use from 1901 to 1903 at Owens College, Manchester, where she worked with William Perkin, Jr.,[110] her research resulting in four substantial papers.

Smith returned to Bangor as Assistant Lecturer in Education and Assistant Lecturer in Organic Chemistry, where she collaborated with Orton on reaction mechanisms, five papers resulting from her research. In 1914, she relinquished her position at Bangor, being succeeded by May Leslie[106] (see Chap. 5). For the next 3 years she held an appointment as Lecturer in Science at the Maria Grey Training College, London.

The Maria Grey Training College played a crucial role in the history of women in education, being the first teachers' training college for women. It was founded by Maria Shirreff Grey, who, in 1871, had initiated the Women's Educational Union, the parent of the Girls' Public Day School Company (GPDSC schools, see Chap. 1).[111] As there was a shortage of trained women to teach in the schools, the Union organized the Teachers' Training and Registration Society for women. The Society then opened its own training college in 1877, which in 1885 was named the Maria Grey Training College. Thus, the College, to be incorporated into the West London Institute of

Higher Education in 1976, served a key role in the success of the GPDSC schools.

Perhaps answering the wartime need for women chemists in industry (see Chap. 12), Smith left the Training College in 1917 to become a research chemist at Messrs Cooper's Laboratory. With the War ended and the male chemists returning to resume their posts, she returned to teaching in 1920, becoming Principal at a private school in Ilkley, Yorkshire — a position she held until her death in 1924.

Commentary

As in England, we find that in Scotland and Wales mentors played an important role for women chemistry students: James Irvine at St. Andrews; Alexander McKenzie at Dundee; Frederick Soddy at Aberdeen and Glasgow; James Walker at Dundee and Edinburgh; George Henderson at the Royal Technical College, Glasgow; and Kennedy Orton at Bangor. Of note, Walker married a chemist, Annie Sedgwick (see Chap. 11); while Soddy's spouse, Winifred Beilby (see Chap. 11), worked with him after marriage .

There was a fundamental difference between the "big" and "small" universities in that some of those from "small" universities found posts in academia (Christina Miller of "big" Edinburgh, under the mentorship of Walker, being an exception). Women chemistry graduates from St. Andrews, Dundee, and Bangor flourished, and in some cases became the stalwarts of the department, such as Ettie Steele at St. Andrews, Isobel Smith at Dundee, and Alice Smith at Bangor. Some of those who left St. Andrews, Dundee, and Bangor were successful in academia in England. For example, Ishbel Campbell of St. Andrews became a Lecturer at University College, Southampton; Mary Lesslie of Dundee, a Lecturer at Bedford (see Chap. 4); Nellie Walker of Dundee, a Lecturer at Bedford, before returning to Dundee; and Phyllis McKie of Bangor, a Lecturer at Bedford College, then Westfield College, London (see Chap. 12).

Notes

1. Moore, L. (2003). Young ladies' institutions: The development of secondary schools for girls in Scotland, 1833–*c*.1870. *History of Education* **32**(3): 249–272.
2. Avery, G. (1991). *The Best Type of Girl: A History of Girls' Independent Schools*. Andre Deutsch, London, p. 255.
3. Moore, L. (1992). The Scottish universities and women students, 1862–1892. In Carter, J. J. and Withrington, D. J. (eds.), *Scottish Universities: Distinctiveness and Diversity*, John Donald Publishers, Edinburgh, pp. 138–146.
4. Manton, J. (1965). *Elizabeth Garrett Anderson*. Methuen, London, 1965.
5. Roberts, S. (1993). *Sophia Jex-Blake: A Woman Pioneer in Nineteenth Century Medical Reform* (The Wellcome Institute Series in the History of Medicine). Routledge, London.
6. Smart, R. N. (June 1968). Literate ladies — A fifty-year experiment. *St. Andrews Alumnus Chronicle* **59**: 21–31.
7. Watson, W. N. B. (1967–1968). The first eight ladies. *University of Edinburgh Journal* **23**: 227–234.
8. McDonald, I. J. (1967). Untapped reservoirs of talent? Social class and opportunities in Scottish higher education 1910–1960. *Scottish Educational Studies* **1**: 53.
9. Crichton, A. C. (1967). Finishing school for young ladies. *The College Courant (Journal of the Glasgow University Graduates Association)* **19**(38): 19–21.
10. Hamilton, S. (1982). Interviewing the middle class: Women graduates of the Scottish universities *c*.1910–1935. *Oral History* **10**(2): 58–67. Reprinted here and subsequently with permission from the Oral History Society, www.ohs.org.uk.
11. Note 10, Hamilton, p. 62.
12. Note 10, Hamilton, p. 65.
13. Read, J. (1953). Schools of chemistry in Great Britain and Ireland–I: The United College of St. Salvator and St. Leonard, in the University of St Andrews. *Journal of the Royal Institute of Chemistry* **77**: 8–18.
14. Sloan, S. (1910–1911). University Hall, St. Andrews. *The Girl's Realm* **13**: 893.

15. "Selah." (27 February 1890). My erudite lady-friends. *College Echoes: St Andrews University Magazine* **15**: 125.
16. "P.F.F." (1922). Obituary notices of fellows deceased: Thomas Purdie, 1843–1916. *Proceedings of the Royal Society of London, Series A* **101**: iv–x.
17. Read, J. (1953). Sir James Colquhoun Irvine, 1877–1952. *Obituary Notices of Fellows of the Royal Society* **8**: 459–489.
18. (June 1953). *St. Andrews Alumnus Chronicle* **40**; and St. Andrews University, student records.
19. University of St. Andrews, student records; and Anon. (June 1953). *The Alumnus Chronicle* **40**.
20. Anon. (June 1984). Obituary. *St. Andrews Alumnus Chronicle* **75**: 24.
21. Anon. (January 1953). *St. Andrews Alumnus Chronicle* **39**: 12; and Anon. (June 1959). *St. Andrews Alumnus Chronicle* **50**: 17.
22. Robinson, R. (1976). *Memoirs of a Minor Prophet: 70 Years of Organic Chemistry*, Vol. 1. Elsevier, Amsterdam, p. 127.
23. (a) Cookson, R. C. (April 1998). Ishbel Campbell, 1906–1997. *Chemistry in Britain* **34**: 72; and (b) Harris, M. M. (2001). Chemistry. In Crook, J. M. (ed.), *Bedford College: Memories of 150 Years*, Royal Holloway and Bedford College, p. 87.
24. Adam, N. K. and Webb, K. R. (1956). Schools of chemistry in Great Britain and Ireland–XXIII, The University of Southampton. *Journal of the Royal Institute of Chemistry* **80**: 133–140.
25. Personal communication, e-mail, M. B. Hocking, 29 August 2007.
26. University of St. Andrews, student records.
27. Letter, W. H. Bragg to J. C. Irvine, 17 January 1927, Royal Institution Archives. Courtesy of the Royal Institution of Great Britain (RI MS WHB).
28. For information on Karlik, see: Rentetzi, M. (2004). Gender, politics, and radioactivity research in interwar Vienna: The case of the Institute for Radium Research. *Isis* **95**: 359–393.
29. Shafe, M. (1982). *University Education in Dundee 1881–1981*. University of Dundee, Dundee.
30. Reed, J. (1952–1953). Alexander McKenzie (1869–1951). *Obituary Notices of Fellows of the Royal Society* **8**: 207–228.
31. "An Undergraduette." (December 1924). Sheiks and shrieks. *The College: Magazine of University College, Dundee* **22**(1): 15. The

magazine is cited here and subsequently by permission of University of Dundee Archive Services.

32. "Alphabeta." (February 1922). A maiden speaks. *The College: Magazine of University College, Dundee* **19**(2): 4.
33. Anon. (Nov 1940). *The College: Official Magazine of SRC, University College, Dundee* **20**(1): 20. The magazine is cited here and subsequently by permission of University of Dundee Archive Services.
34. University College, Dundee (University of St. Andrews), student records; and Bedford College, staff records.
35. Anon. (December 1915). *The College: Magazine of University College, Dundee* 13.
36. Anon. (1981). Personal news: Deaths. *Chemistry in Britain* **17**: 270; University of St. Andrews, student records; and *Calendars*, University of St. Andrews.
37. (a) University of St. Andrews, student records; (b) (1951). *Old Students and Staff of the Royal College of Science,* 6th ed. Royal College of Science, London; and (c) Anon. (October 1998). Personal news: Deaths. *Chemistry in Britain* **34**: 91.
38. "Lady students." (8 February 1889). *The College: Magazine of University College, Dundee* **1**(2): 64.
39. Anon. (January 1904). *The College: Magazine of University College, Dundee.* New series, 1.
40. Anon. (November 1904). Department notes: Science. *The College: Magazine of University College, Dundee.* New series **2**(1): 19.
41. Anon. (December 1904). Department notes: Science. *The College: Magazine of University College, Dundee.* New series **2**(2): 42.
42. Anon. (February 1906). Department notes: Science. *The College: Magazine of University College, Dundee.* New series **3**(4): 77.
43. Anon. (December 1906). Society notes: Scientific society. *The College: Magazine of University College, Dundee.* New series **4**(2): 37.
44. University of Dundee, student records.
45. Strathdee, R. B. (1953). Schools of chemistry in Great Britain and Ireland–V, The University of Aberdeen. *Journal of the Royal Institute of Chemistry* **77**: 220–231.
46. Moore, L. (1991). *Bajanellas and Semilinas: Aberdeen University and the Education of Women 1860–1920.* Aberdeen University Press, Aberdeen, p. 43; and Moore, L. (1980). Aberdeen and the

higher education of women. *Aberdeen University Review* (163): 286–288.

47. Ramsay, M. P. (November 1906). Women students in Aberdeen University. *The World of Dress & Women's Journal* 11.
48. Page, K. R. (Autumn 1979). Frederick Soddy: The Aberdeen interlude. *Aberdeen University Review* 127–148.
49. Fleck, A. (1957). Frederick Soddy. *Biographical Memoirs of Fellows of the Royal Society* **3**: 203–216.
50. Rayner-Canham, M. F. and Rayner-Canham, G. W. (2000). Stefanie Horovitz, Ellen Gleditsch, Ada Hitchins and the discovery of isotopes. *Bulletin for the History of Chemistry* **25**(2): 103–108.
51. Rayner-Canham, M. F. and Rayner-Canham, G. W. (1997). Ada Hitchins: Research assistant to Frederick Soddy. In Rayner-Canham, M. F. and Rayner-Canham, G. W. (eds.), *A Devotion to Their Science: Pioneer Women of Radioactivity*, Chemical Heritage Foundation, Philadelphia, and McGill-Queen's University Press, Montreal, Quebec, pp. 152–155.
52. Cumming, W. M. (1972). John Arnold Cranston 1891–1971. *Chemistry in Britain* 388.
53. (1963). Chemistry: From retort to grid — 1860–1960. In Simpson, W. D. (ed.), *The Fusion of 1860: A Record of the Centenary Celebrations and a History of the United University of Aberdeen 1860–1960*, Oliver and Boyd, Edinburgh, pp. 304–305.
54. Letter, J. C. W. Humfrey, Admiralty Inspection Officer, Sheffield, to Dr. Desch, Glasgow University, 6 January 1916: "We have written to Miss Hitchins today offering her an appointment." University of Sheffield archives, CHD14/WW/180.
55. File 201, Papers of Professor Frederick Soddy, Bodleian Library, Oxford.
56. Cruickshank, A. D. (1979). Soddy at Oxford. *British Journal for the History of Science* **12**: 277–288.
57. F. Soddy, undated reference for A. F. R. Hitchins, Oxford University Archives.
58. Anon. (1972). Personal news: Deaths. *Chemistry in Britain* **8**: 376.
59. Masson, M. (1966). Early women chemistry students at Aberdeen. In Masson, M. R. and Simonton, D. (eds.), *Women and Higher Education: Past, Present and Future*, Aberdeen

University Press, Aberdeen, pp. 316–318; and University of Aberdeen, student records.

60. (1920). Minutes of meeting. *The Senatus Academicus of the University of Aberdeen*, 18 December 1917, item 2, The Aberdeen University Press Ltd.
61. Anon. (17 February 1981). *Aberdeen Press and Journal*, p. 3, col. 2.
62. Orr, J. B. (ed.) (1930). Members of staff. *The Rowett Research Institute Collected Papers*, Vol. II, Edmond & Spark, Aberdeen, p. vi.
63. (Autumn 1964). Obituary: Marion Richards. *Aberdeen University Review* **40**: 395.
64. Mann, F. G. (1960). William Hobson Mills, 1873–1959. *Biographical Memoirs of Fellows of the Royal Society* **6**: 200–225.
65. Watt, T. (compiler) (1935). *Roll of the Graduates of the University of Aberdeen 1901–1925*. Aberdeen University Press.
66. Hirst, E. L. and Ritchie, M. (1953). Schools of chemistry in Great Britain and Ireland–VII, The University of Edinburgh. *Journal of the Royal Institute of Chemistry* **77**: 505–511.
67. Martindale, L. (1951). *A Woman Surgeon*. Victor Gallancz, London.
68. Ross, M. (1996). The Royal Medical Society and medical women. *Proceedings of the Royal College of Physicians, Edinburgh* **26**: 629–644.
69. Cited in: Todd, M. (1918). *The Life of Sophia Jex-Blake*. Macmillan and Co., London, pp. 293–294.
70. "Sarat Mullick." (13 February 1896). A plea for the rational treatment of women students. *The Student: Edinburgh University* **10** (new ser.): 60–61.
71. Doyle, W. P. (17 August 2007). Thomas Charles Hope, MD, FRSE, FRS (1766–1844). http://www.chem.ed.ac.uk/public/professors/hope.html, accessed 3 October 2007.
72. Cockburn, H. (1874). *Letters Chiefly Connected with the Affairs of Scotland*. Ridgway, London, pp. 137–138, cited in Note 71.
73. Lutzker, E. (1973). *Edith Pechey-Phipson, M.D.: The Story of England's Foremost Pioneering Woman Doctor*. Exposition Press, New York.
74. Cited from Note 3 in Hamilton, S. (1983). The first generations of University women 1869–1930. In Donaldson, E. (ed.), *Four*

Centuries: Edinburgh University Life 1869–1983. University of Edinburgh, Edinburgh, p. 114.

75. "J.W." (1924). Obituary notices of Fellows deceased: Alexander Crum Brown, 1838–1922. *Proceedings of the Royal Society of London, Series A* **105**: i–v.
76. Roberts, S. (1993). *Sophia Jex-Blake: A Woman Pioneer in Nineteenth Century Medical Reform*. Routledge, London, pp. 90–104.
77. (a) Chalmers, R. A. (1993). A mastery of microanalysis. *Chemistry in Britain* **29**: 492–494; (b) Mango, K. (Spring 2003). Dr. Christina Miller: A beacon of knowledge and strength. *Hearing Health* 16–21; and (c) Anon. (7 August 2002). University of Edinburgh, news release.
78. Kendal, J. (1932–1935). Sir James Walker. *Obituary Notices of Fellows of the Royal Society* **1**: 537–549.
79. (a) Weigel, H. (1978). Elizabeth E. Percival, F.R.S.E. *Carbohydrate Research* **66**: 7–8; and (b) Manners, D. J. (May 1998). Betty Percival 1906–1997. *Chemistry in Britain* **34**: 61.
80. Hirst, E. L. (1951). Walter Norman Haworth, 1883–1950. *Obituary Notices of Fellows of the Royal Society* **7**: 372–404.
81. Percival, E. and McDowell, R. H. (1967). *Chemistry and Enzymology of Marine Algal Polysaccharides*. Academic Press, London.
82. Cook, J. W. (1953). Schools of chemistry in Great Britain and Ireland–VIII, The University of Glasgow. *Journal of the Royal Institute of Chemistry* **77**: 561–572.
83. Cranston, J. A. (1954). Schools of chemistry in Great Britain and Ireland–X, The Royal Technical College, Glasgow. *Journal of the Royal Institute of Chemistry* **78**: 116–124.
84. Mackie, J. D. (1954). *The University of Glasgow, 1451–1951: A Short History*. Jackson, Glasgow.
85. Anon. (1902). Women students at the Scottish University. *King's College Magazine, Ladies' Department* (16): 11–16.
86. Letter, Miss Melville to Miss Sheavyn, 15 March 1924, University of Glasgow archives, Special Collections 233/2/6/2/17.
87. Melville, F. (1949). Queen Margaret College. *The College Courant (Journal of the Glasgow University Graduates Association)* **1**(2): 99–107.

88. Anon. (1921). Certificates of candidates for election at the ballot to be held at the ordinary scientific meeting on Thursday, February 17th. *Proceedings of the Chemical Society* 14–15; and University of Glasgow, student records.
89. Irvine, J. C. and Simonsen, J. L. (1944). George Gerald Henderson, 1862–1942. *Obituary Notices of Fellows of the Royal Society* **4**: 491–502.
90. Sutherland, M. M. J. (1945). Obituary notices: Forsyth James Wilson, 1880–1944. *Journal of the Chemical Society* 723–724.
91. University of Glasgow Archives, DC306/4/1.
92. *Minute Book 1918–1925* of the Glasgow University Alchemists' Club, University of Glasgow Special Collections, DC306/1/1.
93. Anon. (December 1925). Laboratory gossip. *The Alchemist (Glasgow University)* **1**(2): 12.
94. (Mrs.) Grant, V. A. (November 1927). Letters to the editor: Protest from an Alchemist's mother. *The Alchemist (Glasgow University)* **3**(1): 16.
95. "Nitrogen Iodide." (February 1929). Chemistry women. *The Alchemist (Glasgow University)* **4**(4): 63–64.
96. (2001). *Who, Where and When: The History and Constitution of the University of Glasgow*. The University of Glasgow, p. 15. We thank Venora Skelly, Assistant Archivist, University of Glasgow, for additional biographical information.
97. Rayner-Canham, M. F. and Rayner-Canham, G. W. (1997). ... And some other women of the British group. In Rayner-Canham, M. F. and Rayner-Canham, G. W. (eds.), *A Devotion to Their Science: Pioneer Women of Radioactivity*, Chemical Heritage Foundation, Philadelphia, and McGill-Queen's University Press, Montreal, Quebec, pp. 159–160.
98. Fleck, L. (1963). Early work in the radioactive elements. *Proceedings of the Chemical Society* 330.
99. Lykknes, A., Kragh, H. and Kvittingen, L. (2002). Ellen Gleditsch: Pioneer woman in radioactivity. *Physics in Perspective* **6**: 126–155.
100. Soddy, F. and Pirret, R. (1910). The ratio between uranium and radium in minerals. *Philosophical Magazine* **20**: 345–349; a subsequent paper was: Pirret, R. and Soddy, F. (1911). The ratio between uranium and radium in minerals II. *Philosophical Magazine* **21**: 652–658. It is noteworthy that, in this second paper, Soddy gave Pirret's name priority.

101. (a) Cuthbertson, D. P. (31 August 1968). Mary Andross. *Chemistry and Industry* 1190; and (b) University of Glasgow, student records.
102. Evans, W. G. (1990). *Education and Female Emancipation: The Welsh Experience, 1847–1914*. University of Wales Press, Cardiff.
103. Magnus, L. (1923). *The Jubilee Book of the Girls' Public Day School Trust 1873–1923*. Cambridge University Press, Cambridge, p.174.
104. McCann, J. E. (1972). *Thomas Howell and the School at Llandaff*. D. Brown & Sons, Cowbridge.
105. Zimmern, A. (August 1898). Women at the universities. *Leisure Hour* 433–442.
106. Angus, W. R. (1954). Schools of chemistry in Great Britain and Ireland–XI University College of North Wales, Bangor. *Journal of the Royal Institute of Chemistry* **78**: 291–298.
107. Williams, J. G. (1985). *The University College of North Wales: Foundations 1884–1927*. University of Wales Press, Cardiff, p. 308.
108. (a) "H.K." [probably Harold King, a former student of Orton's] (1931). Obituary notices: Kennedy Joseph Previté Orton. *Journal of the Chemical Society* 1042–1048; and (b) "F.D.C." (1930). Obituary notices: Kennedy Joseph Previté Orton, 1872–1930. *Proceedings of the Royal Society of London, Series A* **129**: xi–xiv.
109. (1961). *Record of the Science Research Scholars of the Royal Commission for the Exhibition of 1851, 1891–1960*. The Commissioners, London; (December 1900). University College of North Wales, Bangor. *Magazine* 38; and University of Wales, Bangor, Archives, student records. E. W. Thomas, Archivist, is thanked for supplying a copy of this document.
110. "J.F.T." (1931). Obituary notices: William Henry Perkin, 1860–1929. *Proceedings of the Royal Society of London, Series A* **130**: i–xii.
111. Ellsworth, E. W. (1979). *Liberators of the Female Mind: The Shirreff Sisters, Educational Reform, and the Women's Movement*. Greenwood Press, Westport, Connecticut.

Chapter 8

Hoppy's 'Biochemical Ladies'

The University of Cambridge was a contradiction: on the one hand, women were barred from formal undergraduate degrees; on the other hand, women scientists were to be found in many of the University's research laboratories. There were women researchers in both the Balfour Biological Laboratory for Women[1] and the Cavendish Laboratory for Physics.[2] In fact, we have shown elsewhere that the Cavendish even attracted women physicists from other countries, including a Canadian, Harriet Brooks (1902–1903); an American, Fanny Cook Gates (1905); and from Bulgaria, Elizabeth Kara-Michailova (1935–1939).[3] Nevertheless, of all the research schools, the biochemical research group of Frederick Gowland Hopkins[4] stands out as exceptional in its high proportion of women researchers.

Women and Biochemistry

The question arises as to why women were attracted to, and flourished in, the field of biochemistry. It could be argued that biochemistry appealed to women scientists by its relevance to the understanding of living processes; however, there were other fields in which women scientists clustered and gained recognition, such as crystallography,[5] radioactivity,[3] and astronomy.[6]

Margaret Rossiter has contended that the fields in which women made up a significant proportion were often the new rapid-growth areas, where the demand for personnel was so great that there was less strident objection to the hiring of women.[7] Biochemistry certainly fitted this paradigm, as Robert Kohler described:

> Biochemistry is one of those fascinating but problematic "new sciences" that have appeared with some regularity in the history of science. It came quite suddenly on the scene in the early years of this century, with a new name and intimations of new insights into the nature of life processes.[8]

Mary Creese has suggested that the argument was feasible for biochemistry:

> The entry paths and entry qualifications of its practitioners were not well defined. This lack of prestige, due to the slowness of academic chemists to recognize the full power and potential of research in the field, offers one explanation for its [biochemistry's] relative openness to women.[9]

Though such explanations have some validity, we have provided evidence that the role of mentor was another important factor.[10] Margaret Rossiter has shown that this factor played a role in the high number of women in the biochemistry department of Yale University between 1896 and 1935. The Yale women biochemists were all former students of Lafayette Mendel.[11] In fact, of the 124 students who completed doctoral degrees with Mendel, 48 were women, and Rossiter concluded that the personality and supportiveness of Mendel were major factors in this exceptional percentage.

Frederick Gowland Hopkins

In Britain, many women chemists veered towards biochemistry for their careers, joining such organisations as the Lister Institute, London (see Chap. 2), and the Rowett Institute in Aberdeen (see Chap. 7). Nevertheless, it was the Cambridge group under Hopkins that provided a unique community of talented women biochemists.

The history of modern biochemistry[12] is inextricably linked with the name of Hopkins, Nobel Laureate, though his encouragement of women to enter biochemical research has been less widely recognised.[13] Creese has been one of the few to note this facet of his character, commenting:

> At the time when there were practically no women research workers in any of the other university departments at Cambridge, Hopkins gave them places in his, despite the criticism which this brought him. Even in the 1920s and 1930s, when, as a Nobel laureate with a world-wide reputation he received hundreds of applications for places in his laboratory, nearly half of the posts in his Department went to women scientists.[9]

Why was Hopkins so supportive of women scientists? We find an interesting parallel with a mentor of women researchers in radioactivity, Ernest Rutherford. Both Rutherford and Hopkins were very close to their respective mothers. Rutherford wrote frequently and endearingly to his mother in New Zealand throughout his life,[14] while the Hopkins archives contain several items of correspondence with his mother that show an exceptionally strong mother–son relationship.[15] In fact, the first 10 years of Hopkins' life was spent alone with his mother.[16] Such strong maternal bonds may well account for supportive attitudes towards women students.

But there is more to creating a women-friendly environment than simply opening the doors to women. One of these factors was the personal style and attitude to their research students. In the case of Hopkins, Dorothy Moyle (see below), one of his former students, contended that Hopkins provided valuable moral support, and that he regarded his students as fellow researchers rather than as underlings in a research empire.[17] Another of his former students, Malcolm Dixon, explained:

> Hopkins was one of the kindest and most lovable of men. ... He never made an unkind or irritable remark, though he could be critical on occasion, it was always with a courtesy that left no sting. He had great charm of manner and was invariably courteous, even to the least important of us. He was always ready to talk over our work and ideas, and somehow contributed to make us feel by the way that he listened to us that our ideas were extremely interesting and our work important.[18]

This nonaggressive environment might have arisen from Hopkins' own experiences at school, where he was repelled by the rampant bullying.[19] Another mentor of women scientists, crystallographer William Henry Bragg (see Chap. 9), was also renowned for the uncompetitive and collaborative environment that he fostered. He, too, had been similarly affected by bullying at his school[20] and, interestingly, had a close relationship with his mother.

Both Mendel and Hopkins engendered a social cohesiveness in their biochemical research students that was remarkable. Students of Hopkins ("Hoppie" or "Hoppy", as he was known) set up "Hoppie Societies" wherever they were, and held periodic meetings. For example, the Hoppie Society of America met from 1934 until (at least) some time in the 1940s, and there was even a Hoppie Club of Leningrad.[21] Not only was there a camaraderie among Hopkins' students *per se*, but there seemed to be a specific friendship amongst the women biochemists. This is particularly apparent from obituaries, where the obituary of one woman biochemist was often written by another woman biochemist.

Contributing to the socialisation, Hopkins believed that the tearoom was one of the two most important rooms in the research building (the other being the departmental library).[17] Throughout the 50 years of Hopkins' time at Cambridge, there were weekly or fortnightly tearoom meetings of his group at which research work would be presented.

The second aspect, that of Hopkins' personality, is summarised best by Joseph Needham and Dorothy Moyle:

> ... he was a living embodiment of the Confucian maxim that one should behave to everybody as one receiving a great guest. The humblest laboratory assistant or the youngest research worker was always sure of a welcome, and a hearing much longer than he was likely to get from any other scientific man of the same standing or generation. Hopkins had faith in people. Colleagues were known to remark lightly, "All Hoppy's geese are swans," but they forget that there is an induction process by which certain geese may be turned into swans if given the hormone of encouragement.[17]

Another aspect of making a group women-friendly was to demonstrate enthusiasm for the subject. Vivienne Gornick interviewed a large number of women scientists and she commented:

> Each of them had wanted to know how the physical world worked, and each of them had found that discovering how things worked through the exercise of her own mental powers gave her an intensity of pleasure and purpose, a sense of reality nothing else could match.[22]

This portrays exactly the environment in the Hopkins group, as Dixon remarked: "... To work there was to feel the thrill and sense of adventure in penetrating into the secrets of living matter, and life was never dull."[18]

Some Women Biochemists at Cambridge

Women undergraduates at Cambridge largely led separate and sheltered lives in Newnham or Girton (see Chap. 6). However, for graduate work, women had to enter the colder climate of a male-dominated laboratory. The physiologist and biochemist William Bate Hardy[23] recalled the atmosphere in the 1880s and 1890s: "At that time women were rare in scientific laboratories and their presence by no means generally acceptable — indeed, that is too mild a phrase. Those whose memories go back so far will recollect how unacceptability not infrequently flamed into hostility."[24]

Thus, it is particularly striking that biochemistry at Cambridge proved to attract a significant number of women graduates, as we will show below. In fact, at least 60 women biochemists worked in the biochemistry unit during Hopkins' "reign," including some from overseas, such as Kamala Bhagwat (Mrs. Sohonie).[25] Bhagwat recalled her arrival at Cambridge in 1937:

> I applied for admission to his [Hopkins'] laboratory, although it was already full. Then the unexpected happened — a kind scientist already working in the laboratory offered me the day-time use of his bench while he would work at night. Professor Hopkins accepted this solution and I was admitted to this great laboratory on 18 December 1937 — the happiest and proudest day of my life.[26]

We have chosen here to document five women biochemists whose contributions, in our view, were particularly interesting.

Edith Willcock

Though Edith Gertrude Willcock[27] had the briefest sojourn of any of the women biochemists, it was the work she performed with Hopkins that was to become a classic in biochemistry. Willcock was born in Albrighton, and she attended the King Edward VI High School for Girls (KEVI) in Birmingham.[28] Thus, it was not surprising (see Chap. 1) that Willcock proceeded to Newnham, which she attended from 1900 to 1905. She completed Parts I and II of the Natural Science Tripos; but as mentioned in Chap. 6, women in that period were not formally awarded Cambridge bachelor's degrees.

During her undergraduate studies, Willcock undertook research with Hardy on the oxidation of iodoform, initiated by radiation from radium. Of more significance, on her own she published a paper on the effects of radium on animal life — probably one of the first studies that showed the damaging effects of exposure to radioactive elements.

Willcock had been inspired by the lectures of Hopkins, recalling that the impression left on her mind was "less a statement of information than a realisation of the existence of vast unexplored tracts and the unfolding of immense opportunities for research."[29] Resulting from this fervour, Willcock became a Bathurst Research Student (1904–1905) and a Newnham Research Fellow (1905–1909) with Hopkins, and it was her work with him between 1905 and 1909 for which she should be remembered.

At the time (1906), it was believed that diets were complete as long as they contained the appropriate chemical functional units. In particular, it was known that the indol unit was required for certain biological functions. Willcock and Hopkins were the first to show that diets had to contain specific molecules — in this case, tryptophane — and that other indol-containing compounds would not function in its place. This initial study focused Hopkins' attention on the essentiality in diets of certain amino acids and led him to formulate his hypothesis of "accessory food factors," later to be named "vitamins."[30] Willcock, another of the "Steamboat Ladies" (see Chap. 6), received a D.Sc. from Trinity College, Dublin, on the basis of her research.

At that time, over 60% of the women who read science at Cambridge remained single,[31] but Willcock was one of the minority who married. Her spouse was Cambridge zoologist, John Stanley Gardiner[32]; and upon their marriage in 1909, in accordance with general expectations of the time, Willcock discontinued her research. During the First World War, she was a local consultant for the British Ministry of Agriculture on the raising of rabbits and poultry, and she wrote leaflets on the subject for public distribution; in addition, she became an advisor on oyster culture. Willcock had several other interests, being a recognised watercolour artist, singer, and author of a popular children's book, *We Two and Shamus*, published in 1911. She had two daughters and travelled widely with them and her husband. She died on 8 October 1953.

Muriel Wheldale (Mrs. Onslow)

Muriel Wheldale[33] was one of only two genetics researchers in the first decade of the 20th century who performed plant breeding experiments *and* investigated the corresponding chemistry of flower pigments of the plants (the other being Erwin Bauer in Germany).[34] To indicate the importance of her contributions, J. B. S. Haldane[35] used her research to conclude that genes control the formation of large molecules such as pigment molecules.[36]

Born on 31 March 1880 in Birmingham, Wheldale was the only child of John Wheldale, a barrister. Like Willcock, Wheldale attended KEVI and then went to Newnham, where she obtained First Class in both parts of the Natural Sciences Tripos, specializing in botany. In 1903, Wheldale joined the research group of William Bateson[37] as a Bathurst Research Student (1904–1906) and a Newnham Research Fellow (1909–1914). With the revival of Gregor Mendel's theories of heredity at the beginning of the 20th century, Bateson had become one of Mendel's most ardent champions. It was Bateson who devised the term "genetics" to describe the field, and he surrounded himself with a group of enthusiastic research students, many of them women.[38]

Wheldale's particular niche was the study of the inheritance of flower colour in antirrhinums, her research proving to be among the most widely recognised of Bateson's group.[39]

She published a full factorial analysis of flower colour inheritance in antirrhinums in 1907. In a reference on behalf of Wheldale, Bateson commented:

> The problem of colour inheritance in *Antirrhinum*, which she set out to solve, proved to be far more complex than was expected, and the solution she proposed (*Proc. Roy. Soc.* vol. 79, B, 1907) is entirely her own work. There is every reason to believe that it is correct and I regard the paper as one of considerable value.[40]

This landmark publication was the first of a flurry of research papers on the linkage between the inheritance of genetic factors and the production of the pigments, the anthocyanins. Her research culminated in the writing of the classic monograph, *The Anthocyanin Pigments of Plants*.[41] In the preface of the monograph, she commented:

> Herein lies the interest connected with anthocyanin pigments. For we have now, on one hand, satisfactory methods for the isolation, analyses and determination of the constitutional formula of these pigments. On the other hand, we have the Mendelian methods for determining the laws of their inheritance. By a combination of these two methods we are within reasonable distance of being able to express some of the phenomena of inheritance in terms of chemical composition and structure.[42]

By 1913, Wheldale's fame was such that she was one of the first three women (the other two being Ida Smedley and Harriette Chick; see Chap. 2) to be elected to the Biochemical Club,[43] the forerunner of the Biochemical Society. She was also awarded a Prize Fellowship in 1915 by the British Federation of University Women for her scientific research.

Wheldale became convinced that the future lay not with the genetic aspects of botany, but with the biochemical basis of plant pigment synthesis. To this end, she briefly attended the University of Bristol to strengthen her background in biochemistry. Though Bateson encouraged her to join his new group at the John Innes Horticultural Institution, she decided her future lay elsewhere.

With Wheldale's departure, Bateson lost not only one of the most gifted members of his research team, but also his leading artist, as he noted in the same reference as above:

> As an artist in colour she has extraordinary skill. If she leaves us, her loss will be a serious one, for she is the only person I know who can reproduce the colours of flowers in such a way as to be an exact record. In respect of accuracy and appreciation of what scientific colouring should be, the quality of her painting far exceeds that of any professional whose work I know.[44]

In 1914, Wheldale was fortunate to be invited to join the research group of Hopkins. The same anthocyanin pigments that she had followed genetically with Bateson, she studied chemically with Hopkins. This proved a fruitful line of research that cemented her reputation as a leading chemical geneticist.[45]

Following from her biochemical studies, she wrote a second edition of *The Anthocyanin Pigments of Plants*.[46] This was no mere reprint of the first edition, but a complete revision in light of the tremendous biochemical advances of the previous decade. As she noted:

> Since the appearance of the first edition the publications of greatest value on the subject of anthocyanin pigments have been in connection with the chemistry and biochemistry of these substances. This later work has now been included, and the present state of our knowledge of the significance of the pigments in relation to plant metabolism has, as far as possible, been indicated.[47]

In 1916, she was introduced to Huia Onslow, second son of the 4th Earl of Onslow. Onslow had had a diving accident, which had paralysed him from the waist down. Nevertheless, he began a research programme in chemical genetics from a laboratory constructed in his home. At their first meeting together, they had tea followed by a "certain amount of Mendelian discussion."[48] Over the following years, Wheldale spent much of her time helping Onslow with his research, particularly that on the origins of the iridescence of some butterflies, moths, and beetles (though she was never acknowledged in his publications). In 1919, they were married and, until his death in 1922, Wheldale became the conduit between Onslow and the Hopkins group.[49]

Wheldale's devotion to research was balanced by an enthusiasm for teaching. From 1907 until its closure in 1914, she was a Demonstrator in Physiological Botany at the Balfour Biological Laboratory for Women.[1] Then, from 1915 until 1926, she held the position of Assistant in Plant Biochemistry under Hopkins; and in 1926, she was promoted to University Lecturer — one of the first women to hold this rank at Cambridge. Her pedagogical interest led her to write *Practical Plant Biochemistry*[50] in 1920, followed by the text, *The Principles of Plant Biochemistry*,[51] the first volume of which was published in 1931. The latter book was written at her house in Norfolk, one of her two favourite places, the other being the Balkans where she spent her holidays. Unfortunately, she died a year later before completing the second volume.

Wheldale's work on flower pigments and genetics was continued by Rose Scott-Moncrieff (Mrs. Meares).[52] Scott-Moncrieff had arrived in 1926 as a new graduate from Imperial College, and Wheldale had persuaded Scott-Moncrieff to work on the isolation of pigments, particularly the pigment of the magenta *Antirrhinum* that had first fascinated Wheldale in 1914. Working initially under the direction of Haldane, Scott-Moncrieff soon became the link person to the Oxford research on plant pigments. The Oxford group was led by Robert Robinson, with much of the anthocyanin research being performed by his co-researcher spouse, Gertrude Walsh (Lady Robinson), as we will see in Chap. 11.

Marjory Stephenson

Of all "Hoppy's Ladies," it is Marjory Stephenson[53] who is most deserving of recognition. The biochemical historian Robert Kohler has shown that the field of bacterial biochemistry was, in large part, defined by the work of Stephenson.[54] Stephenson was born on 24 January 1885 at Burwell, a village near Cambridge, and it was at Cambridge that she was to spend most of her life. Her mother was Elizabeth Rogers, and her father, Robert Stephenson, was a farmer.

Stephenson was the youngest of four children, being over 8 years younger than the next youngest sibling, one of the two boys, and 15 years younger than the other girl, Alice May. She herself remarked on this influence: "Owing to position in my family, almost an 'only child' and somewhat of a little prig, I acquired a childish interest in science from my beloved governess and later from my father."[55] One of the common features amongst woman scientists was being an only child (or eldest child); thus, what seemed to Stephenson a curse at the time — being an "almost only child" — would probably determine, in part, her future career.

Stephenson was educated by a governess until the age of 12, at which point she received a scholarship to attend the Berkhampsted High School for Girls. It was her mother who insisted that Stephenson obtain a university education, and that Newnham was the appropriate place. She attended Newnham from 1903 until 1906, taking the Part I Natural Sciences Tripos in chemistry, physiology, and zoology.

After leaving Newnham, she would have liked to have studied medicine; but lacking the financial resources, she took teaching positions in domestic and household science instead for the next 5 years, including at King's College for Women, in Kensington (see Chap. 3). The *King's College Magazine, Women's Department*, reported:

> The Cookery Side of the Domestic Science staff has a distinct acquisition in Miss Marjory Stevenson [*sic*] from the Gloucester School of Domestic Science. She has taken the Natural Science tripos (chief subject chemistry) and also a first class Diploma in Cookery; a combination of certificates which seems at the present moment to have been achieved only by herself. Unfortunately, owing to present limitations, a good deal of her King's College work will be at the Clapham Housewifery School, though with our students only; she will not be able to spend much time in our own small Kitchen Laboratory.[56]

Her first saviour was Robert Henry Aders Plimmer.[57] Plimmer had studied chemistry under William Ramsay at University College, London (UCL); then in 1899, Plimmer decided that his interest lay in the chemistry of living organisms. Returning to UCL in 1907 as Assistant Professor in Physiological Chemistry, Plimmer invited Stephenson in 1911 to teach advanced classes in the biochemistry of nutrition and to join his research group. It was somewhat ironic, then, that in the same year, Plimmer, co-founder of the Biochemical Club, proposed that only men should be eligible for membership.

As a result of her research on fat metabolism and on diabetes, Stephenson was awarded a Beit Memorial Fellowship in 1913; however, she relinquished the Fellowship upon the outbreak of war, running soup kitchens in France and then supervising a nurses' convalescent home in Salonika. For this war work, she was awarded an M.B.E. In 1919, she took up her Beit Fellowship again, moving to Cambridge to work with Hopkins, probably on the advice of Plimmer, as Hopkins and Plimmer had worked together as co-editors on a series of biochemistry monographs.[57] Hopkins was to be the greatest influence on Stephenson's career. After the expiry of the Fellowship, she worked on annual grants from the Medical Research Council (MRC) until 1929, when she obtained a permanent post with the MRC.

It was Hopkins who encouraged Stephenson to develop her own interests, and she chose chemical microbiology. She explained the reasons for her choice in the preface of her book, *Bacterial Metabolism*, first published in 1929:

> Perhaps bacteria may tentatively be regarded as biochemical experimenters; owing to their relatively small size and rapid growth variations must arise very much more frequently than in more differentiated forms of life, and they can in addition afford to occupy more precarious positions in natural economy than larger organisms with more exacting requirements.[58]

Also in the preface, Stephenson noted the indirect method by which the metabolism of bacterial cells was studied:

> ... we are in much the same position as an observer trying to gain an idea of the life of a household by a careful scrutiny of the persons and material arriving at or leaving the house; we keep accurate records of the foods and commodities left at the door, patiently examine the contents of the dustbin, and endeavour to deduce from such data the events occurring within the closed doors.[58]

Characteristically, Stephenson acknowledges the contribution of Hopkins:

> Finally, and especially, my thanks are due to Professor Sir Frederick Gowland Hopkins, at whose suggestion the book was written and to whose influence alone I owe the incentive to think on biochemical matters.[58]

The book was highly regarded, with a second edition published in 1938, a third in 1949, and a paperback reprint appearing in 1966.

Stephenson's studies of bacteria were immediately fruitful. At the time, studies had shown that bacteria possess enzymes similar to those in other organisms. It was Stephenson who first separated out a pure cell-free enzyme from bacteria (in this case, lactic dehydrogenase from *E. coli*). Next, with Leonard Stickland, she characterised the enzyme hydrogenase from bacteria in a mixed bacterial culture obtained from the River Ouse, which was at that time polluted with waste from a sugar factory. She showed that the bacteria use hydrogenase to produce methane and hydrogen sulphide. Her next challenge was a study of adaptive enzymes. These are enzymes not needed under normal conditions, but synthesised by the bacterium in response to some external influence, such as a change in the growth medium.

With the onset of the Second World War, Stephenson studied acetone-butyl alcohol fermentation as a means of synthesis of industrial solvents. Of greater importance was her work on pathogenic bacteria and her contributions to the MRC Committee on Chemical Microbiology. After the War, she studied the bacterial synthesis of acetylcholine in sauerkraut, while her last years were spent on an investigation of nucleic acids in bacteria and of their breakdown by enzymes within the cells.

Stephenson was largely responsible for the founding of bacterial chemistry as an autonomous branch of biochemistry; hence, it was appropriate that she should have received many honours. The most valued recognition was the election in 1945 as one of the first two women Fellows of the Royal Society (see Chap. 2). Stephenson died on 12 December 1948.

Dorothy Jordan Lloyd

Dorothy Jordan Lloyd[59] had the biological sciences in her genes: her father, George Jordan Lloyd, was a distinguished Professor of Surgery at the University of Birmingham, while her paternal grandfather had been a Lecturer in Anatomy at the Old Mason College (the predecessor of the University of Birmingham). She was born on 1 May 1889 in Birmingham, and by the age of 12 had decided to become a scientist. Jordan Lloyd followed the same path as both Willcock and Wheldale, attending KEVI and then Newnham College, entering the latter in 1908.

In 1912, Jordan Lloyd completed the Part II Tripos in zoology and obtained a B.Sc. (London), but she found she was more interested in the functional and dynamic side of biology than in the structural studies. Following her interest, when she took up research as a Bathurst Student (1912–1914), her topic was the physicochemical studies of proteins under Hardy, who had earlier mentored Willcock. Then in 1914, Hopkins invited her to join his group as a Newnham Research Fellow. With the outbreak of war, Jordan Lloyd worked on culture media for meningococcus, one of the anaerobic pathogens involved in trench diseases, and on the causes and prevention of "ropiness" in bread. For this research, she was awarded a D.Sc. (London) in 1916.

By 1921, her original work on proteins led to an invitation to join the British Leather Manufacturers' Research Association, a move which, according to Marjory Stephenson, "probably robbed this country of a distinguished professor of biochemistry."[59(a)] As a final contribution to her original research field, Jordan Lloyd wrote the classic work *The Chemistry of Proteins*, published in 1926, for which Hopkins wrote the introduction (a second edition, co-authored with Agnes Shore, appeared in 1938).[60]

It was Jordan Lloyd who turned leather manufacture from a craft industry into a scientific process. In 1927, she was appointed Director of the Association, a post in which she served until her death in 1946. For a women to head such a large scientific organisation was an amazing accomplishment for the period. Her predecessor, Robert Pickard,[61] commented in his obituary of her:

> Among scientific women of her time she perhaps was unique in that to the successful management of a large research organization she added the capacity to delegate its routine administration whilst retaining control, and at the same

time also to play a leading and personal part in the actual research.[59(b)]

In 1939, Jordan Lloyd was awarded the Fraser Muir Moffatt Medal by the Tanners' Council of America for her contributions to leather chemistry. Her change in career direction did not diminish her productivity, and over her lifetime she authored or co-authored over 100 scientific papers, together with planning and contributing to the three-volume *Progress in Leather Science, 1920–45*, the classic textbook of leather technology.

A keen mountaineer, Jordan Lloyd achieved the distinction of making the first ascent and descent in one day of the Mittellgi ridge of the Eiger. Her mountain-climbing exploits led to her election to the Ladies' Alpine Club. Later, she adopted competition horse riding as her main interest, which caused her friend Stephenson to comment: "in both those pursuits she excelled and certainly the element of danger encountered in them was not without its attraction to her."[59(a)] She died of pneumonia on 21 November 1946.

Dorothy Moyle (Mrs. Needham)

Culminating her career with a study of the biochemistry of muscle contraction, Dorothy Moyle[62] showed that marriage did not have to end a woman scientist's career. One of the three daughters of John Moyle, a civil service clerk in the Patent Office, and Ellen Daves, Dorothy Mary Moyle was born on 22 September 1896 in London. She attended school at Claremont College, Stockport, and Manchester High School for Girls before being accepted to Girton in 1915. In 1918, as a fourth-year scholar, Needham started research with Hopkins. Her M.A. followed in 1923, and in the next year she received the Gamble Prize for her essay on the correlation of structure, function, and chemical constitution in the different types of muscle.

The year 1924 was also the year of her marriage to Joseph Needham; and for the remainder of her life, part of her research was done independently and partially with her spouse. Probably the most important of her studies was that to elucidate the chemical changes by which glucose is broken down in the cell to yield energy (anaerobic glycolysis). This research established that two molecules of ATP were required for each molecule of glucose degraded. It was for this work that Moyle was elected Fellow of the Royal Society in 1948.

There was another side to Moyle that was equally important: her left-wing political activism. During the 1930s, the Cambridge biochemists and crystallographers (see Chap. 9) comprised a high proportion of the Cambridge Scientists' Anti-War Group, and Moyle and Needham were among the leaders.[63] Though no list of members has survived, a letter of protest against the militarisation of research was signed by 79 faculty members, research workers, and students, of whom 12 (15%) were women (one of whom was Stephenson) even though women comprised only 7% of the personnel.

As Gary Werskey has remarked: "Why women scientists were often drawn to the Left is not an easy question. Though they belonged to the first post-suffragette generation, they received little encouragement, either from the Popular Front or from wider social forces to develop an explicitly feminist perspective."[63] He hypothesised that some women became outraged by the malnutrition issue, and this was certainly true for Moyle, as she had worked hard in the malnutrition campaigns. Moyle's political activities, however, were largely overshadowed by those of Needham. In fact, none of the women scientists were given recognition, as Werskey concluded: "Nevertheless it was the men who publicly dominated the scientific Left, especially as its leading theoreticians. Hence, neither for the first nor the last time, the political contributions of such women would be almost completely 'hidden from history'."[63]

Despite her 1930s pacifism, at the beginning of the Second World War, Moyle became a research worker for the Ministry of Supply (Chemical Defence), studying the biochemical effects of war gases. Then, in 1943, Needham, whose other main interest was the history of science in China, became Scientific Counsellor of the British Embassy in Chungking. Moyle joined him as Associate Director of the Sino-British Science Cooperation Office. It was in China that she contracted tuberculosis, from which it took her until the 1950s to recover fully.[64]

After the War, Moyle returned to teaching and research at Cambridge, never with a permanent position, just subsisting on research grant after research grant from 1945 until 1962. She later commented on her situation:

> Looking back over my 45 years in research I find it remarkable ... that although a fully qualified and full-time investigator ... I simply existed on one research grant after another, devoid of position, rank, or assured emolument ... [I]t was calmly assumed that married women would be supported financially by their husbands, and if they chose to work in the laboratory all day and half the night, it was their own concern.[65]

After retirement in 1963, she began writing her classic work, *Machina Carnis: The Biochemistry of Muscle Contraction in Its Historical Development*,[66] which was published in 1971. The book had a dedication page to "F.G.H. mentor and friend." Then, for many years she assisted Mikulás Teich with *A Documentary History of Biochemistry 1770–1940.* Unfortunately, Moyle's health deteriorated, as her former student, Jennifer Williams, described:

> I last saw "Dophi", as she was known at her ninetieth birthday party in Cambridge in 1986. ... Although I had worked as her assistant for some years, she did not know who I was, as she was then suffering from Alzheimer's Disease. Only a few years earlier we had played Scrabble in the sunshine of her Grange Road garden. She had beaten us all.[67]

Moyle died on 22 December 1987.

Commentary

With Hopkins as administrator, the easygoing atmosphere allowed women researchers such as Stephenson and Moyle to follow their own intuition and direction. In return, Hopkins was revered throughout the Institute. Such adoration did not extend outside; for example, the high marriage rate among the workers was regarded as scandalous at the time.[68] When Hopkins finally retired in 1943, the "Hoppy regime" came to an end. In a letter written in 1948, Stephenson described how life had changed:

> Gone are the days of Hoppy's 5.30 talks when he used to shut his eyes and draw on his memories of the biochemists belonging to the "turn of the century": I enjoyed science then and was ever so happy but at best I could never have stood the pace the modern young scientist must march at.[69]

Mark Weatherall and Harmke Kamminga studied the "Hoppy tradition" that Hopkins' colleagues had enunciated.[68] As much as the loose administrative style, it was also about his warm and supportive personality as we described earlier. In another letter, Stephenson highlighted the difference under the regime which followed Hopkins' departure: "I am worried about the Department; it is beginning to disintegrate owing to the absence of a real Professor. People come here to work and no-one knows they are here or what anyone else is working on or whom to discuss his problems or difficulties with."[70] The collegial life under the benign influence of Hopkins was no more.

Notes

1. Richmond, M. L. (1997). "A Lab of One's Own": The Balfour Biological Laboratory for Women at Cambridge University, 1884–1914. *Isis* **88**: 422–455.
2. Gould, P. (1997). Women and the culture of university physics in late nineteenth-century Cambridge. *British Journal for the History of Science* **30**: 127–149.

3. Rayner-Canham, M. F. and Rayner-Canham, G. W. (1997). *A Devotion to Their Science: Pioneer Women of Radioactivity.* McGill-Queen's University Press, Montreal, Quebec.
4. Dale, H. H. (1948/1949). Frederick Gowland Hopkins, 1861–1947. *Obituary Notices of Fellows of the Royal Society* **6**: 115–145.
5. Julian, M. M. (1990). Women in crystallography. In Kass-Simon, G. and Farnes, P. (eds.), *Women of Science: Righting the Record*, Indiana University Press, Bloomington, Indiana.
6. (a) Warner, D. J. (1979). Women astronomers. *Natural History* **88**: 12–26; (b) Dobson, A. and Bracher, K. K. (1992). A historical introduction to women in astronomy. *Mercury* **21**: 4–15; and (c) Lankford, J. and Slavings, R. L. (1990). Gender and science: Women in American astronomy, 1859–1940. *Physics Today* **43**: 58–65.
7. Rossiter, M. W. (1980). "Women's work" in science, 1880–1910. *Isis* **71**: 381–398.
8. Kohler, R. E. Jr. (1973). The enzyme theory and the origin of biochemistry. *Isis* **64**: 181–196.
9. Creese, M. R. S. (1991). British women of the 19th and early 20th centuries who contributed to research in the chemical sciences. *British Journal for the History of Science* **24**: 275–306.
10. Rayner-Canham, M. F. and Rayner-Canham, G. W. (1996). Women's fields of chemistry: 1900–1920. *Journal of Chemical Education* **73**: 136–138.
11. Rossiter, M. W. (1994). Mendel the mentor: Yale women doctorates in biochemistry, 1898–1937. *Journal of Chemical Education* **71**: 215–219.
12. Teich, M. (1965). On the historical foundations of modern biochemistry. *Clio Medica* **1**: 414–457.
13. Weatherall, M. and Kamminga, H. (1992). *Dynamic Science: Biochemistry in Cambridge 1898–1949.* Cambridge Welcome Unit for the History of Medicine, Cambridge.
14. Wilson, D. (1983). *Rutherford: Simple Genius.* MIT Press, Cambridge, Massachusetts.
15. Personal papers, F. Gowland Hopkins, Cambridge University Library.
16. Stephenson, M. (1948). Obituary notice: Frederick Gowland Hopkins (1861–1947). *Biochemical Journal* **42**: 161–169.
17. Needham, J. and Needham, D. M. (1949). Sir F. G. Hopkins' personal influence and characteristics. In Needham, J. and

Baldwin, E. (eds.), *Hopkins and Biochemistry*, Cambridge University Press, Cambridge, pp. 113–119.

18. Dixon, M. (1949). Sir F. Gowland Hopkins, O.M., F.R.S. *Nature* **160**: 44.
19. Hopkins, F. G. (1949). Autobiography of Sir Frederick Gowland Hopkins. In Needham, J. and Baldwin, E. (eds.), *Hopkins and Biochemistry*, Cambridge University Press, Cambridge, pp. 3–25.
20. Caroe, G. M. (1978). *William Henry Bragg 1862–1942: Man and Scientist*. Cambridge University Press, Cambridge.
21. Reports of these societies and clubs were sent to Hopkins, and the items are to be found in Hopkins' archives, Cambridge University Library, Cambridge.
22. Gornick, V. (1983). *Women in Science*. Simon & Schuster, New York, p. 15.
23. "F.G.H." [Hopkins, F. G.] and "F.E.S." (1934). William Bate Hardy, 1864–1933. *Obituary Notices of Fellows of the Royal Society* **1**: 326–333.
24. Hardy, W. B. (1932). Mrs. G. P. Bidder. *Nature* **130**: 689–690.
25. Haines, M. C. (2001). Sohonie, Kamala née Bhagwat. *International Women in Science: A Biographical Dictionary to 1950*. ABC-CLIO, Santa Barbara, California, p. 291.
26. Sohonie, K. (1982). Women in Cambridge biochemistry. In Richter, D. (ed.), *Women Scientists: The Road to Liberation,* Macmillan, London, p. 19.
27. Teich, M. with Needham, D. M. (1992). *A Documentary History of Biochemistry 1770–1940*. Leicester University Press, Leicester, pp. 314–316.
28. White, A. B. (ed.) (1979). *Newnham College Register, 1871–1971*, Vol. I: 1871–1923, 2nd ed. Newnham College, Cambridge, p. 33.
29. Cited in: Stephenson, M. (1949). Sir F. G. Hopkins' teaching and scientific influence. In Needham, J. and Baldwin, E. (eds.), *Hopkins and Biochemistry*, Cambridge University Press, Cambridge, p. 34.
30. Kamminga, H. and Weatherall, M. W. (1996). The making of a biochemist. I: Frederick Gowland Hopkins' construction of dynamic biochemistry. *Medical History* **40**: 269–292.
31. MacLeod, R. and Moseley, R. (1979). Fathers and daughters: Reflections on women, science and Victorian Cambridge. *History of Education* **8**: 321–333.

32. Forster-Cooper, C. (1947). John Stanley Gardiner, 1872–1946. *Obituary Notices of Fellows of the Royal Society* **5**: 541–553.
33. (a) Rayner-Canham, M. F. and Rayner-Canham, G. W. (April 2002). Muriel Wheldale Onslow (1880–1932): Pioneer plant biochemist. *The Biochemist* **24**(2): 49–51; (b) Creese, M. W. (2004). Onslow, Muriel Wheldale (1880–1932). *Oxford Dictionary of National Biography*, Oxford University Press, http://www.oxforddnb.com/view/article/46433, accessed 4 Nov 2004.
34. Olby, R. (1989). Scientists and bureaucrats in the establishment of the John Innes Horticultural Institution under William Bateson. *Annals of Science* **46**: 497–510.
35. Pirie, N. W. (1966). John Burdon Sanderson Haldane, 1892–1964. *Biographical Memoirs of Fellows of the Royal Society* **12**: 218–249.
36. Glass, B. (1965). A century of biochemical genetics. *Proceedings of the American Philosophical Society* **109**: 227–236.
37. "J.B.F." (1927). Obituary notices of Fellows deceased: William Bateson, 1861–1926. *Proceedings of the Royal Society of London, Series B* **101**: i–v.
38. Richmond, M. L. (2001). Women in the early history of genetics: William Bateson and the Newnham College Mendelians, 1900–1910. *Isis* **92**: 55–90.
39. Richmond, M. L. (2007). Muriel Wheldale Onslow and early biochemical genetics. *Journal of the History of Biology* **40**: 389–426.
40. Bateson letters, Innes collection, cited in Note 38, Richmond, p. 82.
41. Wheldale, M. (1915). *The Anthocyanin Pigments of Plants*, 1st ed. Cambridge University Press, Cambridge.
42. Note 41, Wheldale, p. v.
43. Goodwin, T. W. (1987). *History of the Biochemical Society 1911–1986*. Biochemical Society, London.
44. Bateson letters, Innes collection, cited in Note 38, Richmond, pp. 82–83.
45. Lawrence, W. J. C. (1950). Genetic control of biochemical synthesis as exemplified by plant genetics — Flower colours. *Biochemical Symposia* **4**: 3–9.
46. Onslow, M. W. (1925). *The Anthocyanin Pigments of Plants*, 2nd ed. Cambridge University Press, Cambridge.
47. Note 46, Onslow, p. vii.
48. Onslow, M. (1924). *Huia Onslow: A Memoir*. Edward Arnold, London, p. 162.

49. "F.G.H." [Hopkins, F. G.]. (1923). Obituary notice: Victor Alexander Herbert Huia Onslow. *Biochemical Journal* **17**: 1–4.
50. Onslow, M. Wheldale. (1920). *Practical Plant Biochemistry*, 1st ed. Cambridge University Press, Cambridge; 2nd ed. (1923).
51. Onslow, M. (1931). *The Principles of Plant Biochemistry: Part I.* Cambridge University Press, Cambridge.
52. Scott-Moncrieff, R. (1981). The classical period in chemical genetics: Recollections of Muriel Wheldale Onslow, Robert and Gertrude Robinson and J. B. S. Haldane. *Notes and Records of the Royal Society, London* **36**: 125–154.
53. (a) Robertson, M. (1949). Marjory Stephenson, 1885–1948. *Obituary Notices of Fellows of the Royal Society* **6**: 563–577; (b) Mason, J. (1996). Marjory Stephenson, 1885–1948. In Shils, E. and Blacker, C. (eds.), *Cambridge Women: Twelve Portraits*, Cambridge University Press, Cambridge, pp. 113–135.
54. Kohler, R. E. (1985). Innovation in normal science: Bacterial physiology. *Isis* **76**: 162–181.
55. Stephenson, M. Personal Records, Royal Society, cited by: Mason, J. (1992). The admission of the first women to the Royal Society of London. *Notes and Records of the Royal Society of London* **46**(2): 284–285.
56. Anon. (1910). College notes. *King's College Magazine, Women's Department* (40): 5.
57. Lowndes, J. (2004). Plimmer, Robert Henry Aders (1877–1955). *Oxford Dictionary of National Biography*, Oxford University Press, http://www.oxforddnb.com/view/article/35543, accessed 21 Dec 2007.
58. Stephenson, M. (1930). *Bacterial Metabolism: Monographs on Biochemistry*. Longmans, Green & Co., London.
59. (a) "M.S." [Stephenson, M.]. (1947). Lloyd D. J. *Newnham College Roll Letter* **54**; (b) Pickard, R. (18 January 1947). Lloyd D. J. *Chemistry and Industry* **47**; (c) Theis, E. R. (1947). Lloyd D. J. *Journal of the American Leather Chemists Association* **42**: 40–41; and (d) Phillips, H. (2004). Lloyd, D. J. (1889–1946). *Oxford Dictionary of National Biography*, Oxford University Press, http://www.oxforddnb.com/view/article/34565, accessed 22 January 2008.
60. Lloyd, D. J. (1926). *The Chemistry of Proteins.* J. & A. Churchill, London.

61. Kenyon, J. (1950). Robert Howson Pickard, 1874–1949. *Obituary Notices of Fellows of the Royal Society* **7**: 252–263.
62. (a) Teich, M. (2003). Dorothy Mary Moyle Needham. *Biographical Memoirs of Fellows of the Royal Society* **49**: 353–365; (b) Needham, J. (November 1988). Dorothy Needham. *The Caian* 128–131; and (c) Coley, N. G. (2004). Needham [née Moyle], Dorothy Mary (1896–1987). *Oxford Dictionary of National Biography*, Oxford University Press, http://www.oxforddnb.com/view/article/56147, accessed 4 Nov 2004.
63. Werskey, G. (1988). *The Visible College: A Collective Biography of British Scientists and Socialists of the 1930s.* Free Association Books, London, pp. 219–231.
64. Letter, D. Needham to D. Singer, 17 December 1950, Wellcome Institute Archives.
65. Needham, D. (1982). Women in Cambridge biochemistry. In Richter, D. (ed.), *Women Scientists: The Road to Liberation*, Macmillan, London, p. 161.
66. Needham, D. (1971). *Machina Carnis: The Biochemistry of Muscle Contraction in Its Historical Development.* Cambridge University Press, Cambridge.
67. Williams, J. (11 January 1988). Dr. D. M. Needham. *The Independent.*
68. Weatherall, M. W. and Kamminga, H. (1996). The making of a biochemist. II: The construction of Frederick Gowland Hopkins' reputation. *Medical History* **40**: 415–436.
69. Letter, Marjory Stephenson to Sydney Elsden, 9 November 1948, Newnham College Archives.
70. Letter, Marjory Stephenson to Sydney Elsden, 13 January 1947, Newnham College Archives.

Chapter 9

Women Crystallographers

In the previous chapter, we described how the Cambridge group of F. Gowland Hopkins was a haven for women biochemists. Here, we will see that X-ray crystallography, in part at Oxford, was largely hospitable to women scientists — though with one crucial exception.

The Early History

The characteristic angles between the faces of each crystalline mineral was a subject of study in the late 19th century, and one of the more famous researchers in the field was Henry Alexander Miers.[1] Miers, who had been Keeper of Mineralogy at the British Museum, was elected to the Waynflete Chair of Mineralogy at Oxford in 1895, where he set up a laboratory at the Oxford Museum. The Oxford Museum had been built between 1855 and 1860 at the instigation of Henry Acland, who was concerned that Oxford lacked any facilities for the promotion of the sciences. In the latter part of the 19th century, the Museum housed astronomy, geometry, experimental physics, mineralogy, chemistry, geology, zoology, anatomy, physiology, and medicine. It was Miers who transformed the mineralogy department into a small but productive centre for research in crystallography.[2]

Though there were fewer women science students at late 19th-century Oxford than at Cambridge or the London colleges, women were certainly to be found in mineralogy and astronomy at Oxford, as Janet Howarth noted: "Research-oriented professors who were not overburdened with undergraduate teaching tended to welcome women as collaborators or assistants — over thirty are recorded as working in the Museum Department and Observatory before 1914."[3]

Mary (Polly) Porter

There were two women who contributed to the work of Miers' group: Florence Isaac and Mary (Polly) Winearls Porter.[4] Isaac was Miers' main collaborator on the study of the growth of crystals, co-authoring five publications. However, it was Porter who provided the link to the first generation of women X-ray crystallographers.

Porter, born on 26 July 1886, had little in the way of formal education, as her father was a travelling correspondent for *The Times*. When Porter was 15, her mother became ill in Rome, requiring a lengthy stay in the city. While there, Porter became a collector of fragments of Roman artefacts that she found at classical sites in the city. Her brothers were impressed with her potential and suggested to their parents that she should be given a formal education, but they refused.

Upon the family's return to England, Porter became fascinated by the Corsi collection of antique marbles in the Oxford Museum. Miers was impressed by the visits of this earnest young woman, and he gave her the task of identifying exhibits and translating the catalogue from Italian. He encouraged Porter to attend Oxford as a student, but again her parents were opposed. After working with Miers from 1905 to 1907, she undertook research in crystallography with Alfred Tutton in London for the 1910/1911 year. Porter obtained a position for a year as mineral cataloguer at the National Museum, Washington, DC, and the following year as mineral cataloguer at Bryn Mawr College, Pennsylvania. She recrossed the Atlantic to undertake research at the University of Heidelberg, Germany, for the 1914/1915 year before returning to the Oxford Museum in 1916.

Despite a lack of formal education, Porter was appointed to the Mary Carlisle Fellowship at Somerville College in 1919. Though her career commenced in the time of classical crystallography, she quickly embraced and contributed to the new era of X-ray crystallography in which the atomic arrangements within crystals could be determined. Her research was published in such academically prestigious journals as the *Proceedings of the Royal Society*, *Acta Crystallographica*, and the *Mineralogical Magazine*, and she also co-edited all three parts of the classic work *The Barker Index of Crystals*. She was elected as a Member of Council of the Mineralogical Society of Great Britain from 1918 to 1921, and again from 1929 to 1932. She continued with her research until 1959, and died on 25 November 1980 at the age of 94.

The Braggs and X-ray Crystallography

X-ray crystallography is one branch of physical science whose origins can be exactly defined.[5] In 1912, the theoretician Max Laue persuaded the experimental physicists Walter Friedrich and Paul Knipping to test his hypothesis that crystals diffract X-rays. William Henry (W. H.) Bragg,[6] Professor at the University of Leeds, and his son, William Lawrence (W. L.) Bragg,[7] then at the University of Cambridge, read of the success of Friedrich and Knipping's experiment. They quickly saw the potential for the determination of crystal structures. W. H. Bragg devised and constructed the Bragg X-ray spectrometer for crystal structure analysis, and shortly afterwards W. L. Bragg deduced the relationship between the diffraction pattern and the atomic spacing (Bragg's law). This was the start of X-ray crystallographic methods for the determination of structures.

The First World War interrupted their studies, but with its end, the Braggs decided to divide the crystal world between them. W. H. Bragg chose to work on organic structures and also quartz, while W. L. Bragg was allotted inorganic substances (except quartz). About this time, they both moved: W. H. Bragg to University College, London (UCL), and then to the Royal Institution, London[8]; while W. L. Bragg obtained a faculty appointment at the University of Manchester, a position that he acquired as successor to Ernest Rutherford, the latter having moved to Cambridge.

The Braggs and Women Researchers

The research groups of both Braggs contained a remarkable number of women.[9] Of W. H. Bragg's 18 students, 11 were women: Ellie Knaggs (see below), Grace Mocatta, Kathleen Yardley (see below), Natalie Allen, Thora Marwick, Lucy Pickett, Helen Gilchrist (see Chap. 7), Berta Karlik, M. E. Bowland, Ida Woodward, and Constance Elam (Mrs. Tipper).

W. L. Bragg's first research student was Lucy Wilson, from Wellesley College, Massachusetts, and he, too, had other women researchers working with him over the years: Elsie Firth, Helen Scouloudi, and P. Jones. Maureen Julian, herself a former researcher with Kathleen Yardley, has shown that women have contributed throughout the "family tree" of crystallographers, many of the Bragg students (both male and female) themselves taking on women students when they acquired academic positions.[10]

Why was crystallography such an attractive field for women? One viewpoint is that the Braggs and their crystallographic heirs provided a women-friendly environment. Anne Sayre, spouse of crystallographer David Sayre, is convinced that, at least in crystallography, it was the nonaggressive and friendly attitudes of the supervisors that were so vital to the encouragement of women:

> There is something in the ancient history of crystallography that is hard to isolate but nevertheless was there, that I can best describe as modesty. I have often wondered how much the Braggs were responsible for the unaggressive low-key friendly

> atmosphere that long prevailed in the field (and no longer seems to very much). Somehow the first and second and a few of the third generation crystallographers consistently conveyed an impression of working for pleasure, for the sheer joy of it — the idea of competition didn't seem to emerge very strongly until the 1960s or so. Uncompetitive societies tend to be good for women.[11]

W. L. Bragg's personality can be gleaned from this comment by one of his former students, W. M. Lomer: "[W. L.] Bragg was a gentle man and a gentleman. He never embarrassed anyone and my every contact with him was a pleasure. When he met me showing my fiancée around the little museum ... [he] was so very pleasant that to this day my wife will hear nothing against him. And nor will I."[12]

A less charitable view of why women chose crystallography has been taken by Franklin Portugal and Jack Cohen, who used the laborious nature of crystallography to explain the number of women in the field:

> Since the high speed computer had not yet been invented, the business of calculating data was a very laborious occupation and smart fellows who could find other things to do would generally do them, unless they were absolutely dedicated to the business of X-ray crystallography. Is it possible that these first class women got to be X-ray crystallographers because they were willing to do this work?[13]

This parallels the arguments used to account for the very high proportion of women in astronomy during the late 19th and early 20th centuries.[14] Margaret Rossiter has shown that the rise in women's participation in astronomy corresponded to the change from the active work of observing (men's work) to the passive role of classifying the thousands of photographic plates (women's work).[15] In fact, it was the first of the women astronomer-assistants, Williamina P. Fleming, who commented that women's superior patience, perseverance, and method made such activities particularly suitable for women in science.[16]

John Lankford and Ricky Slavings, in their studies of women in astronomy, added: "In evaluating women [for astronomy], male scientists tended to focus on their [women's] ability to do routine work. Indeed, some recommendations read as if they were descriptions of machines."[17] Yet the women in this chapter did not see their work as dull and routine — to them, the research was the exciting focus of their lives.[18]

Ellie Knaggs

Like Porter, Isabel Ellie Knaggs[19] was another woman scientist to bridge classical and modern crystallography; but whereas Porter was at Oxford, Knaggs was at Cambridge. Knaggs was born on 2 August 1893 in Durban, South Africa. She was educated at North London Collegiate School and Bedford College, going up to Girton in 1913 and completing her studies there in 1917. Arthur Hutchinson[20] at the Mineralogical Laboratory, Cambridge, hired her as a Research Assistant, and she worked with him until 1921. That year, she moved to Imperial College, where she undertook research leading to a Ph.D., which she completed in 1923.

From 1922 until 1925, Knaggs was a Demonstrator in Geology at Bedford; then, in 1925, she was awarded a two-year Hertha Ayrton Fellowship, which she chose to take at the Davy Faraday Laboratory of the Royal Institution to work under W. H. Bragg. When the Fellowship ended in 1927, she was invited to join the staff of the Royal Institution.

Among the many crystal structures she determined was that of the explosive cyanuric triazide, as W. H. Bragg described in a letter that he wrote to the Editor of the journal *Nature*:

> I send you a short note which I hope you may see fit to publish in Nature. I would like you to know however that my writing

> it has something to do with an attempt to do a little act of justice to the lady mentioned in the note, Miss Knaggs. She has been working for some time on an extraordinary substance, cyanuric triazide. It is one of the highly explosive nitrogen compounds. In the recent Faraday Society discussion — see Nature, May 26 — reference was made to her preliminary results without mention of the source from which they had come. It was an accident, of course: there is no question of any unfairness. But this is Miss Knaggs' magnum opus so far and she is naturally disappointed. I have thought I might put matters straight by writing you the short note to which I have referred.[21]

According to Helen Megaw (see below), "Ellie Knaggs was a kind and gentle person, rather shy. She attended scientific meetings, but did not put herself forward."[19(a)] Knaggs' reticence is probably a major reason why her name is rarely mentioned among the pioneer women of crystallography. She had many publications in a wide range of journals, from *Proceedings of the Royal Society* through the *Mineralogical Magazine* to the *Journal of the Chemical Society* and the *Journal of Physical Chemistry*. Knaggs also co-authored *Tables of Cubic Crystal Structures* with Berta Karlik and Constance Elam.

In addition to her research at the Davy Faraday Laboratory, she was advisor on crystallography to Burroughs Wellcome, the pharmaceutical company. During the Second World War, it was noted in the *Girton Register* that she "carried out investigations involving knowledge of crystal structure at the request of various government departments."[19(b)] Some of her later research involved collaboration with Kathleen Yardley (see below). After retirement, Knaggs was elected Visitor to the Royal Institution for the 1963–1966 and 1974–1977 years. In 1979, she moved to Australia to join relatives there, and it was in Sydney that she died on 29 November 1980 at the age of 87.

Kathleen Yardley (Mrs. Lonsdale)

The most famous woman to work with W. H. Bragg was Kathleen Yardley.[22] Yardley was born on 28 January 1903 in Newbridge, County Kildare, Ireland, the youngest of 10 children. When she was 10 years old, her parents separated, and her mother took the younger children to live in a small town just east of London. She attended the County High School for Girls, Ilford; but as it offered little science, she attended classes in physics, chemistry, and advanced mathematics at the High School for Boys. Though she excelled at almost every academic subject and was a good gymnast, her first love was mathematics.

At the age of 16, Yardley's exceptional talents resulted in the offer of a place at Bedford College.[23] After 1 year, she switched to physics, and as Yardley recalled: "I well remember the opposition I met with from the headmistress of the secondary school that I attended when I told her of my intention to specialize in physics."[24] The headmistress argued that Yardley was a fool to think that she would be able to compete in a "man's field." Yardley, however, had rejected the alternative of a mathematics degree, as she believed mathematics would only lead to a teaching position, while physics offered the opportunity of experimental research.[25]

In 1922, at the age of 19, Yardley was awarded an honours B.Sc. degree. W. H. Bragg was one of her B.Sc. oral examiners and, impressed by the fact that she had higher university marks than anyone in the previous 10 years, he offered her a position in his laboratory at UCL. She accepted his offer, commenting: "He inspired me with his own love of pure science and with his enthusiastic spirit of enquiry and at the same time left me entirely free to follow my own line of research."[26] The following year, she moved with his group to the Royal Institution.

While working on her M.Sc. in 1924, Yardley constructed a set of 230 space-group tables, mathematical descriptions of the crystal symmetries that became vital tools for crystallographers.[27] Three years later, in 1927, she received a D.Sc. degree for her work on ethane derivatives, and that same year she married fellow physics student Thomas Lonsdale. Yardley considered giving up research to become a traditional wife and mother, but Lonsdale argued that he had not married to obtain a free housekeeper and that she should keep on working. They shared the shopping for food, while she specialised in devising dinners that could be prepared in 30 minutes.

The Lonsdale family moved to Leeds in 1927, where Thomas Lonsdale commenced work for the British Silk Research Association. Upon her departure from the Royal Institution, she wrote to W. H. Bragg: "I should like to take this opportunity of thanking you again for all the help you have given me in so many ways. I feel sure that it will be difficult to find a place where I shall be as happy in my work as I was at the Davy Faraday."[28]

The work at Leeds between 1927 and 1929 on the structure of benzene[29] was to provide Yardley with international recognition. Auguste Kekulé had proposed in 1865 that benzene had a six-member ring structure, and this was generally accepted by chemists,[30] but the question remained as to whether the ring was planar or puckered. The Braggs had already shown that the carbon atoms in diamond could be pictured as forming six-member puckered rings, and benzene was expected to possess the same puckered structure. As benzene was a liquid at room temperature, Yardley used hexamethylbenzene, a solid that could be obtained in large single crystals. She was able to show that hexamethylbenzene was planar, and hence benzene itself was also likely to be planar. Although her results contradicted W. H. Bragg's belief in a puckered ring, he was enthusiastic in his praise of her work.

Between 1929 and 1934, she had three children. This was a difficult time for the family, as Kathleen Yardley had no formal position and, to make matters worse, in 1930 Thomas Lonsdale's job was terminated. The family returned to London, where Thomas Lonsdale obtained an appointment working at the Testing Station of the Ministry of Transport.

From 1929 until 1931, Yardley worked at home on the calculations needed to solve her next crystal structure, that of hexachlorobenzene.

In 1931, W. H. Bragg wrote Yardley an enthusiastic letter, telling her that he had obtained funds on her behalf to cover the cost of a full time home help so that Yardley could return to the Royal Institution. However, when she arrived, she found that all the X-ray equipment was in use. Discovering a large old electromagnet, she began studies on the diamagnetism of aromatic compounds. After completion of this research, she measured the lattice constants of natural and synthetic diamonds. Her precise work on the diamond structure became so renowned that the discoverers of a rare hexagonal form of diamond announced that it would be named "Lonsdaleite" in her honor.[31]

W. H. Bragg obtained a grant or fellowship for Yardley each year thereafter until his death in 1942, after which she was appointed a Dewar Fellow at the Royal Institution with Henry Dale.[32] It was her isolation after Bragg's death that caused her to apply for her first academic appointment. In 1946, at the age of 43, she finally obtained such a position as Reader in Crystallography at UCL, where her career had started almost 25 years earlier. Initially, most of her time was filled in the role of editor-in-chief of the *International Tables for X-ray Crystallography*.[33] Three years later, she was promoted to Professor of Chemistry and Head of the Department of Crystallography. Her reputation continued to grow, and, as we discussed in Chap. 2, she was one of the first two women to be elected a Fellow of the Royal Society, having been nominated by W. L. Bragg,[34] while in 1966 she was elected the first woman President of the International Union of Crystallography. Among her later work was the study of boron–nitrogen analogues of carbon species, including the identification of a graphite-like form of boron nitride containing alternating boron and nitrogen atoms in arrays of hexagonal rings.

Having lived under the path of air attacks on London in the First World War and having seen the horrific fireball of a bomb-carrying Zeppelin shot down in flames, Yardley became a pacifist. In a letter to Archibald V. Hill,[35] she expanded on her beliefs:

> It was certainly not the imperative of a religious upbringing that made me a pacifist. I like to believe it was commonsense. I came of a military family and have a naturally pugnacious character, the more violent manifestations of which I have to keep continually under control. My religious teaching as a child was of the orthodox kind that rubber-stamps any war that happens to be going. I don't believe in sitting down and being walked on; I believe in non-violent resistance because I believe it to be the only form of resistance that can be really effective, in that it does not perpetuate the evil it aims at eliminating.[36]

Both Yardley and Lonsdale became Quakers, and during the war, she was jailed for 1 month for refusing to register for war duties.[23] From her personal experiences in jail, and as an extension of her beliefs, she became active in prison reform and served on many prison boards. Her indignation at the extensive nuclear testing by the Soviet Union, the United States, and Great Britain caused her to write the book *Is Peace Possible?*,[37] the foreword of which contains her comment that the book was "written in a personal way because I feel a sense of corporate guilt and responsibility that scientific knowledge should have been so misused."[37]

In these later years, she relied more and more on Lonsdale to help with the tremendous amount of correspondence she received. He had always been very supportive of his famous spouse, and Yardley identified the need for such a relationship in her comments on women scientists: "For a woman, and especially a married women with children, to become a first class scientist she must first of all choose, or have chosen, the right husband. He must recognise her problems and be willing to share them. If he is really domesticated, so much the better."[38]

Yardley was very fortunate in her choice of husband. After her death, Thomas Lonsdale wrote to W. L. Bragg:

> When the apple fell on Newton's head someone gathered it and the other windfalls and made a pie for his dinner, thats [sic] my job now a bit, it always has been. Hilton [a prominent mathematician of the time] told me that a Professor of Maths is lucky if in his life he has one student who can see a whole branch of maths as a structure, they know the text book but they don't need it because they go on to build, where the math doesn't yet exist they invent it, for him, Kathleen was that student. Even before we were married I knew she had one of the most powerful intellects of the time. ... Most of my working life has been spent in road engineering, I only know enough about her work to realise its importance and value and how fortunate I have been to have been associated with it, 'in getting Newton's dinner'.[39]

Crystallography was Yardley's life, and she began a study of the crystalline nature of kidney stones, bladder stones, and gallstones when she was nearly 60 years old.[40] She developed an interest in stones after the Chief Medical Officer of the Salvation Army had told her of small children in hot, dry, Third World countries suffering from them.[41] When she fell ill, the first medical diagnosis assumed that she had acquired tropical malaria from her world travels in pursuit of stones and their causes. When cancer of the bone marrow was identified and she was told that her remaining time was short, she began 13-hour workdays to complete a book on human stones, the first draft being completed a few weeks before she died in 1971. Ten years later, in recognition of her contributions to crystallography, the then-chemistry building at UCL was named the Kathleen Lonsdale Building.

The Influence of J. D. Bernal

Though the Braggs had been the "founding fathers" of X-ray crystallography, it was John "Sage" Desmond Bernal[42] who provided the link among the majority of women crystallographers. In addition to a shared passion for crystallography, most of the women espoused the same left-wing political ideals as Bernal and were members of the Cambridge Scientists Anti-War Group (CSAWG). As an example, both Dorothy Crowfoot and Helen Megaw joined Bernal in signing the protest against the Incitement to Disaffection Bill[43] we described in Chap. 8 as having been signed by the biochemists, Marjory Stephenson and Dorothy Moyle.

Bernal was a student at Cambridge, where he became fascinated with the different types of spatial arrangements of atoms.[42(a)] In 1923, he showed a theoretical paper he had written during his vacation to Arthur Hutchinson, Lecturer in Mineralogy. Hutchinson was so impressed that he contacted W. H. Bragg at UCL, where Bragg obtained funding for Bernal. Bernal moved back to Cambridge in 1927 and, notably, his first student was a woman, Nora Martin, who was also a member of the CSAWG. Martin did not stay long with Bernal, marrying her former classmate, W. A. "Peter" Wooster, and continuing her research with him.[44]

Dorothy Crowfoot (see below) was a member of the Bernal group from 1932 to 1934, and she described the very pleasant working atmosphere, particularly the convivial lunches:

> Every day, one of the group would go and buy fresh bread from Fitzbillies, fruit and cheese from the market, while another made coffee on the gas ring in the corner of the bench. One day there was talk about anaerobic bacteria on the bottom of a lake in Russia and the origin of life, another, about Romanesque architecture in French villages, or Leonardo da Vinci's engines of war or about poetry or printing. We never knew to what enchanted land we would be taken next.[45]

When Hutchinson retired in 1931, he had recommended the foundation of a Chair of Crystallography, obviously with Bernal in mind as an occupant; however, the proposal was rejected. Thus, in 1937, when Bernal was offered the Chair of Physics at Birkbeck College, he accepted it and spent the rest of his life at this small constituent college of the University of London, whose prime role was the part-time education of working people. Such an environment matched well with his Communist Party affiliation. At Birkbeck, he continued his policy of taking on women researchers, including Käthe Schiff, Rosalind Franklin (see below), Winifred Booth Wright (see below), and Shirley V. King.

Bernal had been asked whether he ran his laboratory on communist principles, to which he retorted that it was more on feudal lines, where the researchers would spend about half their time on his projects and half on their own.[46] He also continued the social life, as his former student, Alan Mackay, recalled:

> The lab was very socially and politically conscious, a wide spectrum of views being expressed every teatime. Bernal had also thought about the social organization and had brought the institution of the tea club with him from Cambridge so that everyone met twice a day with considerable regularity.[46]

Winifred Booth Wright

One of the goals at Birkbeck was the determination of the crystal structure of small, biologically interesting molecules, and it was Winifred Booth Wright who solved the structure of the first of these: the tripeptide, glutathione.[47] Wright was born on 20 March 1916 in Liverpool and was educated at Holly Lodge Girls' High School, Liverpool.[48] She entered Newnham in 1935, and after completing the degree requirements in 1938, she joined J. C. Lyons & Co. at Cadby Hall. Initially appointed as a Research Assistant, Wright was later promoted to Research Physicist and Physical Chemist (though her application for Fellowship in the Chemical Society listed her as an Analytical Chemist). It was from 1950 to 1958 that she undertook part-time research with

Bernal, which led to the solution of the glutathione structure and her Ph.D. from the University of London. In 1963, she was promoted to Research Management with J. C. Lyons.

Helen Megaw

The careers of Bernal and another of his early students, Helen Dick Megaw,[49] crossed twice: Megaw, born in 1907, worked with Bernal at Cambridge in the early years; and then rejoined him for a short while at Birkbeck. Megaw was born in Dublin, Ireland, initially attending the Alexandra School, Dublin, before being sent to Roedean when her parents moved to Belfast. Initially unable to afford Girton, she spent the 1925/1926 year at Queen's University, Belfast, before transferring to Girton, where she completed a B.Sc. degree in 1930, specialising in chemistry, physics, and mineralogy. She then commenced research towards a Ph.D. in mineralogy and petrology with Bernal on the structure of ice, a topic which required growing crystals in narrow capillary tubes at low and very exact temperatures, and then studying the X-ray diffraction patterns.[50] Her pioneer work on the structure of ice resulted in the Glaciological Society naming an island in the Antarctic the Helen D. Megaw Island.

Bernal suggested that Megaw should try to solve the crystal structure of a clay mineral, hydrargillite, a form of aluminium hydroxide. Bernal, like the US scientist Linus Pauling, held the theory that life had started on clay minerals; hence, the determination of the crystal structure of a clay would advance the study of biogenesis. From this beginning, the crystallographic studies of mineral structures was to be Megaw's research focus for the rest of her active life.

Megaw was awarded a Hertha Ayrton Research Fellowship, which enabled her to spend the 1934/1935 year as a postdoctoral researcher with Herman Mark at the University of Vienna, and the following year with Francis Simon at the Clarendon Laboratory, Oxford. Between 1936 and 1943, she taught at Bedford High School and then at Bradford Girls' Grammar

School, spending her school holidays doing research back at Cambridge. For the next 2 years, she held the position of Crystallographic Scientist in the Materials Research Laboratory, Philips Lamps, Mitcham, London.

Then, in 1945, Megaw was appointed Assistant Director of Research in Crystallography at Birkbeck College. It would seem likely that Bernal had instituted the appointment as a means to get his former protégée back into academic research life. If so, the move succeeded brilliantly, with Megaw being enticed back to Cambridge in 1946 as Assistant for Experimental Research in Crystallography at the Cavendish Laboratory while also being appointed as Fellow, Lecturer, and Director of Studies in Physical Science at Girton.

This period of her life proved to be the most productive, resulting in a stream of research papers together with several books, including the classic, *Ferroelectricity in Crystals.*[51] Megaw's former postdoctoral student, Michael Glazer (who had previously been a graduate student of Kathleen Yardley), commented:

> Helen's impact on ferroelectricity was profound, especially in the early days. She brought to the subject a visual aspect embodied in the crystal structures of ferroelectrics, thus showing how this important effect arose. Her book, *Ferroelectrics in Crystals*, the first book on the subject, was for a long time just about the only book on the subject and became almost a "bible" for those working in ferroelectricity.[52]

Megaw received the Roebling Medal of the Mineralogical Society of America in 1989, the first woman to receive this award. In her award address, Megaw remarked:

> Much has been said about the difficulties of women in science, but I would like to say explicitly that I at least was never or rarely aware of discrimination. Perhaps I was lucky, in that everyone who advised me on my education and guided my

career assumed that women should be given the same opportunities as men.[52]

Yet this belief is contradicted by her early career descent to Assistant Mistress of a high school, for had she been male, an academic position would have followed a postdoctoral fellowship. And had Bernal not "rescued" her, she would never have made the major contributions to mineral crystallography that she did, nor garner the fame and recognition she deserved.

Even after formal retirement in 1972, Megaw remained active, dividing her time between her retirement home in Ballycastle, Ireland, and her desk in the Department of Mineralogy and Petrology at Cambridge. During this period, she continued active research and wrote another landmark book, *Crystal Structures: A Working Approach.*[53] She died in Ballycastle on 26 February 2002, aged 94 years.

Dorothy Crowfoot (Mrs. Hodgkin)

The most famous of all women crystallographers was Dorothy Crowfoot,[54] the Nobel Prize winner, who determined the crystal structures of penicillin, vitamin B_{12}, and insulin. She was born on 12 May 1910 in Cairo, Egypt, where her father, John Winter Crowfoot, an Oxford graduate, supervised Egyptian schools and administered ancient monuments for the British government. Her mother, Grace Mary "Molly" Hood, was a self-taught expert in botany who had written a book on the flora of the Sudan. In 1914, with the outbreak of war, Crowfoot and her two younger sisters were left in England in the care of a nursemaid while their parents returned to the Middle East. Crowfoot believed this background led to her quiet, independent character. Even after the end of the First World War, the children stayed in England while their parents returned from the Middle East on annual visits.

Crowfoot became interested in chemistry at an early age. At age 10, she tried growing crystals of copper(II) sulphate and alum; while at age 13 when visiting the Sudan, she tried analysing local minerals. Back in England, she set up a small attic laboratory where she continued her chemical analyses. In her notebook, she recorded each of her experiments, including: "I had a violent nosebleed and thought — a pity all this good blood should go to waste so I collected it in a test-tube and used it to make haematoporphyrin."[55]

Of particular importance for her future career was the children's book she was given, *Concerning the Nature of Things* by W. H. Bragg, which discussed crystallography, together with a text on biochemistry, which she read with the help of the *Encyclopaedia Britannica*. Crowfoot attended the Sir Thomas Leman School, Beccles, and to enable her to gain a formal knowledge of science, she was given permission to join science courses taken by boys.[56] This experience had the secondary benefit of exposing her to an all-male environment at an early age, so that later in life she seemed comfortable in such circumstances.

In 1926, Crowfoot took the Oxford Senior Local Examination, a national university entrance test, and attained the highest grade for any female student that year. Lacking the Latin and another science required for admission to Oxford, she had to spend a year studying these subjects before she could enter Somerville. At this time, her father was a Director of the British School of Archaeology in Jerusalem, and she visited him, becoming entranced by the Byzantine mosaic floor patterns; in fact, analysing mosaic patterns had a strong parallel to her later work on crystal symmetry.[57]

Near the end of her first year at Oxford, Crowfoot attended T. V. Barker's lectures in the Crystallography Department, and Barker became an early mentor of Crowfoot. Both Barker and her Chemistry tutor, F. M. Brewer, encouraged her to undertake X-ray crystallography in her final year. Polly Porter was another influence, as Crowfoot noted in her recollections:

> Before that [her decision to stay at Oxford] Polly Porter advised me to go to Germany, to work for a few months

> in the laboratory of old Professor Viktor Goldschmidt, a particular friend of hers. He had designed a two-circle goniometer for measuring crystals — Polly bought one of these for Oxford – and also devised a good method for drawing crystals which I learnt. ... — but my time was too short for that.[58]

Wishing to broaden her experience in crystallography, Crowfoot accepted an offer to join Bernal's group at Cambridge. There, she studied early preparations of vitamin B_1, vitamin D, and several of the sex hormones. Finances were a problem, though, and only a substantial gift of money from her aunt enabled her to survive.

During her first year at Cambridge, Crowfoot was offered a post as tutor at Somerville. To entice her back to Oxford, the University agreed that she could spend a second year at Cambridge before taking up the position. Taking Bernal's advice, Crowfoot accepted the offer.

Once back in Oxford, Crowfoot began an X-ray crystal analysis of cholesterol derivatives; but at every opportunity, she would return to visit the sparkling intellectual life at Cambridge, a contrast to the loneliness she found at Oxford, exacerbated by her exclusion from the Senior Alembic Club (see Chap. 6).

In 1937, Crowfoot acquired her first graduate student, Dennis Riley. Riley had invited Crowfoot to address the Junior Alembic Club and, having become excited by her research, insisted that he wished to work with her. His request was not looked upon favourably by fellow chemists. Riley noted:

> This, at the time, was quite revolutionary and several eyebrows were lifted. Here was I, a member of a prestigious college [Christ Church], choosing to do my fourth year's research in a new borderline subject with a young female who held no university appointment but only a fellowship in a women's college.[59]

Also in 1937, Crowfoot married the historian Thomas L. Hodgkin, and over the following years she had three children. It was only during her pregnancies that she obtained relief from the rheumatoid arthritis which had first struck in the early 1930s, resulting in inflamed, tender, and painful joints in her hands. Hodgkin decided early in the marriage that Crowfoot was the more creative of the two, and it was he who looked after the children in the evenings while she returned to the laboratory; Hodgkin also took them to appointments such as the dentist, and on outings such as to the zoo.

To add complication to their later lives, Hodgkin became active in African studies, which required them to have a residence in Ghana as well as England. Despite her work commitments, Crowfoot always found time for her children, being able to switch between roles as mother and scientist with total ease. Fortunately, Crowfoot's social life improved, particularly through a friendship with Max Perutz, another of Bernal's former students, who was still at Cambridge.[60] Perutz would bring his wife and children to Oxford, where the two families would walk beside, and sometimes swim in, the Isis (the local name for the Thames).[61]

Throughout her career, Crowfoot picked projects that were always just beyond the currently accepted limits of feasibility, and her research on the structure of cholesterol was one such example. This molecule had long been a puzzle to organic chemists, and it was Crowfoot and another of her students, Harry Carlisle, who determined the actual bonding arrangement — the first time that X-ray crystallography had been used to deduce an organic structure in which the atomic arrangement was unknown. This work was completed in the early years of the Second World War.

In 1942, Crowfoot embarked upon the molecular structure of penicillin. Structural studies were essential to help in the synthesis of penicillin, a task which was of vital importance during the war. Fortunately, she had met biochemist Ernst Chain[62] some years earlier in the streets of Oxford, and he had promised to provide her with crystals of the antibiotic. Even to start the project was difficult, for it was not realised in those early days that penicillin could adopt different packing arrangements, depending upon the conditions of crystallisation.

To help in the work, Crowfoot had one graduate student, Barbara Rogers Low. About the middle of the Second World War, they acquired night-time use of an IBM punched card machine in an evacuated government building. Low helped write the first three-dimensional computer program and punched the data onto cards. The structure was finally completed in the summer of 1945; but of equal importance, Crowfoot had attacked the problem by a variety of different crystallographic techniques, broadening the options available for structure determination.

Crowfoot's research group steadily expanded in the postwar era, and it included workers from around the world; one undergraduate student was Margaret Roberts (Mrs. Thatcher), who later changed her career direction to become Prime Minister. Crowfoot decided to limit the group size to 10 researchers so as to maintain the interactive environment of the group. A Rockefeller Foundation investigator commented that the lab was

> … kept under good strong scientific discipline by their gentle lady boss who can outthink and outguess them on any score. A lovely small show reflecting clearly the quality of its director … She conducts the affairs of her small laboratory on a most modest, almost self-deprecatory scale.[63]

Though Crowfoot had made important contributions to science, at the end of the war her rank at Oxford was still only that of Tutor. Deeply in debt, she realised that most of her male colleagues had university positions as well as research appointments, so she asked Cyril Hinshelwood, Professor of Physical Chemistry,[64] to help her acquire a better position. With his help, she was appointed as a University Lecturer and Demonstrator in 1946.

In 1948, a researcher with the Glaxo pharmaceutical company gave Crowfoot some deep red crystals of vitamin B_{12} that he had just obtained. It had been in 1926 that raw liver was shown to cure the disease of pernicious anæmia, and the crucial compound had been identified as this vitamin. Before the vitamin could be synthesised, it was essential to determine the chemical structure. With 93 nonhydrogen atoms, many crystallographers regarded the task as impossible. Crowfoot disagreed. Over the following 6 years, the group grew more and larger crystals while they took a total of 2500 X-ray photographs. Having acquired the data, the next task was the analysis, which was far beyond the normal calculating facilities of the time.

A visit to Oxford by Kenneth Trueblood from the University of California at Los Angeles (UCLA), USA, was to make this step possible. At UCLA, Trueblood had programmed one of the first high-speed electric computers for crystallographic calculations, and he had free computing time on the machine. He offered to help, and it was arranged that Crowfoot would send him the data, which he would run through the machine and then mail back the results.[65] In 1956, the structure was finally determined, the largest established up to that time.

Crowfoot was not promoted to Reader until 1957, and, even then she was not provided with modern lab facilities until the following year. The academic pinnacle of success, an endowed chair, was offered to her in 1960, but it was provided by the Royal Society, not the University of Oxford. Worldwide recognition of her work on the determination of the structures of biochemically important molecules came in 1964 with the Nobel Prize in Chemistry. However, indicative of the attitude towards women scientists, the news was announced by the *Daily Mail* newspaper as "Nobel Prize for British Wife."[63]

Some crystallographers felt that Crowfoot should have received the Nobel Prize earlier. For example, Perutz, who was co-recipient of the Nobel Prize for chemistry in 1962, commented:

> I felt embarrassed when I was awarded the Nobel Prize before Dorothy, whose great discoveries had been made with such fantastic skill and chemical insight and had preceded my own. The following summer I said as much to the Swedish crystallographer Gunner Hagg when I ran into him in a tram in Rome. He encouraged me to propose her, even though she had been proposed before. In fact, once there had been a news leak that she was about to receive the Nobel Prize, but it proved false; Dorothy never mentioned that disappointment to me until long after. Anyway, it was easy to make out a good case for her; Bragg [W.L.] and Kendrew [Perutz's Nobel co-recipient] signed it with me, and to my immense pleasure it produced the desired result soon after.[66]

The determination of the structure of the protein insulin was Crowfoot's third major project. This was the culmination of 30 years of work, since her first X-ray photograph of the compound had been taken in 1935; and as early as 1939, she had published a report of X-ray measurements on wet insulin crystals. As techniques had improved over the years, she had kept returning to this particular molecule, but it was only technical advances in the 1960s that made the solution finally possible. When the results were published in 1969, the researchers were listed in alphabetical order, showing her willingness to share the credit and her egalitarian attitude towards all of the research workers. In addition, she had one of her young postdoctoral fellows give the first lecture on the structure so that the glory was again shared rather than focused on herself.

Crowfoot continued the philosophy of the Braggs and of Bernal that science was a social activity. One of her later students,

John H. Robertson, recalled the family atmosphere of the Crowfoot group and particularly the afternoon tea:

> Each member of the community took his or her turn, weekly, to provide the little cakes that went with the afternoon's cup of tea. When anyone had a birthday, or a new baby, or anything comparable to celebrate, it was, by unwritten rule, that person's duty to provide a large iced cake, free, for that occasion. Each person had his or her own desk, of course, but everyone knew, at least in outline, what everyone else was doing. All the problems, and everything that was going on, were interesting. Mutual assistance was frequent; animated, even heated discussions were normal. The motivation was the interest of the subject. Everyone worked hard. It came naturally to do so.[67]

Like Yardley, Crowfoot had a very strong sense of social responsibility: four of her mother's brothers had been killed in the First World War, and her mother had been a strong supporter of the League of Nations. As a result of her beliefs, Crowfoot became a member of the CSAWG and, during the Second World War, chose to demonstrate her pacifist beliefs by her work on the penicillin structure which she felt had the potential of saving lives.[68]

After the Second World War, Crowfoot became a member of the Science for Peace organisation. Membership in this organisation caused her to be denied a visa to attend a meeting in the United States during 1953. For the next 27 years, she had to obtain a special entry permit from the Attorney General to attend scientific meetings in the United States. Only in 1990, at the age of 80, did the State Department relent and approve a visa application.

It is difficult to overstate the challenges that Crowfoot faced in determining the structures of these very complex molecules. As one of her contemporaries, Jack D. Dunitz, commented:

> Dorothy had an unerring instinct for sensing the most significant structural problems in this field, she had the audacity to

> attract these problems when they seemed well-nigh insoluble, she had the perseverance to struggle onward where others would have given up, and she had the skill and imagination to solve these problems once the pieces of the puzzle began to take shape.[69]

Crowfoot formally retired in 1977, but she continued to be active in science until her death on 30 July 1994.

Rosalind Franklin

Whereas Crowfoot's life was one of success, Rosalind Elsie Franklin[70] died without recognition of her contributions to X-ray crystallography and to science as a whole. Franklin was born in London, the second of five children of an affluent Jewish family. Both her father, Ellis Franklin, and her mother, Muriel Waley, had family backgrounds as social activists. Ellis Franklin, a banker, taught science as a volunteer at working men's clubs and helped Jews fleeing from Nazi Germany, while many of her aunts were practicing socialists and women's rights workers. One of her uncles had even been jailed for his suffragette activities.

Her mother concisely described those attributes of Franklin's character that were to lead to antagonisms later in life:

> Her affections both in childhood, and in later life, were deep and strong and lasting, but she could never be demonstrably affectionate or readily express her deeper feelings in words. This combination of strong feeling, sensibility and emotional reserve, often complicated by an intense concentration on the matter of the moment, whatever it might be, could provoke either a stony silence or a storm.[71]

During her early years, Franklin felt discriminated against because she was female and, as she recalled, her childhood was a battle for recognition. She detested dolls, much preferring sewing, carpentry, and Meccano construction sets. She was fortunate to be educated at St. Paul's Girls' School, a strong academic institution that excelled in the teaching of physics and chemistry. Astronomy was her favourite hobby, and by the age of 15 she had decided to become a scientist. Franklin took and passed the entrance examinations for the University of Cambridge, where she planned to study physical chemistry; however, her father refused to pay for her. He had once wanted to be a scientist, and he insisted that he would have been delighted for a son to follow such a path, but in his view, daughters should only consider volunteer work rather than a full-time career. Outraged at this attitude, Franklin's mother, together with her favourite aunt, offered to pay; in the family crisis that followed, Ellis Franklin finally relented. As Sharon McGrayne commented: "his approval was grudgingly given and resentfully received."[72]

Franklin went up to Cambridge in 1938. With the start of the Second World War in 1939, the senior scientists were given war-related research and their time with students dwindled. With the lack of supervision, Franklin, working in the Cavendish Laboratory, enjoyed the opportunity for independent work; however, she did not get a first class degree. Her supervisor in Physical Chemistry, Fred Dainton,[73] provided the reason:

> ... Miss Franklin was certainly capable of getting a First because she had the qualities of pertinacity and penetration which were rather exceptional. But I did not expect her to do so because she was very selective in the use of her time, going to depth in the things which interested her and neglecting the other parts of the subject.[74]

After graduation in 1941, Franklin accepted a position with the future Nobel Laureate, Ronald Norrish,[75] exploring gas-phase chromatography of organic mixtures. Franklin and Norrish had an abrasive relationship, which was due to personality problems on both sides. She had been used to working very much on her own rather than having a supervisor telling her what to do, while Norrish objected to Franklin's belief in sexual equality among scientists.[76] Franklin resigned from the position after 1 year in order to contribute to the war effort by doing research for the British Coal Utilization Research Association (BCURA). Her studies at BCURA contributed significantly to the understanding of coal structure and the effect of heating coals.[77] She wrote up some of this work as her Ph.D. thesis, the degree being awarded in 1945.

During the war, Franklin became friends with the French metallurgist, Adrienne R. Weill, a former worker with Marie Curie and Irène Joliot-Curie at the Institut Curie; and after the war, she wrote to Weill: "If ever you hear of anybody anxious for the services of a physical chemist who knows very little about physical chemistry, but quite a lot about the holes in coal, please let me know."[76] Weill suggested Marcel Mathieu, a former student of W. H. Bragg, and Mathieu invited her to France in 1947 to work as a researcher in the Laboratoire Centrale des Services Chimique de l'État. While in Paris, she worked on the growth of graphite crystals with Mathieu's student, Jacques Méring, with whom she had a very positive relationship. Méring introduced her to the crystallographic techniques which were to become her field of expertise.

The time in France seemed to have been the happiest of her life. Franklin took science very seriously, and in the laboratory she was intense and reserved, but outside she sparkled in the chic social life of Paris. Her co-workers were also friends with whom she had lunch at local bistros, went to dinner parties, and joined for skiing and mountain-climbing vacations. She particularly loved mountains and went on frequent long hikes and bicycle trips. Holidays, in fact, were one of her greatest joys, and they were planned in meticulous detail. As her mother remarked:

> Not for her the lazy holidays basking in the sun. Sunshine and warmth she loved, but her holidays must be full of movement and never idle. Long walks of up to twenty miles a day; climbing,

> swimming; visits to picture galleries and ancient buildings, these were her delight. She was an eager stimulating companion, but those who travelled with her must possess a zest and energy comparable to her own.[78]

After 3 years in Paris, Franklin accepted an offer in 1951 to work at King's College, London. Physicist John Randall,[79] who had formed an interdisciplinary group of physicists, chemists, and biochemists, needed an expert crystallographer to analyse X-ray photographs of DNA taken by Raymond Gosling, a graduate student. Having heard of Franklin's skills, Randall offered her the job, noting: "This means that as far as the experimental X-ray effort is concerned, there will be at the moment only yourself and Gosling, together with the temporary assistance of a graduate from Syracuse, Mrs. Heller."[80]

When Franklin arrived, Randall's second-in-command, Maurice Wilkins,[81] was away, and this proved to be a major problem, for Franklin assumed that she would be working completely independently while Wilkins, on his return, assumed that Franklin had been hired as a technical assistant to produce data for other members of the structured group to analyse. This was the start of an unhappy relationship between them.

Adding to the source of conflicts, McGrayne contends that the social life at King's was far different to Franklin's previous experience:

> A number of women scientists worked on the staff, but they were not allowed to eat in the men's common room; women ate outside the lab or in the student's cafeteria. After work, the men visited a male-only bar for beer and shoptalk; women were not invited. As a result, the men talked science casually among friends while the women operated in a more formal office atmosphere.[82]

This would not be surprising, considering King's reluctance to admit women earlier in the century (see Chap. 3).

This perspective is different to the views of some of the other women who worked with Randall during that time period.[83] For example, Honor Fell,[84] another member of the group, considered that much of the social interaction took place in after-lunch coffee gatherings in one or another research lab. Franklin's "brusque manner" and her "overriding passion" for science meant, according to Fell, that Franklin never developed friendships, even with the 8 women of the 31-member team. Whatever the cause, exclusion from social interaction not only meant a loss of companionship, but it also prevented her from appreciating the significance of the hierarchical nature of Randall's group.

Mainstream crystallography was then concerned with finding precise atomic locations in crystals of pure simple substances. However, the purpose of X-ray studies of complex materials, such as coal and DNA, was to gain a more general view of the atomic arrangements. Over the 2 years that she spent at King's, Franklin established that there were two forms of DNA, A and B, the form depending upon the humidity, a point that had confused other researchers in the field. Franklin developed techniques to produce the most high-resolution X-ray photographs of DNA taken up to that time. She showed that the phosphate groups were on the outside of the DNA molecule and that hydrogen bonding played an important role.

Gosling enjoyed working with Franklin, and he commented on her strong personality and ebullient, argumentative character. These attributes did not endear her to Wilkins, as Wilkins himself was "shy, passive, indirect."[85] As a result, the relationship between the two, which had never been positive, deteriorated into active dislike. Delia Simpson (see Chap. 6), who had been one of Franklin's Lecturers at Cambridge, became aware of Franklin's situation, as Simpson's obituarist described:

> Delia showed great concern for the well-being of her students … This solicitude continued after students had left Cambridge. An example of this was the way in which, when Delia became aware of the problems that Rosalind Franklin was having at

> King's College, she made a special point of calling in for a chat whenever she was in London.[86]

During this period, James Watson and Francis Crick were constructing their models of DNA at the University of Cambridge. Wilkins developed a friendship with the two, confiding in them about his difficulties with Franklin. Without telling Franklin, Wilkins removed the superb X-ray photograph of the B-form and showed it to Watson and Crick, providing the major experimental evidence for the helical structure of DNA. Franklin had summarised her work in a report to the Science Research Council, and this had been distributed to their review committee. A member of this committee passed the report, unethically in the view of many scientists, to Watson and Crick, providing additional essential information.

When Watson and Crick submitted their famous article on DNA structure to the journal *Nature*, the editor contacted both Wilkins and Franklin to ask if they would submit accompanying articles. Franklin had by this time herself deduced the helical nature of DNA, but she had not progressed as far as Watson and Crick. Whether she would have done so in time, or whether Watson and Crick could have solved the structure without Franklin's photograph and report, is still a matter of debate. The three papers appeared sequentially, with the vague comment by Watson and Crick that they had "been stimulated by a knowledge of the general nature of the unpublished experimental results and ideas of Dr. M. Wilkins, Dr. R.E. Franklin, and their co-workers at King's College, London."[87] Franklin was never aware that they had actually seen her photograph and report.

The protein crystallographer, David Harker, identified the key issue:

> And the real tragedy in this affair is the very shady behaviour by a number of people, as well as a number of unfortunate accidents, which have resulted in the transfer of information in an

> irregular way. ... I would never have consciously become involved in anything like this behaviour, especially the transfer of information through a privileged manuscript. And I think these people are — to the extent that they did these things — outside scientific morals as I know them.[88]

By 1953, Franklin felt that relations with Wilkins had deteriorated to the point where she could no longer work at King's. Kathleen Yardley had offered Franklin a position at UCL,[89] but Franklin declined and instead asked Bernal if she could join his group at Birkbeck. Bernal quickly offered an invitation. Randall agreed to the transfer of her Fellowship to Bernal, on the condition that she did not continue working on DNA (at that time, it was not uncommon in Britain for specific research fields to be the "property" of particular research groups). Bernal arranged for Franklin to be the head of her own research group, and he suggested that she look at the structure of the tobacco mosaic virus, a topic that he had started and abandoned years earlier. By working with both Bernal and Mathieu, Franklin was now doubly scientifically related to the Braggs.

In the 5 years at Birkbeck, Franklin and her research group determined the structural features of the virus, showing it to be hollow-cored, not solid as supposed by microbiologists, and she was able to determine surface features of the virus. During this period, her team was the world leader in the structure of viruses and, for the first time, she was able to develop positive collaborations with other research groups. Of equal or greater importance, Franklin was back in a pleasant working environment. However, there were still moments of frustration, such as the refusal (at the time) of the British Agricultural Research Council to fund any project that had a woman directing it; fortunately, the US Public Health Service provided her with adequate funding instead.

In 1956, Franklin experienced extreme pain, which was diagnosed as ovarian cancer. Three operations and experimental chemotherapy followed, but without success. During these last years, Franklin had become friends with Crick and his French wife, Odile, showing no animosity for the DNA dispute. While convalescing with them, Crick had suggested that she move her group to Cambridge, and this she decided to do. Realising that she was terminally ill, Franklin started a dangerous individual project, the structure of the live polio virus. This work was so hazardous that after her death the project was halted. She died on 16 April 1958, at age 37.

Four years later, the Nobel Prize was awarded to Crick, Watson, and Wilkins, for the DNA work. The Nobel Prize is only awarded to living scientists and to no more than three people in any category. Having died, Franklin was no longer eligible, but the controversy focused upon the lack of acknowledgement of her work during their Nobel Lectures in Stockholm.

Many scientists argued that the popular account of the discovery of DNA, *The Double Helix*[90] by Watson, minimised Franklin's contributions and painted a very negative picture of her personality. As her biographer, Sayre, has commented, everyone is entitled to their own perception of another individual, but the picture of Franklin was to generate a totally misleading image which would reinforce the negative stereotype of women scientists. Sayre added:

> "Rosy" [Watson's nickname for Franklin] was less an individual than ... a character in a work of fiction. ... If Rosalind was concealed, the figure which emerged was plain enough. She was one we have all met before, not often in the flesh, but constantly in a certain kind of social mythology. She was the perfect, unadulterated stereotype of the unattractive, dowdy, rigid, aggressive, overbearing, steely, "unfeminine" bluestocking, the female grotesque we have all been taught to fear or to despise.[91]

In fact, her closest collaborator at Birkbeck, Aaron Klug, was so incensed at the belittling of her role that he wrote a formal protest, published in *Nature*.[92] A former colleague of Franklin's, A. J. Caraffi, wrote to thank Klug for his defence of Franklin:

> I would like to say how very much I have appreciated your paper to *Nature* on Rosalind Franklin and DNA. After the journalistic distortions to which her rightful place in the history of science has been subjected, it is most satisfying to one's sense of justice to see published a dignified, objective, and properly documented account of her signal contribution. It brings out her absolute integrity and illustrates admirably the thorough principles applied by great experimentalists to the evaluation of their proofs. There could be no better counter to the titillations of all the offensive "Rosie" nonsense that was bound to attract selection and emphasis by ignorant hacks [journalists].[93]

It is appropriate to give her friend, Klug, the final comment. He wrote a letter to P. Siekevitz of the New York Academy of Sciences:

> However, if she is to be honoured, it should not be as a "woman in science" but for her crucial contributions in sorting out the A and B forms, establishing that the phosphates were on the outside and determining the helical parameters which were used by Crick and Watson in their study. ... There is also, inevitably, a fair amount of discussion as to whether she would have solved the structure on her own. One can only guess, but my view, as stated, is that she would have done so eventually ...[94]

Dorothy Wrinch

The later work of Dorothy Maud Wrinch[95] impinged directly on that of Bernal and Crowfoot, and for this reason her biography is included here even though she was not a member of the

Bernal "family." Wrinch was the single child of Hugh Edward Hart Wrinch, a mechanical engineer, and Ada Minnie Souter. Born in Rosario, Argentina, she grew up in southern England. From Surbiton High School, she won a scholarship in 1913 to Girton, graduating in 1916 with the highest ranking of the year in mathematics. After obtaining an M.A. from Cambridge in 1918, Wrinch accepted a position as Lecturer in Mathematics at UCL, which she held until 1921 when she returned to Girton as a Research Scholar.

In 1922, Wrinch married John William Nicholson, Director of Studies in Physics and Mathematics at Balliol College, Oxford; and she moved to Oxford in 1923, becoming tutor in mathematics to the five women's colleges of the university. The position was annually renewed until 1927, when it finally became a long-term appointment.

Wrinch was a prolific and versatile scientist. Between 1918 and 1932, she published 16 papers on scientific methodology and the philosophy of science, together with 20 papers on aspects of pure and applied mathematics. Though this versatility can be looked upon as the mark of a true scholar, it also had its negative aspects, as Pnina Abir-Am has pointed out:

> Moreover, her prolific output tended to be unfocused because of her splitting her research effort between mathematics and philosophy, a legacy of her youthful infatuation with her teacher and friend Bertrand Russell. Even within mathematics she vacillated between the pure and applied fields, which were separated by conflicting ideologies. This was a legacy of her collaboration with her husband, a mathematical physicist, and her father, an engineer. As a result, Wrinch lacked a strong disciplinary reputation, despite her substantial talents in all of the above mentioned fields. All this was further compounded by the gender problem which limited women's opportunities at Oxbridge and by a class problem ... [Wrinch had come from a middle-class environment and was now in an upper-class milieu].[96]

Wrinch had a child, Pamela, in 1927, and her experiences of combining academic duties with marriage and motherhood led to her writing *The Retreat from Parenthood* under the pseudonym of J. Ayling.[97] Unfortunately, her life was to be suddenly changed by the permanent institutionalisation of her husband as a result of alcoholism in 1930.

The sudden reduction in income, in addition to posing a major financial problem, caused Wrinch to reassess her direction. At this point in time, biological architecture had begun to fascinate her — the application of mathematical topological techniques to the interpretation of biological molecular structures. Wrinch herself commented:

> I had, however, long had a consuming interest in physiology and chemistry, and I had always hoped to find specialists in this field with whom I could develop certain ideas. It proved impossible to arrange such a collaboration, since the mathematical point of view was difficult to link up with the point of view of the professional chemist. At this time, then, it became clear to me that I had but two choices, either to abandon the attempt to develop these ideas or to undergo apprenticeship in chemistry sufficiently extensive to enable me to formulate the ideas in a form suitable for development by specialist workers. I chose the latter course and spent a year's leave of absence from Oxford on the continent of Europe beginning an apprenticeship in many different laboratories.[98]

Though Wrinch's diversity of interests was a problem in some ways, it resulted in her becoming one of the five founder members of the Biotheoretical Gathering. This small, elite group of British scientists focused on redefining the relationships among scientific disciplines and between science and philosophy. Wrinch's contribution dealt with applications of topology to experimental embryology. In the mid-1930s, she developed a new theory of protein structure, a theory that combined ideas of mathematical symmetry with the concept of a covalent bond between the CO of one amino acid and the NH of another amino acid. The cyclic structure that resulted from pairs of these bonds, she referred to as a "cyclol bond."[99] The series of cross-links proposed by Wrinch would result in sheets that could fold to give closed geometric figures. According to her calculations, some of these structures would contain $72n^2$ amino acids, where n was a small integer. Fortuitously, two researchers found egg albumin to contain 288 or $72(2)^2$ amino acids.[100] This, and many other pieces of evidence, seemed to point to the correctness of Wrinch's model of proteins; and many scientists, particularly Irving Langmuir, supported her proposal.

At the same time, there were opponents. For example, neither Dorothy Crowfoot nor J. D. Bernal believed in the cyclol hypothesis. Crowfoot commented:

> She [Wrinch] was a friend of ours, and as a mathematician was very anxious to become acquainted with biology and biochemistry and to contribute to the problems on which we were working ... We were friends of hers, and had helped to develop her theories, but we did not believe in them, and that was our trouble.[101]

Linus Pauling was another critic, and the exchanges between Pauling and Wrinch became acrimonious and often personal. Though neither side was untarnished by the dispute, it is undeniable that Wrinch clung to her general cyclol hypothesis of proteins long after it had been discredited by X-ray crystallographic evidence.[102] Crowfoot delivered her verdict in the affair as follows:

> Gradually and relentlessly, after the war, chemical and crystallographic work proved that protein molecules did not have the cyclol structure. Yet many of Dorothy Wrinch's geometrical instincts and deductions were, in general terms, correct. The structures of protein molecules are not based on simple parallel arrays of peptide chains, yet the packing of molecules

within the insulin unit cell is octahedral in character, there are patterns of close packed units over closed surfaces in many spherical viruses, even the 'cyclol' link itself sometimes exists in the ergot alkaloids. This pleased her – she found it very hard to give up the cyclol theory as a whole.[103]

By this time, Wrinch and her daughter had moved to the United States, spending the 1940/1941 year in the Chemistry Department of Johns Hopkins University. Partially because of Pauling's attacks, Wrinch had a difficult time finding a full-time position; but with the help of Otto Glaser of Amherst College, whom she later married, Wrinch was given a simultaneous Visiting Professorship at Amherst, Smith, and Mount Holyoke Colleges for a year. She then obtained a research position at Smith, where she worked with graduate students, lectured, and continued her research for 30 years until her retirement in 1971.

Though her name is always linked with the cyclol controversy, she maintained a stream of publications covering many fields, giving her a total of 192 in all. In chemistry, she turned her attention towards developing techniques for the interpretation of X-rays of complex crystal structures. This work culminated in the monograph *Fourier Transforms and Structure Factors*.[104] Following retirement, Wrinch moved to Woods Hole, where she had given summer lectures during her years at Smith College. She died there in 1976, 10 weeks after her daughter Pamela had died tragically in a fire.

Commentary

Unusual for their time, both Yardley and Crowfoot managed to raise a family as well as devote long hours in the laboratory. This was, to a large extent, a result of very progressive marriages, with the husbands performing some of the "wifely" duties (whilst hired help undertook other home-making activities). Franklin followed the more common single pattern.

Though she expressed a preference for marriage, not wishing to remain a "spinster professor," she felt that children would have interfered with the research to which she wanted to devote her life. Hilary Rose, sociologist of science, has suggested that the "single and sexually threatening" image contributed to Franklin's problems, compared to the safe and respectable perception of Crowfoot and Yardley as wives and mothers.[105] It can also be argued that Franklin's abrupt manner did not assist with her acceptance into British scientific society.

Crystallography seems to provide one of the clearest examples of the positive and supportive role of supervisors, particularly female role models. Among the many of subsequent "generations" has been W. H. Bragg's crystallographic "granddaughter," Mary Truter[106]; W. L. Bragg's "granddaughter," Carolina MacGillavry[107]; and Crowfoot's "daughter," Jenny Pickworth (Mrs. Glusker).[108] Crowfoot, particularly, was the crystallographic "mother" and "grandmother" of many crystallographers around the world, including India.[109]

Another of Crowfoot's "descendants," Judith Howard, crystallographer at the University of Durham, has speculated that: "... one of the reasons why women continue to be attracted to crystallography is because so many early women crystallographers were so good."[110] She added that in 1995 about one third of the British Crystallographic Association's 800 members were women.

Notes

1. (a) Holland, T. H. and Spencer, L. J. (1943). Henry Alexander Miers, 1858–1942. *Obituary Notices of Fellows of the Royal Society* **4**(12): 368–380; (b) Porter, M. W. (ed.) (1973). *Diary of Henry Alexander Miers, 1858–1942*. Department of Geology and Mineralogy, Oxford.
2. Vincent, E. A. (1994). *Geology and Mineralogy at Oxford 1860–1986: History and Reminiscences*. The Author, Oxford.
3. Howarth, J. (1984). 'Oxford for Arts': The natural sciences, 1880–1914. In Brock, M. G. and Curthoys, M. C. (eds.), *The*

History of the University of Oxford, Volume VII, Nineteenth Century Oxford, Part 2, Clarendon Press, Oxford, p. 469.

4. (a) "A de V." [de Villiers, A.] (1980). Mary Winearls Porter, 1886–1980. *Somerville College Report and Supplement* 32–33; (b) Haines, M. C. (2001). Porter, Mary (Polly) Winearls. *International Women in Science: A Biographical Dictionary to 1950*, ABC-CLIO, Santa Barbara, California, p. 253.
5. (a) Ewald, P. P. (1962). The beginnings. In Ewald, P. P. (ed.), *Fifty Years of X-ray Diffraction*, International Union of Crystallography, Utrecht, The Netherlands, pp. 6–80; (b) Gasman, L. D. (1975). Myths and X-rays. *British Journal for the History of Science* **26**: 51–60; (c) Speakman, J. C. (1980). The discovery of X-ray diffraction. *Journal of Chemical Education* **57**: 489–490.
6. (a) Andrade, E. N. da and Lonsdale, K. (1943). William Henry Bragg, 1862–1942. *Obituary Notices of Fellows of the Royal Society* **4**: 276–300; (b) Caroe, G. B. (1978). *William Henry Bragg 1862–1942: Man and Scientist*. Cambridge University Press, Cambridge.
7. Phillips, D. (1979). William Lawrence Bragg. *Biographical Memoirs of Fellows of the Royal Society* **25**: 75–143; Hunter, G. K. (2004). *Light Is a Messenger: The Life and Science of William Lawrence Bragg*. Oxford University Press, Oxford.
8. Julian, M. M. (1986). Crystallography at the Royal Institution. *Chemistry in Britain* **22**: 729–732.
9. Julian, M. M. (1990). Women in crystallography. In Kass-Simon, G. and Farnes, P. (eds.), *Women of Science: Righting the Record*, Indiana University Press, Bloomington, Indiana, pp. 335–383.
10. Note 9, Julian, p. 342.
11. Letter, Anne Sayre to Maureen Julian, cited in: Note 9, Julian, pp. 339–340.
12. Lomer, W. M. (1990). Blowing bubbles with Bragg. In Thomas, J. M. and Phillips, D. (eds.), *Selections and Reflections: The Legacy of Sir Lawrence Bragg*, Royal Institution of Great Britain, London, p. 117.
13. Portugal, F. H. and Cohen, J. S. (1977). *A Century of DNA: A History of the Discovery of the Structure and Function of the Genetic Substance*. MIT Press, Cambridge, Massachusetts, p. 267.

14. (a) Warner, D. J. (1979). Women astronomers. *Natural History* **88**: 12–26; (b) Dobson, A. and Bracher, K. K. (1992). A historical introduction to women in astronomy. *Mercury* **21**: 4–15; (c) Fraknoi, A. and Freitag, R. (1992). Women in astronomy: A bibliography. *Mercury* **21**: 46–47.
15. Rossiter, M. W. (1980). 'Womens work' in science, 1880–1910. *Isis* **71**: 381–398; also in Rossiter, M. W. (1982). *Women Scientists in America*. Johns Hopkins University Press, Baltimore, Maryland, pp. 53–57.
16. Fleming, W. (1893). A field for 'women's work' in astronomy. *Astronomy and Astrophysics* **12**: 688–689.
17. Lankford, J. and Slavings, R. L. (1990). Gender and science: Women in American astronomy, 1859–1940. *Physics Today* **43**: 58–65.
18. Lonsdale, K. (1970). Women in science: Reminiscences and reflections. *Impact of Science on Society* **20**: 45–59.
19. (a) Megaw, H. D. (1982). Obituaries: Dr. Isabel Ellie Knaggs. *Girton College Newsletter* 30; (b) Butler, K. T. and McMorran, H. I. (eds.) (1948). *Girton College Register, 1869–1946*. Girton College, Cambridge, p. 680.
20. Smith, W. C. (1939). Arthur Hutchinson, 1866–1937. *Obituary Notices of Fellows of the Royal Society* **2**: 483–491.
21. Letter, W. H. Bragg to Sir Richard Gregory, 3 July 1934, W. H. Bragg Archives, The Royal Institution. The note was published as: Bragg, W. H. (1934). Structure of the azide group. *Nature* **134**: 138.
22. (a) Julian, M. M. (1982). Profiles in chemistry: Kathleen Lonsdale 1903–1971. *Journal of Chemical Education* **59**: 965–966; (b) Julian, M. M. (1981). X-ray crystallography and the work of Dame Kathleen Lonsdale. *Physics Teacher* **3**: 159–165.
23. Rice-Evans, P. (2001). Physics. In Crook, J. M. (ed.), *Bedford College: Memories of 150 Years*, Royal Holloway and Bedford College, pp. 265–266.
24. Note 18, Lonsdale, pp. 54–55.
25. Crowfoot, D. M. C. (1975). Kathleen Lonsdale. *Biographical Memoirs of Fellows of the Royal Society* **21**: 447–484.
26. Note 25, Crowfoot, p. 449.
27. Lonsdale, K. (1936). *Simplified Structure Factor and Electron Density Formulae for the 230 Space-Groups of Mathematical Crystallography*. G. Bell & Sons, London.

28. Letter, K. Lonsdale to W. H. Bragg, 16 November 1927, W. H. Bragg Archives, The Royal Institution.
29. Julian, M. M. (1981). Kathleen Lonsdale and the planarity of the benzene ring. *Journal of Chemical Education* **58**: 365–366.
30. (a) Vanderbilt, B. (1975). Kekulé's whirling snake: Fact or fiction. *Journal of Chemical Education* **52**: 709; (b) Wotiz, J. H. and Rudolfesky, S. (1984). Kekulé's dreams: Fact or fiction? *Chemistry in Britain* **20**: 720–723.
31. Frondel, C. and Marvin, U. B. (1967). Lonsdaleite, a hexagonal polymorph of diamond. *Nature* **214**: 587–589.
32. Feldberg, W. S. (1970). Henry Hallett Dale, 1875–1968. *Biographical Memoirs of Fellows of the Royal Society* **16**: 77–174.
33. Henry, N. F. M. and Lonsdale, K. (eds.) (1952). *International Tables for X-ray Crystallography: Symmetry Groups*, Vol. I. Kynoch Press, Birmingham; Kasper, J. and Lonsdale, K. (eds.) (1959). *International Tables for X-ray Crystallography: Mathematical Tables*, Vol. II. Kynoch Press, Birmingham; Lonsdale, K., MacGillavry, C. H. and Reich, G. D. (eds.) (1962). *International Tables for X-ray Crystallography: Physical and Chemical Tables*, Vol. III. Kynoch Press, Birmingham.
34. Mason, J. (1992). The admission of the first women to the Royal Society of London. *Notes and Records of the Royal Society* **46**: 279–300.
35. Katz, B. (1978). Archibald Vivian Hill, 26 September 1886–3 June 1977. *Biographical Memoirs of Fellows of the Royal Society* **24**: 71–149.
36. Letter, Kathleen Lonsdale to A. V. Hill, 7 June 1953, A. V. Hill collection, Churchill Archives Centre, The Papers of Rosalind Franklin, FRKN, University of Cambridge.
37. Lonsdale, K. (1957). *Is Peace Possible?* Penguin, London.
38. Note 25, Crowfoot, p. 474.
39. Letter, Thomas Lonsdale to Sir Lawrence Bragg, 24 May 1971, Lawrence Bragg papers, Royal Institution. For a discussion of the relationship between Kathleen and Thomas Lonsdale, see: Julian, M. M. (1996). Kathleen and Thomas Lonsdale: Forty-three years of spiritual and scientific life together. In Pycior, H., Slack, N. G. and Abir-Am, P. G. (eds.), *Creative Couples in the Sciences*, Rutgers University Press, Brunswick, New Jersey, pp. 170–181.

40. In addition to academic publications, Lonsdale wrote a popular account. See: Lonsdale, K. (1968). Human stones. *Science* **159**: 1199–1207.
41. Letter, Thomas Lonsdale to Mr. Glanville, undated, Lawrence Bragg papers, Royal Institution, London.
42. (a) Brown, A. (2005). *J. D. Bernal: The Sage of Science*. Oxford University Press, Oxford; (b) Hodgkin, D. M. C. (1980). John Desmond Bernal. *Biographical Memoirs of Fellows of the Royal Society* **26**: 17–84.
43. Bell, G. D. H. *et al.* (1934). Correspondence: Scientific workers and war. *The Cambridge Review* **56**: 451.
44. Wooster, W. A. (1962). Personal experiences of a crystallographer. In Ewald, P. P. (ed.), *Fifty Years of X-ray Diffraction*, International Union of Crystallography, Utrecht, The Netherlands, pp. 680–684.
45. Note 42(b), Hodgkin, p. 31.
46. Mackay, A. L. (1995). The lab. *The Chemical Intelligencer* **1**(6): 12–18.
47. Note 42(b), Hodgkin, p. 59.
48. White, A. B. (ed.) (1981). *Newnham College Register, 1871–1971*, Vol. II: 1924–1950, 2nd ed. Newnham College, Cambridge, p. 152.
49. Glazer, A. M. and Kelsey, C. (2006). Helen Dick Megaw (1907–2002). In Byers, N. and Williams, G. (eds.), *Out of the Shadows: Contributions of Twentieth-Century Women to Physics*, Cambridge University Press, Cambridge, pp. 213–221.
50. Note 42(b), Hodgkin, p. 33.
51. Megaw, H. D. (1957). *Ferroelectricity in Crystals*. Methuen, London.
52. Cited in: Crystallography: Helen D. Megaw, http://www.physics.ucla.edu/~cwp/Phase2/Megaw,_Helen@851234567.html, accessed 26 November 1997.
53. Megaw, H. D. (1973). *Crystal Structures: A Working Approach*. W. B. Saunders Co., Philadelphia.
54. (a) Ferry, G. (1998). *Dorothy Hodgkin: A Life*. Granta Books, London; (b) McGrayne, S. B. (1993). *Nobel Prize Women in Science*. Birch Lane Press, New York, pp. 225–254; (c) Farago, P. (1977). Impact: Interview with Dorothy Crowfoot. *Journal of Chemical Education* **54**: 214–216; (d) Perutz, M. F., Dorothy Crowfoot, address delivered at a memorial service in the University Church,

Oxford, 4 March 1995. S. Chandrasekhar is thanked for a copy of the address; (e) Dodson, G. (2002). Dorothy Mary Crowfoot Hodgkin, O.M. 12 May 1910–29 July 1994. *Biographical Memoirs of Fellows of the Royal Society* **48**: 179–219.

55. Cited in: Note 54(a), Ferry, p. 29.
56. Hudson, G. (1991). Unfathering the thinkable: Gender, science and pacifism in the 1930s. In Benjamin, M. (ed.), *Science and Sensibility: Gender and Scientific Enquiry, 1780–1945*, Blackwell, Oxford, p. 275.
57. Note 9, Julian, p. 376.
58. *Crystallography and Chemistry in the First Hundred Years of Somerville College*, a James Bryce Memorial Lecture delivered in Wolfson Hall, Somerville College, on 2 March 1979 by Professor Dorothy Hodgkin, O.M., F.R.S.P. Adams, Archivist, Somerville College is thanked for providing a copy of this unpublished lecture.
59. Riley, D. P. (1981). Oxford: The Early Years. In Dodson, G., Glusker, J. and Sayre, D. (eds.), *Structural Studies on Molecules of Biological Interest, A Volume in Honour of Professor Dorothy Crowfoot*, Clarendon Press, Oxford, p. 17. By permission of Oxford University Press.
60. Blow, D. M. (2004). Max Ferdinand Perutz, 19 May 1914–6 February 2002. *Biographical Memoirs of Fellows of the Royal Society* **50**: 227–256.
61. Perutz, M. (1981). Forty years' friendship with Dorothy. In Dodson, G., Glusker, J. and Sayre, D. (eds.), *Structural Studies on Molecules of Biological Interest, A Volume in Honour of Professor Dorothy Crowfoot*, Clarendon Press, Oxford, p. 6.
62. Abraham, E. (1983). Ernst Boris Chain, 19 June 1906–12 August 1979. *Biographical Memoirs of Fellows of the Royal Society* **29**: 42–91.
63. Cited in: Note 54(b), McGrayne, p. 250.
64. Thompson, H. (1973). Cyril Norman Hinchelwood, 1897–1967. *Biographical Memoirs of Fellows of the Royal Society* **19**: 374–431.
65. Trueblood, K. N. (1981). Structure analysis by post and cable. In Dodson, G., Glusker, J. and Sayre, D. (eds.), *Structural Studies on Molecules of Biological Interest, A Volume in Honour of Professor Dorothy Crowfoot*, Clarendon Press, Oxford, p. 87.

66. Note 61, Perutz, p. 10. By permission of Oxford University Press.
67. Robertson, J. H. (1981). Memories of Dorothy Crowfoot and of the B_{12} structure in 1951–1954. In Dodson, G., Glusker, J. and Sayre, D. (eds.), *Structural Studies on Molecules of Biological Interest, A Volume in Honour of Professor Dorothy Crowfoot*, Clarendon Press, Oxford, p. 73. By permission of Oxford University Press.
68. Note 56, Hudson, p. 281.
69. Dunitz, J. D. (1981). Organic chemistry, X-ray analysis and Dorothy Crowfoot. In Dodson, G., Glusker, J. and Sayre, D. (eds.), *Structural Studies on Molecules of Biological Interest, A Volume in Honour of Professor Dorothy Crowfoot*, Clarendon Press, Oxford, p. 59. By permission of Oxford University Press.
70. (a) Maddox, B. (2002). *Rosalind Franklin: The Dark Lady of DNA*. HarperCollins, New York; (b) Sayre, A. (1975). *Rosalind Franklin and DNA*. W. W. Norton, New York; (c) McGrayne, S. B. (1993). *Nobel Prize Women in Science*. Birch Lane Press, New York, pp. 304–332; (d) Glynn, J. (1996). Rosalind Franklin 1920–1958. In Shils, E. and Blacker, C. (eds.), *Cambridge Women: Twelve Portraits*, Cambridge University Press, Cambridge, pp. 267–282.
71. Franklin, M. (undated). *Rosalind*. Butler & Tanner, Frome, p. 5.
72. Note 54(b), McGrayne, p. 308.
73. Gray, P. and Ivin, K. J. (2000). Frederick Sydney Dainton, Baron Dainton of Hallam Moors, 11 November 1914–5 December 1997. *Biographical Memoirs of Fellows of the Royal Society* **46**: 86–124.
74. Letter, Lord Dainton to Mansel Davis, 10 March 1982, cited in: Davis, M. (1990). Notes and discussions: W. T. Astbury, Rosie Franklin, and DNA: A memoir. *Annals of Science* **47**: 607–618.
75. Dainton, F. and Thrush, B. A. (1981). Ronald George Wreyford Norrish, 9 November 1897–7 June 1978. *Biographical Memoirs of Fellows of the Royal Society* **27**: 379–424.
76. Note 70(b), Sayre, p. 58.
77. Harris, P. J. F. (2001). Rosalind Franklin's work on coal, carbon, and graphite. *Interdisciplinary Science Reviews* **26**: 204–209.
78. Note 71, Franklin, p 12.
79. Wilkins, M. H. F. (1987). John Turton Randall, 23 March 1905–16 June 1984. *Biographical Memoirs of Fellows of the Royal Society* **33**: 492–535.
80. Cited in: Note 54(b), McGrayne, p. 312.

81. Struther, A., Kibble, T. W. B. and Shallice, T. (2006). Maurice Hugh Frederick Wilkins, 15 December 1916–5 October 2004. *Biographical Memoirs of Fellows of the Royal Society* **52**: 455–478.
82. Cited in: Note 54(b), McGrayne, p. 313.
83. Judson, H. F. (1986). Annals of science: The legend of Rosalind Franklin. *Science Digest* **94**: 56–59 and 78–83.
84. Vaughan, J. (1987). Honor Bridget Fell, 22 May 1900–22 April 1986. *Biographical Memoirs of Fellows of the Royal Society* **33**: 236–259.
85. Note 83, Judson, p. 79.
86. Newton, A. A. (1999). Delia Agar, 1912–1998. *Newnham College Roll Letter* 101–103.
87. Watson, J. D. and Crick, F. H. C. (1953). A structure for deoxyribose nucleic acid. *Nature* **171**: 737–738. For a more complete account of the DNA saga, see: Olby, R. (1974). *The Path to the Double Helix*. Macmillan, London.
88. Harker, D., cited in: Note 9, Julian, p. 363. See also: Hubbard, R. (1979). Reflections on the story of the double helix. *Women's Studies International Quarterly* **2**: 261–273; Strasser, J. (1976). Jungle law: Stealing the double helix. *Science for the People* **8**: 29–31.
89. Letter, K. Lonsdale to R. Franklin, 4 January 1955, Churchill Archives Centre, The Papers of Rosalind Franklin, FRKN, University of Cambridge.
90. Watson, J. D. (1968). The double helix. In Stent, G. S. (ed.), *The Double Helix: A New Critical Edition*, Weidenfeld & Nicolson, London. This edition includes additional commentaries, several of which contradict Watson's viewpoint.
91. Note 70(b), Sayre, p. 18.
92. (a) Klug, A. (1968). Rosalind Franklin and the discovery of the structure of DNA. *Nature* **219**: 808–810 and 843–844; (b) Klug, A. (1974). Rosalind Franklin and the double helix. *Nature* **248**: 787.
93. Letter, A. J. Caraffi to A. Klug, 29 August 1968, Churchill Archives Centre, The Papers of Rosalind Franklin, FRKN, University of Cambridge.
94. Letter, A. Klug to P. Siekevitz (New York Academy of Sciences), 14 April 1976, Churchill Archives Centre, The Papers of Rosalind Franklin, FRKN, University of Cambridge.

95. (a) Abir-Am, P. G. (1993). Dorothy Maud Wrinch (1894–1976). In Grinstein, L. S., Rose, R. K. and Rafailovich, M. H. (eds.), *Women in Chemistry and Physics: A Biobibliographic Sourcebook*, Greenwood Press, Westport, Connecticut, pp. 605–612; (b) Julian, M. M. (1984). Dorothy Wrinch and the search for the structure of proteins. *Journal of Chemical Education* **61**: 890–892; (c) Abir-Am, P. G. (1987). Synergy or clash: Disciplinary and marital strategies in the career of mathematical biologist Dorothy Wrinch. In Abir-Am, P. G. and Outram, D. (eds.), *Uneasy Careers and Intimate Lives: Women in Science, 1789–1979*, Rutgers University Press, New Brunswick, New Jersey, pp. 239–280.
96. Abir-Am, P. G. (1987). The biotheoretical gathering, trans-disciplinary authority and the incipient legitimation of molecular biology in the 1930s: New perspective on the historical sociology of science. *History of Science* **25**: 64.
97. Ayling, J. [Wrinch, D.] (1930). *The Retreat from Parenthood.* Kegan Paul, Trench, Trubner & Co., London.
98. Cited in: Note 95(b), Julian, p. 890.
99. Wrinch, D. M. and Langmuir, I. (1939). Nature of the cyclol bond. *Nature* **143**: 49.
100. Crowfoot, D. C. (1976). Obituary: Dorothy Wrinch. *Nature* **260**: 564.
101. Cited in selection 134 in Teich, M. with Needham, D. M. (1992). *A Documentary History of Biochemistry 1770–1940.* Leicester University Press, Leicester, p. 349.
102. Martin, R. B. (1987). Dorothy Wrinch and the structure of proteins. *Journal of Chemical Education* **64**: 1069.
103. Note 100, Crowfoot, p. 564.
104. Wrinch, D. (1946). *Fourier Transforms and Structure Factors.* American Society for X-ray and Electron Diffraction, Cambridge, Massachusetts.
105. Rose, H. (1994). Nine decades, nine women, ten Nobel Prizes: Gender at the apex of science. In Rose, H. (ed.), *Love, Power and Knowledge: Towards a Feminist Transformation of the Sciences*, Polity Press, Cambridge, p. 157.
106. Cruickshank, D. (2005). Milestones: Mary Rosaleen Truter (1925–2004). *International Union of Crystallography Newsletter* **12**(3): 28.

107. (a) Haines, M. C. (2001). MacGillavry, Carolina Henriette. *International Women in Science: A Biographical Dictionary to 1950*, ABC-CLIO, Santa Barbara, California, p. 189; (b) Note 9, Julian, pp. 346–347; (c) Anon. (2007). Carolina Henriette Mac Gillavry, http://en.wikipedia.org/wiki/Carolina_Henriette_Mac_Gillavry, accessed 14 November 2007.
108. Rose, R. K. and Glusker, D. L. (1993). Jenny Pickworth Glusker (1931–). In Grinstein, L. S., Rose, R. K. and Rafailovich, M. H. (eds.), *Women in Chemistry and Physics: A Biobibliographic Sourcebook*, Greenwood Press, Westport, Connecticut, pp. 207–217.
109. Ramaseshan, S. (1996). Reminiscences and discoveries: Dorothy Hodgkin and the Indian connection. *Notes and Records of the Royal Society* **50**(1): 115–127.
110. Cited in O'Driscoll, C. (1996). Minorities and mentors: The X-ray visionaries. *Chemistry in Britain* **32**: 5–8.

Chapter 10

Women in Pharmacy

Though pharmacy is regarded as a distinct profession, many academic women chemists saw pharmacy as one of their few employment options. As we will show, several of the women were also associated with the Chemical Society and/or the Institute of Chemistry, and two of them were signatories to the petition for admission of women to the Chemical Society (see Chap. 2).

Of particular interest, the fight for women's entry into the Pharmaceutical Society, though conducted in an earlier period, shows interesting parallels with the battle for membership of the Chemical Society. Pharmacy also provides a contrast to chemistry in that the Association of Women Pharmacists became a strong focus for women, unlike the short-lived Women Chemists' Dining Club (see Chap. 2).

Society of Apothecaries

Apothecaries had been involved in the production and dispensing of patent medicines from the Middle Ages.[1] There are records of women being registered as apothecary or "chymist & druggist" as far back as the 17th century.[2] Apothecaries were a recognized Guild with its own Court of Examiners to license those who wished to dispense the herbal remedies of the time.

The Court of Examiners' first encounter with a woman candidate occurred in 1860.[3] Elizabeth Garrett[4] (see Chap. 3) had been refused by every medical school to which she had applied on the grounds of her gender. She then focused her attention on becoming an apothecary. Passing the Apothecaries' examination entitled its holder to a place on the Medical Register, which was Garrett's long-held desire. Having taken private lessons with an apothecary, she enquired about the possibility of being examined at Apothecaries' Hall. This request caused great consternation among the members of the Court of Examiners. Their legal counsel advised them that the Apothecaries' Act of 1815 opened the examinations to all persons and as, by British law, women were persons, they could not be excluded. Garrett passed the preliminary examination and then both parts of the professional examination, obtaining her diploma in 1865.

The Council realized that Garrett's success would encourage other women to follow suit. In fact, in 1867, three women passed the preliminary examination. The portents of more women in the pipeline persuaded the Council to act. Before admission to the examination, the candidate had to provide a certificate of attendance at the required lectures. As women were excluded from the lectures at public schools, they had to be tutored privately. Thus, later in 1867, the Court of Examiners announced that they would no longer accept the validity of certificates of attendance at private lectures. This tactic proved an effective means of circumventing the law and blocking the aspirations of subsequent women candidates.

From the middle of the 19th century, the profession of apothecary lost ground to that of the more professionally trained pharmacist. Nevertheless, women found a niche as Apothecaries' Assistants. These assistants were required by any general practitioner or dispensary who dispensed their own medicine. In 1887, Fanny Saward became the first woman to gain admission to the Apothecaries' Assistants' examination. The Assistants' certificate, known as "Apothecaries' Hall," or simply "Hall," was, as Penelope Hunting remarked: "… a post suited to a young lady willing to work as the handmaiden of the doctor for little pecuniary reward."[5] This was certainly a female ghetto; for example, in 1917, 9 men and 233 women registered for the Apothecaries' Assistants' examination. Among those women was Agatha Mary Clarissa Christie, who later gained fame as a writer of detective stories. As the pharmaceutical industry developed during the 1930s, greater and greater scientific knowledge was demanded of the dispensers of medicines. For this reason, the more basic examinations of the Apothecaries' Hall fell out of favour compared with the more lengthy and rigorous studies demanded of those holding qualifications of the Pharmaceutical Society.

Pharmaceutical Society

The profession of pharmacy has been controlled by legal statute for over 150 years.[6] There were two examinations of the profession: the Minor and the Major. The Pharmacy Act of 1852 allowed those who passed the Major to have their names entered on the Register of Pharmaceutical Chemists. The 1864 Pharmacy Act required that successfully passing the Minor was mandatory before being registered as a chemist or druggist. From 1868, anyone wishing to open a pharmacy was required to pass the Major, though those already in business were able to continue to do so without examination.

The Pharmacy Examinations

The Pharmaceutical Society was given the authority to organise the examinations under the terms of the Acts; however, the legal administration of the examinations was quite separate from the regular activities of the Society. Following the 1868 Pharmacy Act, there were three examinations: the Preliminary, the Minor, and the Major. The Preliminary examination was more of a skills test in Latin, French, and arithmetic, and it entitled the successful candidate to be registered as an apprentice. Passing the Minor resulted in the designation as an assistant to a chemist or druggist, while passing the Major allowed the graduate to call themselves a pharmaceutical chemist. The Major examination was described as "decidedly difficult," and it focused on advanced chemistry, *materia medica*, and botany. For pharmacy assistants who had been actively involved in the profession for 3 years, passing a Modified examination was all that was necessary.[7]

By the early part of the 19th century, women pharmacists were to be found all over the country.[8] In fact, when the first compulsory Register of all prastising pharmacists was undertaken in 1869, 215 of the 11 638 registered chemists and druggists were women.[6] Most of the women were continuing a business that had been started by their father or husband who had subsequently died. The first woman to take the examinations required under the 1868 Act was F. E. Potter, who applied in 1869 to take the Modified examination. It was only when the individual appeared to sit the examination was it realized she was a woman — Frances (Fanny) Elizabeth Potter. As the Act made it clear that the Society had a duty to examine all persons, Potter was allowed to take the exam, which she successfully passed. Potter (Mrs. Deacon) was followed 6 months later by Catherine Hodgson Fisher.[6] In 1873, Alice Vickery of Camberwell, Surrey, became the first woman to pass the Minor; and then in 1875, Isabella Skinner Clarke (Mrs. Clarke-Kerr) became the first woman to pass the Major examination and register as a pharmaceutical chemist.

Admission of Women to the Pharmaceutical Society

Though entry to the profession had been easy, admission to the professional body, the Pharmaceutical Society, proved challenging. The Pharmaceutical Society had been founded in 1841.[6] Pharmacy students and graduates were encouraged to join the Society — those who had passed the Preliminary as students, the Minor as associates, and the Major as members. Even though by law the Society was forced to admit women to its examinations, the Society acted on the premise that pharmacy was a male profession and that the Society itself was a male preserve. In fact, the assumption had been that the practice of female relatives taking over a pharmacy would cease once Registration became law. Thus, they were unprepared for the "women problem" to be a regular business item for the Council of the Pharmaceutical Society for the next decade. However, the fact that, in 1868, *The Englishwoman's Review* had pronounced pharmacy as a suitable profession for a woman[9] should have warned the Society members of what was ahead.

The Pharmaceutical Society had its own School of Pharmacy. In 1861, the discovery that a "lady" had acquired a ticket for admission to the lectures of the School caused a flurry of concern. As she was already in possession of the ticket, the Library, Museum and Laboratory Committee of the Society had little choice in the matter, but they added that ladies "must be regarded as attending upon sufference."[6] The "lady" seeking admission had been the aforementioned Elizabeth Garrett, who had wished to use the lectures of the School of Pharmacy to prepare herself for the examinations of the Society of Apothecaries. The following year, the Council of the Pharmaceutical Society rejected a proposal to formalise the admission of women to lectures, thus preventing any other woman from following suit.

The first woman to apply for membership of the Pharmaceutical Society was Elizabeth Leech in 1869.[10] Leech had learned her pharmacy skills from her father, having worked in his shop for 7 years. Following his death, she shared the running of the shop with her brother for 6 years and then on her own for another 9 years. The Lancashire cotton famine had forced her out of business, and she had then become compounder and dispenser of prescriptions at the Munster House Lunatic Asylum, Fulham. Her application noted that she believed that membership in the Pharmaceutical Society would help her resume her business. Fearful that the Council might think she was a troublemaker, she wrote: "I have no wish upon any occasion to interfere with the Council or its meetings. All I want is the Membership."[11] The Council rejected her request a total of three times, the last being in 1872.

Robert Hampson and His Crusade

It is often overlooked how much the progress of women relied on sympathetic men, and the cause of women in pharmacy was no exception.[12] The "white knight" in this case was Robert Hampson.[13] In 1872, Hampson was elected to the Council of the Pharmaceutical Society. In the 1860s, he had been a prominent member of the rival United Chemists and Druggists Society. However, following the passage of the Pharmaceutical Act, he redirected his radicalism, especially women's rights, to the activities of the Pharmaceutical Society. He was joined in his crusade by other, equally radical, new members of the Council.

On 2 October 1872, Hampson proposed to the Council that, since women had been permitted to take the examinations of the Society, they should have been allowed to attend lectures and use the laboratories of the Society's School of Pharmacy. It is interesting that a report of his opening remarks reflected the perceived progress of women that had occurred in the preceding decade:

> In 1862 perhaps the admission of lady students to the classes and laboratory might appear a step fraught with great danger, and tending to revolution; but in the present day, in remembrance of the social and educational changes that had taken place, he could not for a moment assume that the present Council, elected on a much broader basis, would endorse the decision of the predecessors, which was, in fact, most arbitrary, unjust and impolitic.[14]

Hampson contended that, by the use of "person" in the Pharmacy Act of 1868, the framers of the Act did not exclude the possibility of women. His eloquent plea — in this instance — was successful. The Council agreed unanimously to the admission of women to classes; however, as there was limited laboratory space, they were denied access to the laboratories.

This was Hampson's sole early victory. His proposal in November of that year that Elizabeth Leech be admitted as a member was defeated.[15] In December, his motion that women who took the examinations should be eligible for the Society's prizes, certificates, and fellowships was defeated by one vote.[16]

At the Council meeting of February 1873, the names of three women were put forward among the 166 candidates for admission as "registered students" of the Society. These were Rose Coombes Minshull, Louisa Stammwitz, and Alice Marion Rowland (Mrs. Hart). Minshull had attained the highest mark among the 166 who had passed the Preliminary examination, while Stammwitz was about midway in the list. Hart had a certificate from the Society of Apothecaries in lieu of taking the examination. Though Hampson and his allies may have hoped that the list would be approved *in toto* without debate as usually happened, this was not to be. An amendment was moved to reject the names of the women, but the amendment was defeated. A second amendment was then made to defer the decision on the women candidates until after the Council elections; this proposal produced a tie vote, with the chair then casting his vote in support. The decision was therefore postponed and the women barred from admission.

The then-President of the Society, George W. Sandford, articulated his position in the Correspondence pages of the *Pharmaceutical Journal*:

> The only question which came before the Council on the 5th of February was, whether Alice Marion Hart, Louisa Stammwitz, and Rosa Coombes Minshull should be admitted as apprentices *of the Pharmaceutical Society*. Now, Sir, these ladies are utter strangers to me, therefore no personal feeling could influence me in moving that they be not admitted; but as I have always held that the Pharmaceutical Society was intended to be a Society of men, that certain disadvantages would arise from its being a mixed Society of men and women, and that the admission of females as apprentices would be

only a stepping-stone to their admission as members, I felt bound to oppose them on the threshold.[17]

Society for Promoting the Employment of Women

Evidence suggests that Hampson had the active support of one of the women's organisations, the Society for Promoting the Employment of Women (SPEW).[10] Since its founding in 1859, the Society had been fighting for the opening of male-exclusive trades to women. Its *modus operandum* was to find a few suitable, talented women for sponsorship and find employers who would take them as apprentices or paying students, the Society covering their fees. The Society contended that once a few women had made a success in each particular field, it would be much easier for others to follow their path. However, the Society seemed to have little interest in supporting the aspirations of comparatively affluent ambitious women seeking careers in the professions, that is, until Hampson entered the cause of women in pharmacy.

In 1866, the year after gaining her medical licence, Elizabeth Garrett opened St. Mary's Dispensary for Women and Children in Marylebone, London. Over the following years, Garrett willingly accepted students sponsored by the SPEW to be trained as dispensers for hospitals and for medical practitioners. Two of the sponsored young women were Minshull and Stammwitz. After Hampson had persuaded the SPEW to take on the cause of women in pharmacy, it appears that Minshull and Stammwitz were chosen to lead the assault.

Rowland was probably encouraged to apply for admission by her surgeon-husband, Ernest Hart. Hart had shown his support for women's rights by being a founding member of, and major donor to, the London School of Medicine for Women (see Chap. 4).

Success at Last

The early months of 1873 also saw the first series of letters from readers of the *Pharmaceutical Journal* and the *Chemist and*

Druggist on the subject of "Pharmaceutical Women." One correspondent contended that the job was unsuitable for women because of "the common occurrence of prescriptions and remedies dealing with maladies of the most revolting nature" and of the necessity to deal with "subjects which possess the power to appal and disgust the sternest members of the sterner sex."[18] Another writer thought that modest English women would not want to see objects "which are sold generally to men of not the highest moral character."[19] These two writers missed the whole point that the 1868 Pharmacy Act already permitted women to become pharmacists: the only issue was whether women were to be admitted to the professional society.

Minshull, Stammwitz, and Rowland put forward their case in a letter to the *Pharmaceutical Journal*: "All that we ask is to be allowed the same opportunities for study, the same field for competition and the same honours, if justly won."[20] The issue of women's admission was next raised at the Annual General Meeting in May 1873. Hampson and his friends pressed the case of the women pharmacists. As Holloway commented: "The supporters of Hampson's motion dominated the debate, won all the arguments, and were heavily defeated at the end of the day."[21]

Jordan notes that, following the defeat, Hampson's crusade changed tack:

> From that point on explicit appeal to nonconformist, radical and feminist principles vanished from their speeches. They seem to have concluded that demonstrating inconsistency with these discourses left most members unmoved, and that the expedient tactic was to show that refusing admission to women was inconsistent with principles to which all members subscribed.[22]

In the following years, Minshull and Stammwitz (Rowland having been admitted to medical school instead; see below) continued their studies by attending the Society's lectures while using the laboratories of the private South London School of

Chemistry and Pharmacy.[23] Opened in 1870, the School was run by John Muter,[24] a former public analyst, and it was one of the few private schools of pharmacy that admitted women students.

Minshull and Stammwitz were joined in their quest for admission to the Society by Isabella Skinner Clarke. Clarke, too, started her career in Garrett's Dispensary for Women and Children. She passed both the Minor and Major examinations in 1875 (gaining fourth place in the Major). She became a registered Pharmaceutical Chemist, and in early 1876 opened her own chemist's shop in Paddington, following a stint working in Hampson's shop. Upon passing the Major, Clarke applied for Membership of the Pharmaceutical Society in January 1876. She was refused, many of the Council arguing that they were bound by the previous rejections. Her case was raised again in July 1876 and in October 1877. Later that year, Minshull and Stammwitz passed the Minor examinations and requested admission as Associates. They, too, were refused on the grounds of the ruling of the May 1873 Annual Meeting.

At the end of 1877 and in early 1878, the topic of the admission of women rose to prominence once more in the correspondence pages of the *Pharmaceutical Journal and Transactions*. The issue came to a vote again at the Annual Meeting of 1878. After initial confusion as to the voting, defeat was by a narrow margin of 59 to 57. At the following Council meeting, a motion to admit women resulted in an 8–8 tie, with the Chair deciding the matter for the status quo. The next year, Stammwitz and Minshull passed the Major examination. Minshull applied and Clarke reapplied for full Membership of the Pharmaceutical Society. The Council decided the matter had to go to the Annual Meeting. In a rerun of the arguments of the 1878 Annual Meeting, the May 1879 event rejected the women's candidacy by 81 votes to 78.

The issue would not go away. The names of Minshull and Clarke were put forward again at the Council meeting of 1 October 1879, as the *Chemist and Druggist* reported: "Mr. Hampson moved that they should be elected. He thought the question ought never to have been referred to the annual meeting and he urged that it was the duty of the Council to elect all eligible persons irrespective of sex."[25]

The outcome this time was very different. As one of the Council members, Mr. Bottle, a former strong opponent of women's admission, was quoted as saying:

> He should vote for the motion, not with a view to conceding the ladies what Mr. Hampson asserted was their right, but as a matter of courtesy, which he thought they had well earned by passing the examinations, and also with a view to bringing about a peaceful termination to a question which had formed a bone of contention for some years. A prolonged agitation would be infinitely worse than admitting even a dozen women into the society.[26]

At the vote, only Sandford, still the President of the Society, opposed their admission. The opposition had collapsed; this particular battle had been won. Women were now admitted to the Pharmaceutical Society School of Pharmacy. Stammwitz applied and was accepted a year later.

More Challenges

In 1899, pharmacy was recommended as a suitable career for young women by R. Kathleen Spencer in the pages of *The Girl's Own Paper*, though she did add caveats:

> The employment of girls as dispensers becoming much more general both in hospitals and pharmacies, a few words on the subject may be useful to any who seriously contemplate adopting the occupation. In the first place, only girls of education, of average health, and who can afford to give the necessary expenditure of time and money should take up the profession of pharmacy.[27]

Even though it was possible to find appointments with just the Apothecaries' Assistant's examination, Spencer expressed concern that these minimally qualified women would hurt the cause of women pharmacists: "If they do so, by their inexperience they bring the whole question of the employment of ladies in pharmacy into disrepute."[27]

Admission to the School of Pharmacy did not end the challenges for women students: they had to use a separate entrance and were usually ignored by their male colleagues. Nevertheless, the two journals, *Pharmaceutical Journal* and *Chemist and Druggist,* maintained their strong support for the cause of women in pharmacy. In 1891, the *Chemist and Druggist* had an editorial on Lady Pharmacists,[28] reviewing the history of their cause and suggesting that the defeat of the motion for admission in 1879 was the result of negative votes by young male pharmacists who feared the potential competition from women. In that editorial, the journal asked women pharmacists to report on their experiences and, 6 months later, the journal published a fascinating three-and-a-half-page summary of comments by some of the more prominent women pharmacists of the time, including Louisa Stammwitz, Margaret Buchanan (see below), Anne Neve, and Rose Minshull.[29]

Despite the hostile atmosphere at the School of Pharmacy, women performed excellently in their studies, as was noted by a columnist in the *Chemist and Druggist* in 1908:

> The Male Intellect is evidently not equal to the contest with feminine rivals in the class rooms and in examinations. Miss Wren annexes three out of four silver medals which the Pharmaceutical Society contributes annually ... and at the same time establishes her claim to the Periera medal. Miss Neve supplements this demonstration of the superiority of the sex by scooping in exactly the same proportion of the bronze medals awarded in the Minor course. This appropriation of the Society's bullion by two very young ladies ... leaves but a scanty "distribution" of honours ... among the masculine

> majority of competitors. Moreover the ominousness of the event is that it is not merely occasional or accidental. It is just the climax of a consistent progress which has been noted again and again ... Undoubtedly the average of the academic work of the ladies at schools of pharmacy (and not in this country alone) has been much higher than that of the male students ...[30]

Both Gertrude Wren,[31] who had been educated at North London Collegiate School (NLCS), and Grace Neve[32] became recipients of the coveted Periera Medal.

Nevertheless, the columnist was reassured by the following fact: "But there remains the consoling fact that though women have been demonstrating their capabilities in pharmacy for 30 years or more, it is still men who teach them, men who examine them, and men who hand them out their medals."[30]

Employment of Women Pharmacists

In 1904, a column in the *Pharmaceutical Journal* reported upon a large increase in the number of women students. Nevertheless, the author of the column was very pessimistic on the opportunities for women pharmacists: "There is no sign that any considerable proportion of master-pharmacists will ever be willing to employ women behind their counters as assistants."[33] The article concludes that women of "average intelligence and good health" should seek a calling other than pharmacy. This column provoked a flurry of Letters to the Editor of both *Pharmaceutical Journal* and *Chemist and Druggist*, debating whether women were capable of being good pharmacists.[34]

In the *Girl's Realm Annual* of 1908, J. E. Walden of the Westminster College of Chemistry, Pharmacy, and Dispensing[35] wrote an article: "Girls as Chemists: How a Girl May Take Up the Work of Chemistry, with a View to Keeping a Pharmacy, or Becoming a Doctor's Dispenser."[36] Despite the title, the author emphasized the route of the Society of Apothecaries examination leading to the role of "lady dispenser." As to becoming a "girl chemist," in view of the "severity" of the Pharmaceutical Society examination, Walden suggested that girls consider the Apothecary examination as a stepping stone. He commented that:

> The girls who take up the profession are principally doctors' daughters, or other relatives, those who have some means, and yet want something to do. A great many nurses also take it up, for the dispensing certificate is a valuable asset to the trained nurse. It is, however, generally suitable for the well-educated of the middle classes.[36]

Women pharmacists usually ended up in the worst-paid and most demanding areas of pharmacy. By 1908, there were still only 160 women registered as pharmacists, while only about two thirds were practising pharmacists.[6] Of those who had abandoned their careers, some had married (and often had been forced to resign), others had emigrated, and the remainder had changed careers (for example, to medicine). Of those in the profession of pharmacy, over 60% were working in hospitals and institutions; about 20% were in retail pharmacy, most often as assistants; 12% were dispensers to doctors in private practice; a few were employed by wholesale pharmaceutical houses; and still fewer were analysts, lecturers, or researchers.

Those women in institutions or working for medical practitioners were particularly disadvantaged financially, as they were competing with the lowly paid ranks of the male dispensers who held the Assistants' Certificate of the Society of Apothecaries.[37] Women who attempted to practise in the retail trade found that they were not always welcome, as Emily Forster (see Chap. 3), Lecturer at the Westminster College of Pharmacy, described in 1916:

> Where to settle! ... The woman pharmacist has something else to weigh besides expense: it is the question of her sex, and the fact that at present she is a pioneer in her profession, and

must naturally turn to where she thinks an enterprising woman will be respected and her ability made use of rather than to a locality that appears very "Early Victorian" ... The places to avoid are centres, such as cathedral towns, where anything new is looked upon with suspicion, and must stand the test of time before it can be trusted.[38]

Though the hospital work was simply routine dispensing, the women pharmacists did make their mark, as Margaret Buchanan commented: "In such a position the orderliness and attention to detail, the tact and desire to please, which are supposed to be natural to most women, are the most necessary parts of their stock-in-trade, and there have been not a few instances, where women's business capacity and an up-to-date knowledge of drugs and economical methods have led to practical appreciation on the part of committees."[39]

Many male pharmacists still refused to take on women apprentices. One of the major boosts for training women in pharmacy was a single chemist's shop: Number 17, The Pavement, Clapham Common.[40] The pharmacy had been opened by Henry Deane in 1837. Deane had joined the Pharmaceutical Society in 1841, being one of its earliest members, and he was President from 1853 until 1855. The pharmacy was purchased sometime between 1911 and 1914 by Buchanan as a training facility for women. She was joined in 1914 by Agnes Borrowman (see below). One of their student intake of 1915 recalled:

Surely no youngster could have had two more energetic or exacting tutors than the two "Miss B's" — Miss Buchanan and Miss Borrowman — at that time partners in the somewhat decaying business of Deane's of Clapham. I well remember the contagious enthusiasm with which she [Borrowman] tackled the job of putting that old business back on its feet, and how proud we were to be allowed to help and perhaps earn an occasional word of appreciation when we had worked with her from early morning until late into the night.[41]

By 1923, of the 15 young women trained at The Pavement who had studied at the Pharmaceutical Society's School of Pharmacy, 14 had taken prizes and scholarships. As the *Pharmaceutical Journal* noted: "Is there another pharmacy in the country that can beat this record?"[42] Borrowman took over sole proprietorship in 1924. The pharmacy was badly damaged by the nearby impact and explosion of a V-2 rocket in January 1945. At this point, Borrowman converted the business to a limited company with Hilda Wells as Director. Wells had been an apprentice at The Pavement in 1918. The business ceased to be women-owned and women-run in 1958, and it finally closed in 1984.

Association of Women Pharmacists

Membership of the Pharmaceutical Society brought few, if any, benefits. Women pharmacists found themselves isolated and marginalized. Buchanan and Clarke are said to have discussed the problems in Buchanan's kitchen at her home in Gordon Square, and they concluded that a separate organization for women pharmacists was needed.[43] This discussion led to the founding of the Association of Women Pharmacists on 15 June 1905, with membership confined to those who had passed the Minor or Major examination.[44] A letter in the *Chemist and Druggist* reported on the inaugural meeting, noting: "It is hoped that all qualified lady chemists will join the Association now formed, which, on the principle that 'union is strength,' cannot fail to emphasise and bring to the front the latent power of women's work in pharmacy."[45]

The first President was Clarke; the Vice-President, Buchanan; the Treasurer, Hilda Caws[46]; and Secretaries, Georgina Barltrop[47] and Elsie Hooper (see below). Fifty women joined immediately; and by 1912, Buchanan, then President,[48] proclaimed that "practically every woman practicing pharmacy" was a member.[49]

The objectives of the Association were as follows:

1. to discuss the employment of women in pharmacy from an ethical and practical standpoint;
2. to keep a register of all qualified women and their appointments; and
3. to keep a register of members requiring assistants or locum tenens and to put them in touch with members desiring employment.

In addition, the Association's employment bureau compiled a blacklist of poorly paid posts, and put pressure on hospitals and other institutions to improve their salaries and terms of employment.

Women were encouraged to use the Association's employment service; to take up the Association's special insurance and annuity scheme; to participate in the training programme involving the interchange of women apprentices between retail and hospital services; and to start their own businesses, preferably as a joint venture between "two or three women of congenial tastes."[50]

The first public meeting of the Association of Women Pharmacists was held in London on 17 October 1905. One hundred women and 12 men heard Louise Creighton, first president of the Union of Women Workers, speak on "The Present Responsibilities of Women." Within the ranks of the Association were a number of "radical" members and, on 17 June 1911, they marched across the city with more than 40 000 other women to show their solidarity with the Suffragette Movement. The *Chemist and Druggist* expressed its approval: "It was a magnificent demonstration of the organising abilities of women and of the universality of their desire to get their Parliamentary vote. We give two photographs of the small section composed of women pharmacists."[51] In 1918, branches of the Association were established in other British cities and, that same year, the organisation changed its name to the National Association of Women Pharmacists.

With all young able-bodied men called up for military service during the First World War, women were required to take over many positions (see Chap. 12), including that of pharmacist. However, the end of the War brought a flood of returning males expecting (and demanding) to reoccupy their former jobs. Nevertheless, women were able to retain some of the inroads they had made, with 7% of the names on the Pharmaceutical Register being women in 1920.

The need for the Association of Women Pharmacists became apparent once more during the backlash of the 1922–1929 period. For example, in 1922, the *Pharmaceutical Journal* contained a letter from "A Pharmacist":

> It has recently been reported to the Press that in future, the London Hospital intends to restrict its students to men only. It occurs to me that this may be a good lead to the Pharmaceutical Society and also to Colleges of Pharmacy generally. While there are so many male chemists and chemists' assistants unemployed at the present time, it seems the limit of absurdity to flood the business, or profession, with a motley horde of untrained and incompetent surplus females. Is it not practicable to eliminate this undesirable element all together?[52]

Recovery came by 1937, when women made up 10% of the names on the Register, although the first woman President was not elected until 1947.[53] The National Association of Women Pharmacists continues to thrive today.[54]

Early Pioneers

Of the women at the forefront of the battles with the Pharmaceutical Society, Clarke, Minshull, and Stammwitz all pursued successful careers in pharmacy. In 1876, Isabella Clarke[55] established her own business at Spring Street, Paddington, London. She provided the dispensing course at her pharmacy for women medical students at the London School of Medicine for Women (LSMW; see Chap. 4). As a result, Clarke was appointed as Tutor in Pharmacy at the LSMW. Clarke married Thomas Kerr in 1883; she had met him years earlier when they were both students at the South London School of Chemistry and Pharmacy under Muter. After marriage, she sold the Spring Street pharmacy and became Kerr's partner in a pharmacy on Bruton Street, Berkeley Square. She took in women pharmacy students as boarders at her home in Endersleigh Street, which was also used for early meetings of the Association for Women Pharmacists. Following Kerr's death in 1898, Clarke ran a shorthand and typewriting school in Westminster; then during the First World War, in her 70s, she worked daily at the Admiralty. She died in Croydon on 30 July 1926, aged 84 years.

Minshull became head dispenser at the North Eastern Hospital for Children. Dying on 9 May 1905 at the comparatively young age of 58, she was described in an obituary as: "… not by nature a fighter, but a bright and charming little woman of an affectionate nature."[56]

Stammwitz spent 9 years as a dispenser at the New Hospital for Women (an expansion of Garrett's original dispensary). She then went into partnership with Anne Neve, a former apprentice of Isabella Clarke, who had qualified in 1884. They established a successful chemist's shop in Paignton, Devon. Stammwitz and Neve retired together to Croydon, where Stammwitz died in 1916.[57]

Neither Rowland nor the first woman to pass the Minor, Alice Vickery, had any intention of pursuing a career in pharmacy.[10] They had both planned to study medicine, but had been thwarted by the collapse in 1872 of a plan to set up a Ladies' Medical College in London. Thus, both women left pharmacy and enrolled with the LSMW upon its founding in 1874. Vickery became a successful medical practitioner and was also active in radical political causes, including the birth control movement.

Rowland completed her medical studies and worked together with her husband.

The second woman to pass the Major was Lucy Boole (see Chap. 4).[58] After serving her apprenticeship with Clarke, Boole became the first woman in Britain to undertake research in pharmaceutical chemistry. Working with Wyndham Rowland Dunstan,[59] Professor of Chemistry to the Pharmaceutical Society (and a Vice-President of the GPDSC), Boole's 1889 procedure for the analysis of tartar emetic, published in the *Pharmaceutical Journal*, became the official method of assay from that date until 1963. She continued her research in the Pharmaceutical Society's laboratory for some years after her appointment as Lecturer in Chemistry at the LSMW.

Margaret Buchanan

The most important pioneer of women in pharmacy, Margaret Elizabeth Buchanan,[60] born in 1864, was educated at NLCS. Her father was a medical doctor and she began her pharmaceutical training with him, and then later with Clarke and Kerr. She qualified in 1887 and passed the Major a year later, obtaining second place in the Periera competition.

Until taking over proprietorship of the Deane pharmacy (see above), Buchanan had been a hospital pharmacist. Then, following her transfer of ownership of the Deane pharmacy to Borrowman, Buchanan founded the Margaret Buchanan School of Pharmacy for Women at Gordon Hall, for which she was Principal for many years. Her links with the Deane pharmacy, however, did not come to an end: "... and it was arranged that her pupils, all of whom were women, should attend the Clapham establishment [the Deane pharmacy] — three in the morning and three in the afternoon — in order to gain practical experience."[40] Buchanan also taught pharmacy to women students at the London School of Medicine for Women.

Buchanan was the first woman to be elected to the Council of the Pharmaceutical Society. It was no accident, but rather a well-planned endeavour by members of the Progressive Pharmaceutical Club, as Herbert Skinner described in one of her obituaries:

> Things were not moving nicely and strong resistance from the Pharmaceutical Society at Bloomsbury Square was still strongly manifested. ... A few kindred spirits met at the Progressive Pharmaceutical Club including John Humphrey, Hugo Wolff, myself and a few others. We determined to try and bring in one of our club members, Miss Buchanan, and run her as a candidate for the 1918 election. She was a progressive in every sense, but somewhat diffident on the political side. We succeeded in persuading her to stand for the Pharmaceutical Council — the first woman candidate. The consternation that arose in certain quarters is one of those things not easily forgotten.[60(a)]

Perhaps fired by the opposition to the nomination, the Progressive Pharmaceutical Club members threw all of their efforts into the campaign:

> We progressives had to secure Margaret's return [election]... The *C & D.* [*Chemist and Druggist*] helped us magnificently and for successive nights we addressed envelopes, wrote postcards, sent circulars for Margaret E. Buchanan, the woman progressive pharmacist for the Pharmaceutical Council. ... The returns on the election day rewarded us for all we had worked to achieve.[60(a)]

Buchanan had made history. Following her election to the Council of the Pharmaceutical Society, though her interest was in education, she was placed on the Benevolent Fund Committee.

In 1922, Buchanan travelled extensively in Canada as the accredited representative of the Council; and on her return, she submitted a detailed report on the requirements for qualification, registration, and practice in the different Canadian provinces. One result of this document was a reciprocity agreement between the Society and the Province of Ontario.

Buchanan retired from the Council in 1926 due to ill health. She died on 1 January 1940. One of her first students, Elsie Hooper, was quoted in another obituary:

> She was a pioneer of women's work in pharmacy and has trained many of the best women pharmacists. No one who came in contact with her bright intelligent personality could fail to be affected by it. She was a wonderful and inspiring teacher, and laid the foundation of sound pharmaceutical knowledge.[60(b)]

Agnes Borrowman

Born near Melrose, Scotland, in 1881, Agnes Thomson Borrowman[61] spent 4 years completing a pharmacy apprenticeship before joining the Edinburgh pharmacist, William Lyon, as a junior assistant. During her limited spare time, she studied for the Minor at the Edinburgh Central School of Pharmacy. She passed the examination in 1903, at the age of 21.

Woman pharmacists in Scotland often faced discrimination. As was commented by Borrowman's obituarist: "At that time the Edinburgh public had not outgrown its prejudice against women in medicine and pharmacy, and customers often walked out of the shop rather than let a girl serve them, so that it was difficult for a woman pharmacist to get employment."[61] In fact, Lyon would not permit Borrowman to be seen at the front counter; instead, she had to work in the backshop, where "a large range of galenicals [lead compounds]" were made.[61]

For this reason, Borrowman moved south to Runcorn, Cheshire. In addition to pharmacy duties, she undertook her first research. This research — on an arsenic, iron, and quinine mixture — she presented to an Edinburgh Evening Meeting in 1904. It was while at Runcorn that she penned a spirited letter to the *Pharmaceutical Journal*, in which she expressed frustration with the timidity of women pharmacists in pressing their claims to equality of remuneration and parity in treatment with male members of the profession.[62]

In 1906, she moved to Dorking to work for 3 years with J. B. Wilson, an accomplished pharmacist and a cultivator of medicinal plants. She passed the Major examination in 1909, after which she became a research assistant at the Pharmacy School with Henry Greenish, studying components of quinine. Having been recommended by Arthur Crossley,[63] Borrowman subsequently left the School's laboratories and took up a better-paying appointment as Research Chemist in the London laboratory of the Rubber Growers' Association of Malaya and Ceylon.

Borrowman's research activities covered an incredible diversity of topics, as her obituarist in the *Pharmaceutical Journal* noted:

> … the examination of the physical and chemical properties of vulcanized rubber, and of the various processes of vulcanization, with a view to their improvement; soil analysis with the object of increasing the yield of latex, examination of distinctive fungi of the rubber plant, cellulose, paper-making, new processes for production of artificial silk, and examination of possible plant paper-making materials. In this connection Miss Borrowman acquired such a facility in the microscopic examination of fibres that she could tell at a glance the proportion or percentage of different fibres in a given paper. She also experimented extensively with papers, for the detection of forgery.[61]

The obituary went on to note that she also spent four nights per week attending classes at Borough Polytechnic, Chelsea Polytechnic, and the Cass Institute. In her spare time, she read and searched specifications at the Patent Office library.

However, in 1913, following the death of her father, Borrowman had to return to the more remunerative pharmacy trade to support the younger members of her family. She took a year of retail experience at a pharmacy in Slough before joining Buchanan at the Deane pharmacy in 1914, becoming the sole proprietor after the First World War.

Borrowman was active with the Association of Women Pharmacists from its founding. In 1924, she was the first woman to be invited onto the Board of Examiners of the Pharmaceutical Society, an appointment she held until 1937. She was also elected President of the South-West London Chemists' Association. Among other activities, she helped compile *Pharmacopedia* with Edmund White and John Humphrey.[58]

The near-destruction of the Deane pharmacy in 1945 and the years of wartime duties, such as fire watching, that she added to her load took their toll on her health. However, she kept active with the pharmacy until her death on 20 August 1955, aged 74 years. A correspondent for the *Pharmaceutical Journal*, "An Onlooker," commented: "Miss Borrowman was a firm believer in the equality of the sexes but this view applied only to other women — she knew that she herself was more than equal to most men."[64]

Dorothy Bartlett (Mrs. Storey)

Greenish and Crossley seem to have played mentoring roles for the early women research pharmacists. For example, they had both aided Borrowman, while Crossley was thanked by Lucy Boole for his interest in her research. Dorothy Bartlett,[65] too, was a woman pharmacist whose career owed a lot to these two mentors.[58]

Born in 1887 and educated at Streatham Hill High School (a GPDSC school), Bartlett was another of the brilliant early students who won a large number of the prizes of the Pharmaceutical Society during her years at the School. Passing the Major examination in 1911 and obtaining a B.Pharm (London) from King's College in the same year, she undertook research with Crossley from 1911 until 1912.

In 1912, Bartlett was awarded the Burroughs Research Scholarship, followed by the Redwood Research Scholarship, to work with Greenish in the Pharmacognosy Research Laboratories, chiefly engaged in making detailed microscopic studies of commercially available powdered vegetable drugs. Three publications resulted from her work. Bartlett then became a Research Chemist with Burgoyne Burbridges & Co. Ltd. In 1918, she married W. A. Storey and apparently terminated her professional activities. She died on 20 January 1940 in Manchester, aged 53 years.

Hope Winch

The most tragic ending was that of Hope Constance Monica Winch.[66] Born in 1895, Winch had been educated at the Clergy Daughters' School, Casterton, Kirby Lonsdale, Westmoreland. After leaving school, Winch had a year's training in practical pharmacy at the Royal Victoria Infirmary, Newcastle-upon-Tyne, before spending nearly 3 years in dispensing at the hospital. Then, she moved to London and attended the School of Pharmacy, where, among her many honours, she was the fifth woman student to be the Pereira Medallist. She passed her Major in 1917 and obtained her B.Pharm (London) in 1918.

Upon graduation, Winch held the appointment of Demonstrator in the School until 1920, while undertaking research in the School's research laboratories as a Redwood Scholar. In 1920, she was appointed Lecturer in Pharmaceutical Subjects and Botany at the Technical College, Sunderland, where, from 1921, she was Head of the Pharmacy Department. Tragically, on 8 April 1944, Winch fell from about 150 feet while climbing the peak of Scafell and was killed. At the inquest, it was suggested that: "the fall might have been due to a temporary "blackout," or to the rock which she was holding disintegrating through frost and crumbling in her hand."[67] Winch was only 49 years old.

Elsie Hooper (Mrs. Higgon)

Another pioneer stalwart of the Association of Women Pharmacists, Elsie Hooper,[68] born in 1880, passed the Major examination in 1902. Hooper was probably one of the most prominent early women researchers in pharmacy.[58] In 1901, she was the first woman to be awarded the Redwood Research Scholarship; and in 1903, she was the first woman recipient of the Burroughs Research Scholarship.

Under Greenish and T. E. Wallis, Hooper performed research on natural products, presenting some of her work at the 1905 British Pharmaceutical Conference. During the same time period, she was a demonstrator at the Pharmaceutical Society's School of Pharmacy, while in the evenings she studied towards a B.Sc. in botany and chemistry at Birkbeck College.

Following completion of her degree in 1905, Hooper worked on the publication of the first British *Pharmaceutical Codex*, followed by a year with Alfred Kirby Huntington at King's College (see Chap. 3). Then, she joined the staff of E. F. Harrison and worked on the analysis of "secret remedies" for the British Medical Association. Subsequently, for 4½ years, she held a lectureship in chemistry at Portsmouth Municipal College. She was also one of the signatories of the 1909 letter to *Chemical News* (see Chap. 2).

During the First World War, Hooper was an analyst with U.C.A.L. at Cheltenham. After the War, she opened a retail pharmacy in Cheltenham called Ladies Chemists Ltd. In 1920, Hooper left the pharmacy to return to London and assist in the teaching at Margaret Buchanan's School of Pharmacy for Women. Kathleen King joined the staff in 1922, and Hooper and King took over ownership of the School from Buchanan in 1925, renaming it the Gordon Hall School of Pharmacy for Women. In view of the continued prejudice against women pharmacists, Hooper purchased two retail pharmacies in northwest London specifically to provide work experience for the women students.

Hooper remained owner and main teacher at the School until it closed in 1942. She sold one of the pharmacies in 1945 and the other in 1961.[69]

Hooper was the first Secretary of the Association of Women Pharmacists, serving from 1905 to 1908, and she was elected President for the year 1927/1928. In 1911, she was one of the pharmacists demonstrating with the Coronation Suffragette Procession, marching behind the banner: "Women Pharmacists Demand the Vote." There is nothing noted on any later activities, but she did marry Mr. Higgon. Perhaps the marriage occurred somewhere about the time of the closing of the School. Hooper died in Paignton, Devon, on 6 May 1969, aged 89 years.

Nora Renouf

Nora Renouf[58] was born in Jersey in the Channel Islands, though the date is unrecorded. She gained some experience there working in a chemist's shop and decided to follow a career in pharmacy. She moved to London and studied at the Pharmaceutical Society School of Pharmacy, passing the Minor in 1902 and the Major in 1903. Renouf followed Hooper as a recipient of the Redwood Research Scholarship in 1904.

In 1905, Renouf was the first woman and the first pharmacist to be awarded the Salters' Research Scholarship. For the next 2 years, she performed research with Crossley on camphor derivatives, work that resulted in 12 publications. In fact, Renouf was Crossley's leading researcher, as his obituary noted: "... in many of those [communications] from the School of Pharmacy, Miss Nora Renouf was his collaborator."[70] Renouf did not take up the third year of the Scholarship.

Renouf was another founder member of the Association of Women Pharmacists, and she was the Treasurer from 1907 until 1916. She, too, was a signatory of the 1909 letter to *Chemical News* (see Chap. 2). Hooper and Renouf overlapped in their time at the School of Pharmacy, Hooper passing the Major the same year that Renouf passed the Minor. Thus, it seems likely that these two sole pharmacy signatories of the letter knew each other.

To contribute to the war effort, in 1916, Renouf returned to the Channel Islands to work in a hospital. Following the end of the War, Renouf returned to London and, in 1920, was one of the first cohort of women to be admitted to the Chemical Society (see Chap. 2). In her application to the Society, she noted that she was a survey officer at the Fuel Research Board of the Department of Scientific and Industrial Research.[71] Renouf died in 1934.

Ella Caird (Mrs. Corfield) and Elsie Woodward (Mrs. Kassner)

Ella Caird,[72] born in 1894, spent a 3-year apprenticeship at the Buchanan School of Pharmacy for Women between 1911 and 1914. Entering the Pharmaceutical Society School of Pharmacy, she proceeded to accumulate more awards than any other student up to that time. Her obituarist, C. A. Johnson, commented: "As Ella Caird, she gave early warning of her potential at 'The Square', when passing the 'Minor' in 1914, and the Ph.C. in 1915, with top place in every subject — and this was against the formidable opposition of at least two fellow students who were later to become professors in their subjects at the same college."[72] She became the first woman demonstrator at the College and also became an assistant to the Examiners.

In 1920, she collaborated with C. E. Corfield on a study of the fat content of momordica seeds; and in 1921, they were married. Upon marriage, Caird abandoned her promising career and devoted herself to her family, raising two sons, both of whom entered medicine. In 1925, C. E. Corfield went into partnership of the company Harrison & Self, which specialised in pharmaceutical analysis. However, he became increasingly ill and Caird had to take over more and more of the running of the company. This work became even more arduous for Caird in 1940 when, through enemy action, the laboratories were totally destroyed. It fell to Caird and her friend, Elsie Woodward, to organise the rebuilding of the establishment.

Elsie Woodward[73] had become a Demonstrator in Chemistry at the School of Pharmacy in 1919, and it was presumably there that Caird and Woodward met. Woodward had two publications with C. E. Corfield; thus, there was obviously a close connection between the three of them. Upon Woodward's marriage to Mr. Kassner in 1925, she moved to the United States. Woodward returned to England in 1938 and resumed her appointment at the School of Pharmacy, until joining Corfield and Caird in running Harrison & Self. In 1942, Caird and Woodward were made partners in the Company. Then, in 1945, Corfield died.

The business flourished under Caird and Woodward, a success that came as a surprise to some, as Johnson commented: "With the early death of her husband in 1945, she and Mrs. Kassner [another "remarkable woman pharmacist" according to Johnson] took over full control of the firm despite considerable scepticism on the part of many of their professional colleagues."[72] In fact, as the *Chemist and Druggist* noted: "So well did they maintain the cachet that went with the name of Harrison and Self that buyers of a number of important drug firms, including opium, continued to be satisfied only with a certificate endorsed with the name of the firm, which had enjoyed a world wide reputation."[74]

Caird and Woodward retired simultaneously from the company in the 1950s. Woodward died on 5 September 1959. Caird then returned to her pharmacy roots, undertaking part-time work in a hospital pharmaceutical department, first at the North Middlesex Hospital and then at Highlands Hospital. Her longtime friend, Mrs. E. Bradford, commented on Caird's longevity in the hospital service: "When Ella Corfield finally retired from Harrison and Self in the mid 1950's, she found retirement ill suited her so, in her late 60s she returned to pharmacy practice in the hospital service, where she remained until well into her 80s, celebrating her 80th birthday with a huge party at the hospital."[75] She died on 11 August 1986, aged 93 years.

Agnes Lothian (Mrs. Short)

We conclude this chapter with an individual whose scholarship bridged pharmacy and scientific librarianship (see Chap. 13). Agnes Lothian,[76] born in 1903, was the daughter of John Lothian, teacher of pharmacy in both Edinburgh and Glasgow. She took her pharmacy qualification at Herriot-Watt College, Edinburgh, graduating in 1926. She obtained appointments in retail pharmacies, first in Redhill, Surrey, then with Allen & Hanburys in London. In September 1940, she was appointed as a librarian to the Pharmaceutical Society. Following her appointment, she formally qualified as Librarian, being elected Associate of the Library Association in 1944. Her librarian activities were recalled by Desmond Lewis, a later Secretary and Registrar of the Pharmaceutical Society:

> To a student she was always kind and helpful: to others she could be brusque to the point of abrasion. It took courage to invite an opinion on a personal matter, because she could be devastatingly frank. But the formidable front that she adopted concealed an inner kindness that could burst into ripples of laughter.[77]

She developed a special interest in the history of pharmacy, and became a world authority on historical drug jars and mortars. Lothian was the first woman to be elected member of the Academie International d'Histoire de la Pharmacie. In 1967, she married G. R. A. Short, a Fellow of the Pharmaceutical Society.

Two years later, the Council of the Pharmaceutical Society bestowed upon her the title of Emeritus Keeper of the Historical Collection. As Lewis added:

> ... she used her combined knowledge and a limited budget to build the Society's library into the finest of its kind in the world. ... Inevitably, she will be best remembered for her researches into the history of pharmacy and its artefacts. She was recognised as a world authority on drug jars and mortars, and the collections that she built for the Society and the major acquisitions which she made, often against strong opposition of those who resented the expenditure, will remain as her memorial and our heritage.[76]

Lothian died on 13 October 1983, aged 80 years.

Commentary

In several respects, the fight for the admission of women to the Pharmaceutical Society resembled the later battle for admission to the Chemical Society. The two Societies were not only professional bodies, but also served as a "men's club" necessitating all possible measures to exclude the "alien species" known as females. As the Chemical Society had its excluder in Armstrong, so the Pharmaceutical Society had Sandford, with his quote that: "... the Pharmaceutical Society was intended to be a Society of men ...."[17] Likewise, the Chemical Society supporters, particularly Ramsey and Tilden, had their equivalent in the Pharmaceutical Society's Hampson. And, of course, there were a series of young women willing to take on the establishment of the Pharmaceutical Society, just as the 1904 petitioners had taken on the Chemical Society.

There was one notable difference between the two sagas: whereas the Women Chemists' Dining Club, organised by Martha Whiteley and Ida Smedley, never flourished and ceased to exist soon after the death of Whiteley, the Association of Women Pharmacists founded by Buchanan and Clarke lives on until the present day. The difference is reflected in the fact that it was the Royal Institute of Chemistry, not the Chemical Society, which handled professional issues for chemists. Thus, the Women Chemists' Dining Club was solely a social and networking body, while the goals of the Association of Women Pharmacists encompassed professional rights of women pharmacists and provided a continuing reason for its existence even when the generation of the founders had passed on.

Notes

1. Hunting, P. (1998). *A History of the Society of Apothecaries*. The Society of Apothecaries, London.
2. (a) Rawlings, F. H. (1984). Two 17th century women apothecaries. *Pharmaceutical Historian* **14**(3): 7; (b) Anon. (1999). Information please. *Pharmaceutical Historian* **29**(1): 3.
3. Note 1, Hunting, pp. 207–210.
4. Manton, J. (1965). *Elizabeth Garrett Anderson*. Methuen, London.
5. Note 1, Hunting, p. 230.
6. (a) Holloway, S. W. F. (1991). *Royal Pharmaceutical Society of Great Britain 1841–1991: A Political and Social History*. The Pharmaceutical Press, London; (b) Creese, M. (2005). How women pharmacists struggled for recognition before 1905. *Pharmaceutical Journal* **274**: 730.
7. Jordan, E. (1998). "The great principle of English fair-play": Male champions, the English women's movement and the admission of women to the Pharmaceutical Society in 1879. *Women's History Review* **7**(3): 381–409.
8. Burnby, J. G. L. (1990). Women in pharmacy. *Pharmaceutical Historian* **20**(2): 6. See also: Austen, J. (1961). Second period (1800 to 1841). *Historical Notes on Old Sheffield Druggists*, J. W. Northend Ltd., Sheffield.
9. Bernard, B. (1868). Pharmacy as an employment for women. *The Englishwoman's Review* (First series) **1**: 348.

10. Jordan, E. (2001). Admitting a dozen women into the society: The first women members of the British Pharmaceutical Society. *Pharmaceutical Historian* **31**: 18–27.
11. Cited in: Note 6(a), Holloway, p. 262.
12. Strauss, S. (1982). *Traitors to the Masculine Cause: The Men's Campaigns for Women's Rights*. Greenwood Press, Westport, Connecticut.
13. Celebrating women in pharmacy: Robert Hampson, www.rpsgb.org.uk/members/museum/nawpexhib/rhampson.html, accessed 19 May 2006.
14. (5 October 1872). Lady students (from Meeting of the Council, 2 October 1872). *Transactions of the Pharmaceutical Society* **3**: 268.
15. (9 November 1872). Election of members (from Meeting of the Council, 6 November 1872). *Transactions of the Pharmaceutical Society* **3**: 366.
16. (1872) Lady students (from Meeting of the Council, 4 December 1872). *Transactions of the Pharmaceutical Society* **3**: 456.
17. Sandford, G. W. (1 March 1873). Correspondence. *Pharmaceutical Journal* **3**: 698.
18. "H.L." (1 March 1873). Correspondence. *Pharmaceutical Journal* **3**: 699.
19. de Lancy, M. (29 March 1873). Correspondence: Pharmaceutical women. *Pharmaceutical Journal* **3**: 780.
20. Cited in: Note 6(a), Holloway, p. 263.
21. Cited in: Note 6(a), Holloway, p. 264.
22. Note 7, Jordan, p. 397.
23. Earles, M. P. (1965). The pharmacy schools of the nineteenth century. In Poynter, F. N. L. (ed.), *The Evolution of Pharmacy in Britain*, Charles C. Thomas, Springfield, Illinois, p. 88.
24. Hehner, O. (1912). Obituary: John Muter. *The Analyst* **37**: 76–80.
25. Anon. (15 October 1879). The Pharmaceutical Council: Ladies admitted as members. *Chemist and Druggist* **21**: 422.
26. (4 October 1879). Transactions of the Pharmaceutical Society: Meeting of the Council, Wednesday, 1 October 1879. *Pharmaceutical Journal* **9**: 265.
27. Spencer, R. K. (1899). Pharmacy as an employment for girls. *The Girl's Own Paper* **21**: 19.
28. Anon. (12 December 1891). Lady pharmacists. *Chemist and Druggist* **39**: 848.

29. Anon. (30 July 1892). Lady pharmacists. *Chemist and Druggist* **41**: 143.
30. "Xrayser." (4 July 1908). *Chemist and Druggist.*
31. Gertrude H. Wren, educated at North London Collegiate School, was the first woman to win the Pereira Medal, and that same year she also received the Redwood Research Scholarship. Wren was appointed Demonstrator at the School of Pharmacy, but terminated her career upon marriage in 1910. See Shellard, E. J. (August 1982). Some early women research workers in British pharmacy. *Pharmaceutical Historian* **12**(2): 2.
32. The following year, 1909, Grace Neve won the silver medal in the Periera competition. She was also the recipient of the 1909 Burroughs Research Scholarship. See Note 31, Shellard.
33. Anon. (9 July 1904). Pharmacy for women. *Pharmaceutical Journal* **73**: 84.
34. (a) Bedell, M. I. (16 July 1904). Women as pharmacists. *Pharmaceutical Journal* **73**: 107; (b) "Homo." (23 July 1904). Women as pharmacists. *Pharmaceutical Journal* **73**: 135; (c) Oliver, E. (30 July 1904). Women as pharmacists. *Pharmaceutical Journal* **73**: 167; (d) Spencer, R. K. (23 July 1904). Women as pharmacists. *Pharmaceutical Journal* **73**: 167; (e) Borrowman, A. T. (23 July 1904). Women as pharmacists. *Pharmaceutical Journal* **73**: 168; (f) "Homo." (7 January 1905). Lady assistants. *Chemist and Druggist* **66**: 26; (g) "Bismuth." (14 January 1905). Lady assistants. *Chemist and Druggist* **66**: 64; (h) "Darby." (28 January 1905). Lady assistants. *Chemist and Druggist* **66**: 181; (i) Weston, J. H. (11 February 1905). Lady assistants. *Chemist and Druggist* **66**: 258; and (j) "One who has a sister." (25 February 1905). Lady assistants. *Chemist and Druggist* **66**: 320.
35. Kurzer, F. (2007). George S. V. Wills and the Westminster College of Chemistry and Pharmacy: A chapter in pharmaceutical education in Great Britain. *Medical History* **51**: 477–506.
36. Walden, J. E. (1908). Girls as chemists. *Girl's Realm Annual* **10**: 395–398.
37. Jordan, E. (2002). "Suitable and remunerative employment": The feminisation of hospital dispensing in late-ninteenth century England. *Social History of Medicine* **15**: 429–456.
38. Forster, E. L. B. (12 August 1916). The ideal neighbourhood for the woman pharmacist. *Pharmaceutical Journal* **97**: 158.

39. Buchanan, M. E. (23 May 1908). The present position of women in pharmacy. *Pharmaceutical Journal* **80**: 675.
40. Hudson, B. (September 2004). The "petticoat peril" on the pavement: Women's pharmacy history in Lambeth. *Newsletter, National Association of Women Pharmacists* 8.
41. Cited in: Anon. (27 August 1955). Deaths: Miss A. T. Borrowman. *Pharmaceutical Journal* **224**: 155.
42. Anon. (15 December 1923). Personal items: Miss Agnes Thomson Borrowman. *Pharmaceutical Journal* **107**: 625.
43. Jones, D. M. (10 November 1959). Progress of women in pharmacy. *Chemist and Druggist* 42.
44. Symonds, S. (11 June 2005). An event great with possibilities. *Pharmaceutical Journal* **274**: 733.
45. (24 June 1905). Letters in brief: Women pharmacists. *Chemist and Druggist* **66**: 975.
46. Celebrating women in pharmacy: Hilda Caws, www.rpsgb.org.uk/members/museum/nawpexhib/hcaws.html, accessed 19 May 2006.
47. Celebrating women in pharmacy: Georgina Barltrop, www.rpsgb.org.uk/members/museum/nawpexhib/gbarltrop.html, accessed 19 May 2006.
48. (6 March 1909). Winter session of chemists' associations: Association presidents. *Chemist and Druggist* **74**: 377.
49. Women and pharmacy. *Information Sheet.* Museum of the Royal Pharmaceutical Society of Great Britain.
50. Cited in: Note 6(a), Holloway, p. 267.
51. (24 June 1911). Women and the vote. *Chemist and Druggist* **78**: 36.
52. "A Pharmacist." (11 March 1922). Letters to the editor: Women in pharmacy. *Pharmaceutical Journal* **108**: 208.
53. Celebrating women in pharmacy: Jean Kennedy Irvine, www.rpsgb.org.uk/members/museum/nawpexhib/jirvine.html, accessed 19 May 2006.
54. Mason, P. (11 June 2005). Activities of members after 1918. *Pharmaceutical Journal* **274**: 740.
55. (a) Buchanan, M. and Neve, A. (18 September 1926). The late Mrs. Clarke-Kerr. *Pharmaceutical Journal* **110**: 374; (b) Celebrating women in pharmacy: Isabella Clarke, www.rpsgb.org.uk/members/museum/nawpexhib/iclarke.html, accessed 19 May 2006.

56. Anon. (20 May 1905). Deaths: Minshull. *Chemist and Druggist* **66**: 786.
57. Celebrating women in pharmacy: Louisa Stammwitz, www.rpsgb.org.uk/members/museum/nawpexhib/lstammwitz.html, accessed 19 May 2006.
58. Shelland, E. J. (unpublished work). Chapter II, Some outstanding women pharmacists of the late 19th and the early 20th century. In *Women in Pharmacy*. We thank the Archivist of the Pharmaceutical Society for a copy of this manuscript.
59. Henry, T. A. (1950). Wyndham Rowland Dunstan, 1861–1949. *Obituary Notices of Fellows of the Royal Society* **7**: 62–81.
60. (a) Anon. (13 January 1940). Deaths: Buchanan. *Chemist and Druggist* **132**: 27; (b) Anon. (6 January 1940). Deaths: Buchanan. *Pharmaceutical Journal* **144**: 10.
61. Anon. (27 August 1955). Obituary: Miss A. T. Borrowman. *Pharmaceutical Journal* **175**: 155.
62. Borrowman, A. T. (30 July 1904). Letters to the editor: Women as pharmacists. *Pharmaceutical Journal* **73**: 168.
63. "W.P.W." (1928). Obituary notices of Fellows deceased: Arthur William Crossley, 1869–1927. *Proceedings of the Royal Society of London, Series A* **117**: vi–x.
64. Anon. (3 September 1955). An onlooker's notebook: Miss Borrowman. *Pharmaceutical Journal* **175**: 191.
65. Anon. (1941). Obituary. *Journal of the Royal Institute of Chemistry* (Pt. 1): 59.
66. Anon. (11 May 1918). Personal items: Miss Hope Constance Monica Winch. *Pharmaceutical Journal* **100**: 227.
67. Anon. (15 April 1944). Deaths: Winch. *Pharmaceutical Journal* **152**: 161.
68. Anon. (21 June 1969). Deaths — Hooper. *Pharmaceutical Journal* **202**: 714.
69. Celebrating women in pharmacy: Elsie Hooper, www.rpsgb.org.uk/members/museum/nawpexhib/ehooper.html, accessed 19 May 2006.
70. Wynne, W. P. (1927). Obituary notices: Arthur William Crossley. *Journal of the Chemical Society* 3165–3173.
71. (1920). Certificates of candidates for election at the ballot to be held at the ordinary scientific meeting on Thursday, 2 December 1920. *Proceedings of the Chemical Society* **117**: 95.

72. Johnson, C. A. (23 August 1956). Deaths: Corfield. *Pharmaceutical Journal* **237**: 224.
73. Anon. (1960). Obituary: Mrs. Kassner. *Journal of the Royal Institute of Chemistry* **84**: 44.
74. Anon. (4 August 1956). Figures in the pharmaceutical world. *Chemist and Druggist* **166**: 117.
75. Bradford, E. (30 August 1956). Deaths: Corfield. *Pharmaceutical Journal* **237**: 249.
76. Anon. (1983). Obituary: Lothian Short. *Pharmaceutical Historian* **13**(4): 4.
77. Anon. (22 October 1983). Deaths: Agnes Lothian Short. *Pharmaceutical Journal* **231**: 475.

Chapter 11

The Roles of Chemists' Wives

In the late 19th and early 20th centuries, it was argued by some that women who excelled academically should regard it an honour and a duty to devote their lives to knowledge in a similar manner to a religious vocation, thus excluding the possibility of family life. Alice Freeman, President of Wellesley, a women's college in Massachusetts, had pronounced in the 1880s: "Civilization rests upon dedicated [women's] lives, lives which acknowledge obligation not to themselves or to other single persons, but to the community, to science, to art, to the cause."[1]

For a male chemist, marriage was an expectation. Marriage for the male reduced the workload; whereas for a female chemist, marriage meant acquiring additional duties and responsibilities and probably the end of her career. A retired woman Lecturer at the University of Cambridge, when asked whether she regretted not marrying, responded that she would have been glad to marry had she only found someone who would have made a good wife.[2]

Not all of the women academics who gave up their occupations upon marriage did so willingly. The opposition to a married woman's continued employment was harsher during the interwar period (see Chap. 13); for example, Elsie Phare, upon applying in 1931 for a position in English at University College, Southampton, was asked if she was engaged to be married, and she answered truthfully that she was not.[3] She later became engaged to a classics lecturer, at which time she was accused of deceit and told she would be dismissed upon marriage. Phare noted: "We married nevertheless, and I was sacked; married women were unemployable, our income £200 per annum, and for several years we lived largely on lentils."[3]

The Amateur Assistant

Male chemists usually relied upon women for support, the best example being Frederick Donnan, Professor of Chemistry at University College, London (UCL), from 1913 until 1937.[4] It was Donnan's two sisters, Leonora (Nora) and Jane, who were indispensable to him. Nora kept house for Donnan for 38 years; while Jane was his Secretary at UCL, in later years also being responsible for the ICI Research Fellows there.

Kathleen Holland (Mrs. Lapworth)

More often it was the wife who undertook secretarial duties, and the three Holland sisters exemplified this role. Kathleen Holland was one of the three daughters (the others being Mina and Lily) of William Thomas Holland and Florence DuVal.[5] Each of the sisters married a renowned chemist: Mina married William Henry Perkin, Jr. in 1887[6]; Lily married her cousin, Frederic Stanley Kipping, in 1888[7]; and Kathleen married Arthur Lapworth in 1900.[8] Both Mina and Kathleen were childless, while Lily had four children.

In a fictionalized biography, authors Eugene Rochow and Eduard Krahé contend that each sister contributed to the success of her husband. There is strong evidence that it was true for Kathleen. According to Arthur Lapworth's obituarist, Kathleen assisted Lapworth with paperwork, acting as his secretary: "As already mentioned the valuable help of Mrs. Lapworth lightened the load of routine for several years."[9] Likewise, in Perkin's biography, Mina's support was acknowledged: "... those of us whose privilege it was to know the Perkins more intimately are aware how greatly Perkin was helped by her constant sympathy with his ideals and with his work, and with what unobtrusive care she shielded him as far as possible from the minor worries and irritations of life."[10]

Lapworth was joined in 1922 by Robert Robinson to form what was known as the Lapworth–Robinson "golden age" of organic chemistry at Manchester. In a review of the Chemistry Department, George Burkhardt noted that the spouses of Lapworth and Robinson played a prominent role:

> An unusual feature of the life of the School of Chemistry at this time was the presence in it of the wives of both professors. Mrs. Lapworth as her husband's secretary helped him greatly with the detail of the heavy administrative responsibilities in the department. Mrs. (later Lady) Robinson [see below], as an Honorary Research Fellow, worked on long-chain acids in the professor's laboratory. Both took a kindly and active interest in staff and students.[11]

Winifred Beilby (Mrs. Soddy)

An example of an untrained spouse performing laboratory work is provided by Winifred Moller Beilby,[12] spouse of Frederick Soddy (see Chap. 7). Born on 1 March 1885 in Edinburgh, Scotland, she was the only daughter of George Thomas Beilby, an industrial chemist. When she was 14 years old, the family moved to Glasgow, where she was educated at home by a governess.

It was through her father that she met Soddy, for George Beilby had helped finance Soddy's research work. This assistance included a fundraising campaign in 1904 to raise money for equipment needed for radium research. Soddy and Beilby became engaged in 1906 and married in 1908.[13] Beilby wanted a quiet, early morning wedding so that they could go off quickly on their mountain-climbing honeymoon.

Beilby became interested in Soddy's work and helped him considerably, her research on gamma rays emitted by radioactive atoms being published in 1910. Soddy himself commented: "She did quite a bit of work for me at Glasgow. You can read it all in my communications. It was a tedious investigation and she stuck at it, like Marie Curie, I used to say."[14]

Upon Soddy's move to Oxford, Beilby seems to have given up research activity to devote her time to the required social life and spending her free moments painting and gardening. Towards the end of 1935, Beilby fell ill, and she died suddenly in August 1936. Quite possibly, her death was as a result of exposure to lengthy doses of gamma rays during her research work. Her passing came as a grievous shock to Frederick Soddy, and it was the cause of his resignation from his position at Oxford.

Grace Toynbee (Mrs. Frankland)

The most acknowledged of the amateur assistant spouses was Grace Coleridge Toynbee.[15] The youngest of nine children of Joseph Toynbee, pioneer ear specialist, and Harriet Holmes, Toynbee was born on 4 December 1858 in Wimbledon. Initially home-schooled, she also studied in Germany and then spent 1 year at Bedford College, although little else is known of her early life. In 1882, she married Percy Frankland,[16] who was then a Lecturer in Chemistry at the Normal School of Science, South Kensington (later part of Imperial College; see Chap. 3). Subsequently, they had a son.

Percy Frankland, together with his father, Edward Frankland, had set up a private analytical laboratory in London, and it was here that Toynbee commenced her scientific career. Though both father and son were chemists, they had a strong interest in bacteriological problems, particularly those relating to human health. Toynbee's first publication, co-authored with Percy Frankland, was on microorganisms in air.

Percy Frankland was appointed in 1888 as Professor of Chemistry at University College, Dundee, and the institution's magazine, *The College*, reported: "Any notice of Dr. Frankland would be incomplete without some reference to Mrs. Frankland, who has worthily aided and seconded him in his scientific career, and whose achievements are about as well known in the world of science as are those of her husband."[17]

At Dundee, Toynbee continued her research, which included studies of the reactions involved in bacteriological fermentation as a means of synthesising chemical compounds. In 1894, they moved to Birmingham, where Frankland had been appointed Professor of Chemistry. It was during her time at Birmingham that Toynbee added her signature to the 1904 petition for the admission of women to the Chemical Society (see Chap. 2).

In 1894, the Franklands co-authored a book, *Micro Organisms in Water: Their Significance, Identification and Removal*[18]; then Toynbee, on her own, wrote a more popular book, *Bacteria in Daily Life,*[19] published in 1903. It would seem that after the move to Birmingham, Toynbee focused more on science journalism than laboratory research. She was elected a Fellow of the Royal Microscopical Society in 1900, and was one of the first 12 women scientists admitted to the Linnean Society in 1904.

With Frankland's retirement in 1919, they moved to Scotland, where Toynbee died on 5 October 1946, Frankland dying 3 weeks later. Frankland's obituarist, William Garner, noted:

> Probably in few cases have husband and wife collaborated so effectively and enthusiastically in both research and professional work. On one occasion it was said, "Many women in the past have helped their husbands, but Percy Frankland is the first man who had the chivalry to admit it."[20]

The Woman Chemist as Professor's Wife

In the socially restricted world of academic research, university-educated women tended to marry university-educated men — particularly their supervisors or colleagues. Such a fate was mourned by Flora Garry, a graduate of the University of

Aberdeen, in her poem, *The Professor's Wife*, in which the final stanza of the poem reads:

> 'Learnin's the thing', they wid say,
> 'To gie ye a hyste up in life.'
> I wis eence a student at King's.
> Noo I'm jist a professor's wife.[21]

Louisa Cleaverley (see Chap. 2) and Elison Macadam (see Chap. 3) were two of the women chemists who terminated their careers upon marriage to a chemistry Professor, another being Annie Purcell Sedgwick.[22] Sedgwick, the eldest daughter of Lieutenant-Colonel Sedgwick, was born in Cork on 4 December 1871. She had completed the science Tripos at Girton in 1893, and then she joined Norman Collie[23] at UCL as a research student in chemistry. It was at UCL that she met James Walker.[24] Walker was appointed Professor of Chemistry at University College, Dundee, in 1894, and they married in 1897. Sedgwick authored three publications on organic chemistry, but abandoned research in 1905. Walker accepted a Professorship at the University of Edinburgh in 1908, and it was in Edinburgh that Sedgwick died on 7 September 1950.

A small, but slowly increasing number of these women continued with their research after marriage, joining the ranks of the "academic couples."[25] However, the women usually found themselves given secondary status and were overshadowed by their male partners. The writer Sharon McGrayne has commented on the prevalence of the spousal problem:

> The academic countryside was littered with scientific couples studying botany, genetics, chemistry, and other sciences. Professor husbands and their low-ranking, low-paid [or unpaid] wives often worked together for decades ... the women were generally low-level instructors, lecturers, or research

> assistants while their male partners were professors with tenure. A woman had a permanent position only as long as her relationship with the man continued. In case of divorce or disaffection, the woman could be fired.[26]

An interesting contrast of marital outcomes is provided by the Badger sisters, May Badger[27] and Louise Midgely Badger. May continued with her chemistry career after marriage (see Chap. 5), working with Frank Sinnatt[28] at the Faculty of Technology, University of Manchester. Louise Badger had followed in her sister's path, also completing a B.Sc. (Tech.) in Applied Chemistry from the Faculty of Technology, but in 1912. In 1920, Louise was noted as being a researcher at the University of Manchester, then marrying Sinnatt a year later. Unlike May, Louise apparently discontinued research upon marriage, the sole mention of her in Sinnatt's obituary being: "Sinnatt left a widow, Louise Midgely Badger, who, herself a University trained chemist, had been his devoted companion for over 20 years, …."[28]

Marjory Wilson-Smith (Mrs. Farmer)

Once married, some of the women chemists are simply noted as having assisted their husbands, but without having any formal position. The career of Marjory Jennet Wilson-Smith[29] provides an example. Wilson-Smith was born on 19 May 1899, the daughter of a doctor, and was educated at Bath High School and Cheltenham Ladies' College. She entered Royal Holloway College (RHC) in 1918, graduating with a B.Sc. in chemistry in 1921.

From 1921 until 1925, Wilson-Smith was a Demonstrator in Chemistry at the London School of Medicine for Women, though she spent the year 1922/1923 as a research student in the organic chemistry department of Imperial College, another of the research students being Ernest Farmer.[30] In 1925, she accepted an appointment with RHC, but resigned in 1930, as the *College Letter* reported:

> The last loss to be recorded is of Miss M. J. Wilson-Smith (now Mrs Farmer) for two years Assistant Lecturer and Demonstrator in the Chemistry Department. Shortly after the end of term we heard of her engagement to a fellow-scientist at Imperial College. Her marriage followed within a few weeks. Those who remember her record for industry will note, perhaps only with faint surprise, that, according to Rumour, Mrs Farmer has for the last two months done "not a stroke" of scientific work.[31]

According to the obituarist of Farmer, Wilson-Smith worked with Farmer after marriage: "During this period, in 1930, he married Marjorie Wilson-Smith, she was one of his research students and continued for many years to assist him with his researches."[30] However, none of Farmer's later publications list Wilson-Smith as a co-author. We will never know how many more women chemists continued to work upon marriage, but without acknowledgement or recognition. Helena Pycior, Nancy Slack, and Pnina Abir-Am have named such women the "invisible" assistants.[32]

The Independent Wife-Chemist

A small number of married women chemists forged a path independent of their nonchemist spouse. In Chap. 4, we described the life and work of Margaret Seward, who continued university teaching even after marriage and the birth of her son. In crystallography (see Chap. 9), Dorothy Crowfoot and Kathleen Yardley were both very fortunate in their choice of academic husbands who took over much of the parenting role to allow their famous wives to continue research. In biochemistry, Helen Archbold (see Chap. 3) was also a true "independent."

The research relationship between the married biochemists Dorothy Moyle (see Chap. 8) and Joseph Needham could better be described as "autonomous" rather than independent, as their research overlapped to a significant extent. We will see below that the descriptor "autonomous" also fits the research profile of Gertrude Walsh with Robert Robinson. Finally, Muriel Wheldale's collaboration with Huia Onslow (see Chap. 8) might be considered farther along the collaborative spectrum as "semi-autonomous." Nevertheless, most active women spouse-chemists were part of a collaborative couple.

The Collaborative Couples

In contrast to the "invisible" assistants, the contributions in other relationships were widely acknowledged, particularly in co-authorship of publications. Here, we will see four examples of such collaborative couples.

Mildred Gostling (Mrs. Mills)

Daughter of George James Gostling, dental surgeon and pharmaceutical chemist, and Sarah Abicail Aldrich, Mildred May Gostling[33] was born on 15 December 1873 in Stowmarket, Suffolk. She attended Royal Holloway College from 1893 to 1897, obtaining an honours B.Sc. in Chemistry. From there, she spent 1899 to 1900 at Newnham College as a Bathurst research student, working with Henry Fenton[34] on carbohydrate chemistry and co-authoring four publications. In Fenton's obituary, one of Gostling's contributions was described:

> With Miss M. M. Gostling he found that various carbohydrates, in particular fructose, gave a purple colour when dissolved in ether and treated with hydrogen bromide, and this proved to be due to an oxonium salt of a yellow crystalline compound which could be thus obtained in considerable quantity and was shown to be ω-bromomethylfurfuraldehyde.[35]

Gostling returned to Royal Holloway College in 1901, becoming Demonstrator in Chemistry. During her time at Newnham, she had met William Mills.[36] In 1902, Mills was appointed as Head of the Chemistry Department of Northern Polytechnic Institute in North London, and they must have remained in contact, for in 1903 Gostling resigned from her position to marry Mills. She subsequently joined Mills' research group, which also included Alice Bain (see Chap. 7), while Sibyl Widdows (see Chap. 4) had a publication co-authored with him. Gostling's extensive research on dinaphanthracene was published in 1912. She had one son and three daughters — two of the daughters, Mildred Marjorie Mills and Sylvia Margaret Mills, became students at Newnham; while the third, Judith Isobel Emily Mills, became, later in life, a staff member at Newnham. Gostling died on 19 February 1962.

Ellen Field (Mrs. Stedman)

Not to be confused with E. Eleanor Field, Ellen Field together with her spouse, Edgar Stedman,[37] formed the most equal chemistry partnership of those we have studied. Of course, in those days, equal work was not reflected in equal status. Field was born on 29 October 1883 at Greenwich, Kent, the daughter of William Frederick Field, a labourer, and Ellen Bobey.[37] She studied towards a chemistry degree at Goldsmiths' College.

Goldsmiths' College, New Cross, London, had originally been the Royal Naval School, and it had been purchased by the Worshipful Company of Goldsmiths in 1889 and opened as the Goldsmiths' Company's Technical and Recreative Institute in 1891. The intention of the Institute, funded by the Goldsmiths' Company, was the "promotion of the individual skill, general knowledge, health and well being of young men and women belonging to the industrial, working and poorer classes," and between 1907 and 1915 included evening classes towards a B.Sc. (London).[38]

After completing her B.Sc. degree in 1910, Field was hired as a Demonstrator in Chemistry at Goldsmiths'; while in her spare time, she commenced research with George Barger,[39] the Head of the Department of Chemistry. Her work in 1912 on the blue compounds of iodine formed with starch and other substances provided her first publication. While she was a Demonstrator, a young undergraduate student, Edgar Stedman, regularly visited her house for help with his chemistry studies.

Field followed Barger to Royal Holloway College. Then, the following year, when Barger was appointed to the Department of Biochemistry and Pharmacology of the Lister Institute, Field undertook research at Bedford College, resulting in her being awarded an M.Sc. (London). In 1919, when Barger accepted the Chair of Chemistry in Relation to Medicine at the University of Edinburgh, he invited Field to join him, which she did. Barger found the workload of the new Department overwhelming, and Field proposed that he hire the newly graduated Stedman as Lecturer. Barger concurred, and hired Stedman. Field, equally qualified to Stedman, was not appointed Lecturer for another 8 years.

With Stedman having been offered a position, he and Field married. This was to prove an ideal match, as Stedman's obituarist, H. J. Cruft, commented:

> She [Field] had a stimulating, often mildly caustic, sense of humour and was never to be overawed by Edgar with his single-mindedness. She never hesitated to express her views, even if they might be unpopular — an attribute Edgar admired. They enjoyed many activities together such as hill walking, and later gardening and above all the sense of achievement when experiments went well and produced a positive result in the laboratory.[40]

While Stedman worked towards a Ph.D., Field continued research with Barger, studying plant alkaloids and their derivatives, resulting in two more publications: one in 1923 and the other in 1925. With Stedman's Ph.D. granted, Field shifted to research with her husband, studying hæmocyanin between 1925 and 1928 and co-authoring five publications on the subject. As Cruft noted: "This work was primarily carried out by Ellen and for which she had received several grants...."[41]

In the 1930s, Field and Stedman undertook research on acetylcholine and choline esterase, co-authoring 10 publications. Then, in the 1940s, they entered a more controversial study: that of the chemical composition of the cell nucleus, arguing in a series of papers that DNA was not a major constituent of chromosomes. By the 1950s, Field and Stedman had become isolated from the life of the Department, as Cruft described: "He [Stedman] and Ellen at this stage, to a certain extent, retired from the banter of the department coffee room and had their lunch of fruit and sandwiches in two leather chairs in Edgar's almost Dickensian office overlooking the Medical Quadrangle."[42]

Sadly, during the summer of 1953, Field suffered a massive coronary thrombosis and was severely ill for some months. For the rest of her life, she was essentially bedridden, the boredom being a major problem for her. She died on 22 September 1962.

Mary Laing (Mrs. McBain)

Of the four, we have the least information on Mary Evelyn Laing. The daughter of Scottish parents, David Laing and Jeannie McGeogh, she earned a B.Sc. and an M.Sc. in Chemistry at Bristol, the latter qualification being completed in 1919.[43] That year, she was hired as a "scientific collaborator" with the physical chemist James McBain, resulting in nine publications on the structure of soap solutions — five on her own, and the remainder with McBain (one also being co-authored with Millicent Taylor; see Chap. 5).

McBain accepted a position at Stanford University, California, in 1926, and Laing moved with him, becoming a

Research Associate in Chemistry. As Eric Rideal, one of McBain's obituarists, commented:

> On 1 January 1929 he [McBain] was married again to a woman of remarkable ability, Mary Evelyn Laing, who had been on the Science Faculty at Bristol and was a research associate in Chemistry at Stanford. ... They had one son, John Keith McBain. During all his years at Stanford in the interludes of foreign travel and in India, Evelyn McBain was there to help him with his daily need, whilst she herself, apparently without effort, maintained a high standard of social activity in the home.[43(b)]

Laing authored eight papers from Stanford, four under her name alone, which was now given as M. E. L. McBain.

Catherine Tideman (Mrs. LeFèvre)

The highest-profile spouse-collaborator was Catherine (Cathie) Gunn Tideman. Born in Glasgow on 1 November 1909, she went to UCL to study chemistry,[44] and it was there that the physical organic chemist Raymond LeFèvre noticed her:

> Among those taking organic chemistry during 1928–1929 was one who seemed always cheerful, lively, ready with relevant comments on current or local affairs, full of conversational topics, of repartee, energy and vigour; a hockey player of enthusiasm (who had once knocked unconscious an opponent through an accidental head-to-head collision), a tireless dancer, a rider of horses (her grandmother had owned a riding school in Glasgow), not excessively teetotal but convivial with most of the women and men contemporary with her at U.C.L.[45]

A romance soon developed, with the marriage taking place in 1931. Tideman became LeFèvre's long-term research colleague, coauthoring papers and reviews under the name of C. G. LeFèvre.

Much of the careful laboratory work was performed by Tideman, such as that which led to the validation of the molar Kerr constant as a numerical criterion for the discrimination between often closely related molecular configurations.[46] Tideman's son, Ian, was born in 1938; and shortly afterwards, Tideman became a Demonstrator in Chemistry at UCL and also taught chemistry at Queen's College, Harley Street (see Chap. 4). She had a second child, Nicolette, in 1940. As a result of Tideman's research contributions at UCL, she was awarded a D.Sc. (London) in 1960.

The LeFèvres departed for the University of Sydney in 1946, and there, in addition to continuing research on the Kerr effect, Tideman developed broader interests. She became involved with social issues, particularly drug dependency and the drug problem in New South Wales. Later, she became involved in forensic science issues, becoming the first woman elected to the Council of the Australian Academy of Forensic Sciences.[47] She was also heavily involved in programmes to encourage women into science. Tideman died in Sydney on 9 March 1998.

A Contrast in Women Organic Chemists

For much of their later lives, the organic chemists Robert Robinson[48] of Oxford and Christopher Ingold[49] of UCL were in professional competition. The dispute, and the resulting enmity between them, has been described elsewhere[50]; what is usually overlooked is that they each had a wife who continued research after marriage.

Their spouses, Gertrude Walsh (Mrs. Robinson) and Hilda Usherwood (Mrs. Ingold), combined a dedication to organic chemistry with marriage and motherhood. However, the nature of the research relationships with their spouses were quite different, and this almost certainly arose from their personality differences. Walsh, outgoing and strong-willed, developed her own clear research field, even though she was involved in some joint projects. Usherwood, on the other hand, "painfully shy," became very much a collaborator of Ingold. Usherwood's contributions merged into Ingold's productivity, and this contributed to a lower profile of Usherwood compared with Walsh.

Both, however, faced the common problem for married women chemists at the time: that of being perceived as appendages of their husbands with no opportunity or expectation of an independent scientific career. It was their shared love of organic chemistry that allowed them to spend countless hours at the research bench, unpaid for their efforts, while each raising a family.

Gertrude (Gertie) Walsh (Mrs. Robinson)

Of all the chemist spouses, Gertrude Maud Walsh[51] was the most renowned, being deemed worthy of obituaries in two scientific journals. Born on 6 February 1896 in Winsford, Cheshire, Walsh was the youngest daughter of Thomas Makinson Walsh, a representative of a coal merchant, and Mary Emily Crosbie. She was educated at Verdin Secondary School, Winsford, and then obtained her B.Sc. degree in 1907 and M.Sc. in 1908 from Owens College, Victoria University of Manchester. After graduation, she undertook organic chemistry research with Chaim Weizmann (later President of Israel) and taught chemistry at the Manchester High School for Girls.

Robinson, an organic chemist at Manchester, organised a walk for the "Tea Club" researchers through Cheedale and Cowdale in Derbyshire, among the participants being Ida Smedley (see Chap. 2) and Walsh. Torrential rain discouraged the participants from Robinson's subsequent invitations to hikes, except one: Walsh. As Robinson recalled: "The mass invasion of Derbyshire dales was not repeated, but Miss Walsh and I devoted many weekends to rock climbing on the millstone grit edges of the Peak District."[52]

A common bond in mountaineering turned into a personal relationship, and in 1912 they were married. They had two children: a daughter, Marion (born 1921); and an invalid son, Michael (born 1926). Robinson's first post was at the University of Sydney, Australia, where Walsh was employed as an unpaid Demonstrator in Organic Chemistry, the University banning the paid employment of married women. Walsh was very popular among the students, as this commentary notes:

> Mrs. Professor Robinson, who acts as a honorary demonstrator for her husband, and says the things to students that he is not game to say, is an amazing amateur fire brigade, all by herself. Fires are of frequent occurrence in organic chemistry and a rug is specially provided with which to extinguish them. When a blaze arises Mrs. Robby hurls herself into the air, grabbing the rug as she flies, falls upon the conflagration, puts it out regardless of singed hair and eyelashes, and without even waiting to regain her breath, rounds upon the luckless student until he wishes the fire had consumed him too![53]

During her years in Sydney, Walsh developed elegant methods of synthesising both saturated and unsaturated fatty acids. However, her most prolific period was during the 1930s, after Robinson was appointed Professor of Chemistry at the University of Oxford. It was here that she began studying chemical genetics, specifically plant pigments (see section on Muriel Wheldale, Chap. 7).[54] J. C. Smith, who chronicled life in the Oxford chemistry department, commented:

> R. R. [Robinson] took over Perkin's office and laboratory, wisely allotting two benches to his wife. Mrs Robinson was probably the hardest worker of us all. She arrived at 9 am…., left at 1 pm and then put in a 2.15–7 pm afternoon.[55]

Walsh drove around Oxford in her Standard 12 car, as Smith recalled: "Mrs. (Gertrude) Robinson was not a good driver; she not only forgot to wear her glasses, but she sat so low that she peered *through* the steering-wheel. One was often surprised at the positions in which her car stood outside the D. P. [Dyson Perrins Laboratory]."[56]

As Walsh and Robinson did not have a crushing machine to extract the flower pigments, they put the plants in a pile, covered them with boards, and then drove one of their cars backwards and forwards over them.[57] The extractions were usually performed using ether (ethoxyethane) as a solvent, which was often contaminated by highly explosive peroxo compounds. One particular explosion involved Walsh, as Smith remembered:

> In 1938 there was a shattering explosion on the top floor while I was demonstrating below. Senior assistant ("Gertie") Miller and I rushed upstairs and found Mrs. Robinson on the floor, her face covered with blood and pieces of glass. Miller, the unflappable, quickly did all that we dared and quickly took her to the Oxford Eye Hospital. Fortunately there was no glass in the eyes, but many pieces were removed from the face.[58]

Walsh was one of the few wife-assistants to gain some recognition for her work. Robinson, in his autobiography, gave her due credit:

> Further reference to the work of G.M.R. and to incidents in her later life will be found in Volume II of these Memoirs [which was never published]. Nevertheless, I cannot postpone an acknowledgement of the very great help which she gave me at all stages of my career. Looking back, I can see how she subordinated her interests to mine, was always such a ready collaborator in scientific work, and cheerfully followed my chief vacation activity, namely mountaineering.[59]

Robinson's support for Walsh was very public, as Smith recalled, in the context of Walsh's pioneering and often overlooked synthesis of oleic acid[51(b)] and higher fatty acids:

> R. R. admired her confident and cheerful attitude to research-work (and to the rather frequent breakages of glassware). He became embarrassingly enthusiastic over her successes. ... R. R.

startled one of his demonstrators dining with him in Magdelen College, by suddenly asking, "Have you heard about my wife's fatty acids?"[60]

During the Second World War, Walsh became involved in antibiotic synthesis, and she was the first chemist to synthesise a penicillin analogue with antibiotic character. Unfortunately, little was published on her research in this field.[51(b)]

As a woman, Walsh was excluded from the meetings of the Alembic Club — the chemistry society at Oxford (see Chap. 6). So, it was not surprising that she exhibited a determination for sexual equality, as illustrated by the following account from Louis Hunter of the University of Leicester:

> ... A memorable event in this period was the visit of the British Association for the Advancement of Science to Leicester in 1933.... It was the custom at that time for the Sectional Dinner to be open only to male members, and this custom fell particularly unfairly on women chemists. With characteristic energy and hospitality, Mrs. (later Lady) Robinson arranged a dinner party at the same time as the Sectional Dinner, in the same hotel and with the same menu, to which she invited other women chemists as well as wives of the sectional officers and of other prominent members. This bold action finally broke down the practice of restricting the dinner to men, and at all meetings of the British Association subsequent to 1933 the dinners of Section B have been graced by the presence of ladies.[61]

In 1953, the University of Oxford conferred on Walsh an honorary M.A. degree. Tragically, she died of a heart attack the following year.

Hilda Usherwood (Mrs. Ingold)

Whereas Walsh's productivity was maintained throughout her career, that of (Edith) Hilda Usherwood diminished over time.[62]

Usherwood, the eldest daughter of Thomas Scriven Usherwood, a teacher of engineering, and Edith Howarth, showed academic brilliance from an early age, winning a scholarship to North London Collegiate School.[63] She graduated from Royal Holloway College in 1920 with a first-class honours degree in chemistry, and then undertook postgraduate research with Whiteley (see Chap. 3) at Imperial College (IC), from where she obtained a Ph.D. in 1922.

It was at IC that she met Christopher Ingold.[63] One of Ingold's co-workers, Frank Dickens,[64] recalled that, in 1923, Ingold gassed himself with phosgene and it was Usherwood who first rushed to the rescue.[65] Dickens added: "As far as we were concerned that was the first intimation that we had that an engagement was in the offing and, of course, they were married shortly afterwards."[65] Their honeymoon was spent in Snowdonia, though Usherwood preferred walking to climbing.

Marriage did not initially diminish Usherwood's research, and she received the D.Sc. in 1925, resulting from a series of papers on tautomerism published under her name alone. However, a move by Ingold in 1924 did affect her work. Ingold was appointed Professor of Organic Chemistry at the University of Leeds, while Usherwood served as unpaid demonstrator there, in addition to undertaking research. At the time, socialising was the major occupation of the other faculty wives, particularly morning visits to each other, a pastime the shy Usherwood avoided when possible.

From 1925 onwards, most of Usherwood's publications were joint with Ingold. An incident in 1926 illustrates how closely Usherwood and Ingold worked as a team. Robinson's group had repeated Ingold's research on the nitration of *S*-methylthioguiacol, obtaining very different results. On being told this, Usherwood and Ingold went into the laboratory and worked for 72 hours without a break, repeating the disputed reactions, and rapidly submitted a follow-up paper declaring the correctness of Robinson's findings.[66]

Usherwood's productivity declined with the birth of their first child, Sylvia, in 1927 (followed by Keith in 1929 and Dilys in 1932), though she did co-author a paper on the first synthesis of benzyl fluoride. In 1930, Ingold was appointed as Professor of Chemistry at UCL. Usherwood continued to participate in research work at UCL as honorary Research Associate, hiring a "mother's helper" and a cleaner to help her cope with family responsibilities. During the 1930s, she took on the additional task of aiding Jewish refugees from Germany and Austria.

In 1939, UCL's chemistry department was evacuated, mostly to Bangor, and the remainder — including Ingold and Usherwood — to Aberystwyth. No secretarial staff was posted to Aberystwyth, so Usherwood became *de facto* Administrative Assistant and, after teaching herself typing, proceeded to become Departmental Secretary as well. At Ingold's request, Usherwood finally received a salary — for secretarial work.

Following their return to London in 1946, Usherwood was formally appointed as Administrative Assistant to UCL's Chemistry Department, with her last research publication appearing in 1947. The chemistry historian, Joan Mason, recalled overlapping with Usherwood at UCL from 1951 to 1956:

> I found her not especially friendly to me, and realized after a while that she and her friends were anti-Oxbridge as centres of privilege, and extended this to people like me! I counted myself as politically on the left as she was; also we had been to the same school (North London Collegiate, where I overlapped with her daughter Sylvia). But I can well understand the resentment of non-Oxbridge people, particularly high calibre ones such as Ingold, to effortless assumptions of superiority by Oxbridge.[67]

This was a resentment which was probably reinforced by the personal feud of the Ingolds with the Robinsons at Oxford.

After Ingold's death in 1971, Usherwood continued to be active at UCL, including serving as President of the UCL Chemical and Physical Society during the academic year

1976/1977. She died peacefully at the age of 90, leaving her body to the anatomy department of University College, London.

Commentary

For those women chemists who married chemists, the wife often continued to participate in some way, for example, as a secretarial assistant or laboratory worker. There were very few "independents"; while "autonomous" and "semi-autonomous" relationships, where the spouses worked on related or overlapping areas, also occurred. The most common situation was for the woman to become assistant to her husband. To find such occurrences, we had to identify postmarriage publications by the husband that had the wife's name as a co-author. A second avenue was to find mention of the role of the wife in the husband's obituary, where, in a few cases, the husband readily acknowledged the importance of the spousal role. The question remains how many more were "invisibles" — women who contributed significantly to their husband's chemical career, but whose name was not acknowledged in publications nor by the husband's obituarist.

Notes

1. Barnard, J. (1964). *Academic Women*. Pennsylvania State University Press, University Park, Pennsylvania, p. 207.
2. Apter, T. (1993). *Professional Progress: Why Women Still Don't Have Wives*. Macmillan, Basingstoke.
3. Phare, E. E. (26 February 1982). From Devon to Cambridge, 1926: Or, mentioned with derision. *Cambridge Review* 149.
4. Freeth, F. A. (1957). Frederick George Donnan, 1870–1956. *Biographical Memoirs of Fellows of the Royal Society* **3**: 23–39.
5. A fictionalized account of their lives was published as: Rochow, E. G. and Krahé, E. (2001). *The Holland Sisters: Their Influence on the Success of Their Husbands, Perkin, Kipping and Lapworth*. Springer, Berlin.

6. "J.F.T." (1931). Obituary notices: William Henry Perkin — 1860–1929. *Proceedings of the Royal Society of London, Series A* **130**: i–xii.
7. Challenger, F. (1950). Frederick Stanley Kipping, 1863–1949. *Obituary Notices of Fellows of the Royal Society* **7**: 182–219.
8. Robinson, R. (1947). Arthur Lapworth, 1872–1941. *Obituary Notices of Fellows of the Royal Society* **5**: 554–572.
9. Note 8, Robinson, p. 566.
10. Greenway, A. G., Thorpe, J. F. and Robinson, R. (1932). *The Life and Work of Professor William Henry Perkin*. The Chemical Society, London, p. 36.
11. Burkhardt, G. N. (1954). Schools of chemistry in Great Britain and Ireland–XIII The University of Manchester (Faculty of Science). *Journal of the Royal Institute of Chemistry* **78**: 448–460.
12. Letter, Hilda M. Beilby (spouse of Hubert Beilby, Winifred's brother) to Muriel Howorth, 30 October 1957, Soddy collection, Oxford University Archives (OUA).
13. Howorth, M. (1958). Chapter 15: The perfect marriage. *Pioneer Research on the Atom: The Life Story of Frederick Soddy*, New World Publications, London.
14. Note 13, Howorth, p. 168.
15. (a) Cohen, S. L. (2004). Frankland [née Toynbee], Grace Coleridge. *Oxford Dictionary of National Biography*, Oxford University Press, http://www.oxforddnb.com/view/article/62321, accessed 2 October 2004; (b) Creese, M. R. S. (1998). *Ladies in the Laboratory: American and British Women in Science 1800–1900*. Scarecrow Press, Lanham, Maryland, pp. 150–151.
16. Garner, W. E. (1945–1948). Percy Faraday Frankland. *Obituary Notices of Fellows of the Royal Society* **5**: 697.
17. Anon. (16 March 1889). Professor Percy F. Frankland. *The College, Magazine of University College, Dundee* **1**(3): 87.
18. Frankland, P. and Frankland, P., Mrs. (1894). *Micro Organisms in Water: Their Significance, Identification and Removal*. Longmans, Green & Co., London.
19. Frankland, P., Mrs. (1903). *Bacteria in Daily Life*. Longmans, Green & Co., London.
20. Note 16, Garner, p. 700.

21. Garry, F. (1995). *Collected Poems.* Gordon Wright Publishing, Edinburgh, pp. 16–17.
22. Melville, H. W. (2004). Walker, Sir James (1863–1935). *Oxford Dictionary of National Biography*, Oxford University Press, http://www.oxforddnb.com/view/article/36693, accessed 10 Sept 2007.
23. Baly, E. C. C. (1943). John Norman Collie, 1859–1942. *Obituary Notices of Fellows of the Royal Society* **4**: 329–356.
24. Kendal, J. (1932–1935). Sir James Walker. *Obituary Notices of Fellows of the Royal Society* **1**: 537–549.
25. Dyhouse, C. (1998). *No Distinction of Sex? Women in British Universities, 1870–1939.* UCL Press, London, pp. 162–163.
26. McGrayne, S. (1993). *Nobel Prize Women in Science: Their Lives, Struggles, and Momentous Discoveries.* Birch Lane Press, New York.
27. Anon. (1954). Obituary notes: May Badger Craven (Mrs.). *Journal of the Royal Institute of Chemistry* **78**: 105.
28. Egerton, A. C. (1943). Frank Sturdy Sinnatt, 1880–1943. *Obituary Notices of Fellows of the Royal Society* **4**: 429–445.
29. Royal Holloway College, student records.
30. Gee, G. (1952). Ernest Harold Farmer, 1890–1952. *Obituary Notices of Fellows of the Royal Society* **8**: 159–169.
31. (November 1930). *College Letter, Royal Holloway College Association* 16.
32. Pycior, H. M., Slack, N. G. and Abir-Am, P. G. (1987). Introduction. In Pycior, H. M., Slack, N. G. and Abir-Am, P. G. (eds.), *Creative Couples in the Sciences*, Rutgers University Press, New Brunswick, New Jersey, p. 12.
33. (a) White, A. B. (ed.) (1979). *Newnham College Register, 1871–1971.* Vol. I: 1871–1923 (2nd ed.). Newnham College, Cambridge, p. 146; (b) Creese, M. R. S. (1998). *Ladies in the Laboratory: American and British Women in Science 1800–1900.* Scarecrow Press, Lanham, Maryland, p. 265; (c) Mann, F. G. (2004). Mildred May Gostling (1873–1962) in Mills, William Hobson (1873–1959). *Oxford Dictionary of National Biography*, Oxford University Press, http://www.oxforddnb.com/view/article/35030, accessed 4 November 2004.
34. "W.H.M." [Mills, W. H.] (1930). Obituary notices: Henry John Horstman Fenton. *Proceedings of the Royal Society of London, Series A* **127**: i–v.

35. Note 34, "W.H.M.," p. v.
36. Mann, F. G. (1960). William Hobson Mills, 1873–1959. *Biographical Memoirs of Fellows of the Royal Society* **6**: 200–225.
37. Cruft, H. J. (1976). Edgar Stedman, 12 July 1890–8 May 1975. *Biographical Memoirs of Fellows of the Royal Society* **22**: 528–553.
38. Firth, A. E. (1991). *Goldsmiths' College: A Centenary Account.* Athlone Press, London.
39. Dale, H. H. (1940). George Barger, 1878–1939. *Obituary Notices of Fellows of the Royal Society* **3**: 63–85.
40. Note 37, Cruft, p. 533.
41. Note 37, Cruft, pp. 535–536.
42. Note 37, Cruft, p. 543.
43. (a) Taylor, H. (1956). Obituary notices: James William McBain, 1883–1953. *Journal of the Chemical Society* 1918–1920; (b) Rideal, E. K. (1952–1953). James William McBain 1883–1953. *Obituary Notices of Fellows of the Royal Society* **8**: 529–547.
44. Aroney, M. J. and Buckingham, A. D. (1988). Raymond James Wood Le Fèvre. *Biographical Memoirs of Fellows of the Royal Society* **34**: 375–403.
45. Autobiographical notes by R. J. W. Le Fèvre, cited in: Note 44, Aroney and Buckingham, p. 383.
46. Davies, M. (1987). R. J. W. Le Fèvre 1905–1986. *Chemistry in Britain* **23**: 564.
47. Bright Sparcs Biographical Entry: Le Fèvre, Catherine Gunn (1909–1998), http://www.asap.unimelb.edu.au/bsparcs/biogs/P003055b.htm, accessed 25 November 2007.
48. Todd, L. and Cornforth, J. W. (1976). Robert Robinson, 13 September 1886–8 February 1975. *Biographical Memoirs of Fellows of the Royal Society* **22**: 414–527.
49. Shoppee, C. W. (1972). Christopher Kelk Ingold, 1893–1970. *Biographical Memoirs of Fellows of the Royal Society* **18**: 348–411.
50. Saltzman, M. D. (1980). The Robinson–Ingold controversy: Precedence in the electronic theory of organic reactions. *Journal of Chemical Education* **57**: 484–488.
51. (a) Baker, W. (1954). Lady Robinson (Obituary). *Nature* **173**: 566–567; (b) Simonsen, J. L. (1954). Obituary notices: Gertrude Maud Robinson 1886–1954. *Journal of the Chemical Society* 2667–2668.

52. Robinson, R. (1976). *Memoirs of a Minor Prophet: 70 Years of Organic Chemistry*, Vol. 1. Elsevier, Amsterdam, p. 45.
53. Cited in: Williams, T. I. (1990). *Robert Robinson: Chemist Extraordinary*. Clarendon Press, Oxford, pp. 34–35. By permission of Oxford University Press.
54. Scott-Moncrieff, R. (1981). The classical period in chemical genetics: Recollections of Muriel Wheldale Onslow, Robert and Gertrude Robinson and J. B. S. Haldane. *Notes and Records of the Royal Society of London* **36**: 125–154.
55. Smith, J. C. (n.d.). *The Development of Organic Chemistry at Oxford. Part II: The Robinson Era 1930–1955*, unpublished, p. 5. H. Anderson Memorial University is thanked for a copy of this manuscript.
56. Note 55, Smith, p. 46.
57. Williams, T. I. (1990). *Robert Robinson: Chemist Extraordinary*. Clarendon Press, Oxford, p. 67.
58. Note 55, Smith, p. 43.
59. Note 52, Robinson, pp. 55–56.
60. Note 55, Smith, p. 8.
61. Hunter, L. (1955). Schools of chemistry in Great Britain and Ireland–XV The University College of Leicester. *Journal of the Royal Institute of Chemistry* **79**: 15–16.
62. Rayner-Canham, M. F. and Rayner-Canham, G. W. (1999). A tale of two spouses. *Chemistry in Britain* **35**: 45–46.
63. (a) Leffek, K. T. (1997). *Sir Christopher Ingold: A Major Prophet of Organic Chemistry*. Nova Lion Press, Victoria, British Columbia, pp. 58–73, 231–233 (and other pages); (b) Leffek, K. T. (2004). (Edith) Hilda Ingold (1898–1988) in Ingold, Sir Christopher Kelk (1893–1970). *Oxford Dictionary of National Biography*, Oxford University Press, http://www.oxforddnb.com/view/article/34102, accessed 26 Nov 2007.
64. Thompson, R. H. S. and Campbell, P. N. (1987). Frank Dickens. 15 December 1899–25 June 1986. *Biographical Memoirs of Fellows of the Royal Society* **33**: 188–210.
65. Cited in: Note 63(a), Leffek, p. 55.
66. Note 63(a), Leffek, pp. 100–101.
67. Personal communication, e-mail, J. Mason, 14 October 1999.

Chapter 12

Women Chemists and the First World War

The First World War changed everything for women. Universities lost most of their male staff and students, resulting in women being hired as lecturers and leaving women students in the majority. A Newnham student, M. G. Woods (Mrs. Waterhouse), remarked: "the preponderance of women in the classrooms made the salutation of 'Gentlemen' more ridiculous than ever."[1]

The War and Young Women

Up until 1914, women had entered university as part of the "New Woman" movement.[2] Marriage and child-raising were still the preference of the majority, while the unmarried could still be ladies of leisure. A career for a middle-class girl as a necessity was a new dimension. An incredibly blunt article in 1915 by The Editor of *The Girl's Realm* laid out the new reality:

> In the old Victorian days the woman who did not marry was supposed in some way or other to find a niche in the family, where she made herself "generally useful" but did not work, where she was dependent on the money left her by her father, or the generosity of her relations. Those days are passing. Every woman, as well as every man, has in times like these to show "the reason of her existence."[3]

The Editor pointed out that the enormous number of deaths on the battlefields, particularly among the middle-class,[4] made the possibility of marriage by the teen girl readers an unlikely prospect: "I address these remarks principally to girls who are facing life at the opening of their careers. The war has made us all face reality, and, not least of all, the women of the world."[3] The article ended by promising that *The Girl's Realm* would continue to provide information on careers open to young women.

For girls who had taken chemistry at secondary school, their laboratory skills were in demand. A note in *Our Magazine*, the student magazine of North London Collegiate School for Girls, reported: "Gwendolen A. James is working in an analytical laboratory for a year before going on with her Science Degree work. Margarethe Mautner is analysing drugs in a Chemical Laboratory before taking up her medical studies."[5]

At university, women students were uncertain of what was expected of them. At Armstrong College, Newcastle, the women students heard of the need for women munitions workers:

> Perhaps the most thrilling event of the term from the feminine point of view has been the appeal for women munition-makers (workers). After twenty-four distracted and heart-searching hours; after some of us had obtained permission to go and were feeling heroic; after some of us had decided to stay and were feeling "desper'te mean"; and after some of us had defied our obdurate parents and guardians in the names of Patriotism and Legal Majority, and were feeling Shelleyesque, we were informed that our highest patriotic duty was to complete our education![6]

The call for women students to continue their education had come from H. A. L. Fisher in early 1917, representing the Board of Education: "My own view is that for the present women students at the universities should continue their academic courses until such time as they may be called up by the branch of the National Service Department presided over by Mrs. Tennant."[7]

The fear of being "called up," as Fisher had indicated in his communication from the Board of Education, was very real for the Newcastle women students:

> Nowadays the most serenely stay-at-home and selfish of us have been under Zeppelin fire two or three times. When Professor Mawer announced that he had a communication from the Government to read to us the involuntary start and look of dismay by which everyone betrayed the impression that conscription for women had come at last, and the general relief which succeeded his explanation that the Government only meant to warn us not to repeat military information. Yet even among the women, numbers have made it already in the interests of their country. The latest to go are Miss Oliver, Miss Clayburn and Miss Cummings, late candidates for Honours in Chemistry, now chemists to munition works at Middlesboro'.[8]

The war impacted the women at university and also the newly graduated. For example, Somerville College recorded the war work of its former students, three of whom were undertaking chemistry-related duties[9]:

1. Arning, Dorothy — Asst. Loading Chemist, No. 21 Natl. Filling Factory, Coventry.
2. Cudworth, Mary — Analytical Chemist for Mssrs. Richardson & Co., manufacturers of agricultural fertilisers.
3. Markham, Claudia E. — Science Mistress, Bude County School, substitute for a man. Then, Chemist to British Resorcin Manufacturing Co.

In the following sections, we will illustrate the wide range of war duties performed by women chemists.

Women Chemists in Organic Synthesis

At the start of the First World War, nearly all fine chemicals, such as pharmaceuticals and anæsthetics, were produced in Germany and Austria. Thus, it was of the utmost priority to develop a homegrown organic chemical industry. This task took time. Arthur Schuster, Secretary to the Royal Society, wrote to all university chemistry departments: "It has been thought desirable to enlist the voluntary services of the many chemical laboratories connected with the educational institutions of the country, to meet the urgent demand for the immediate supply of certain of these drugs, mainly organic products."[10]

The academic chemical laboratories were turned into miniature factories. It was noted in a Royal Society report of 21 May 1917 that the narcotic β-eucaine was being produced by the University of Glasgow, Bedford College, and Imperial College; arabinose by the University of St. Andrews, University of Leeds, Imperial College, and Borough Polytechnic Institute; and atropine by the University of Liverpool and the University of Dundee.[11] In a more detailed report of December 1917, the production of β-eucaine by Millicent Taylor (see Chap. 5) and her students at Cheltenham Ladies' College was noted.[12]

Women chemists made up a significant proportion of the scientific workforce. Fortunately, the Women's Work Collection of the Imperial War Museum (IWM) has a significant amount of documentary evidence on the wartime women scientists. This useful material was compiled in 1919 by Agnes Ethel Conway of the Women's Work Subcommittee of the IWM. Conway circulated a questionnaire to universities and industries informing them that the Committee was compiling a historical record of war work performed by women for the National Archives. In particular, Conway added: "they [the Subcommittee] are anxious that women's share in scientific research and in routine work should not be overlooked ..."[13] Enough replies were received to provide a sense of the breadth of employment of scientifically trained women during the war.

Had it not been for the significant increase in the number of women taking chemistry degrees during the first 15 years of the 20th century (see Chap. 13), it is apparent from the replies to Conway's enquiries that the British war machine would have faced a significant shortage of chemists. Fortunately, there was a pool of qualified women chemists ready and willing to do their part towards the war effort. As illustration, Kennedy Orton, Professor of Chemistry at the University College of Wales, Bangor (see Chap. 7), commented in a report that: "The demand for young women who have received a training in Chemistry, both for educational and professional work, has increased greatly during the past Session. The demand is far in excess of the supply ..."[14]

The Chemistry Department of the University of Sheffield was one of the major participants in organic synthesis (see Chap. 5). William Palmer Wynne assembled a team of six women chemists, including Emily Turner, Dorothy Bennett, and Annie Mathews, to synthesise β-eucaine.[15] However, the most noteworthy production was that at Imperial College, in the group run by Martha Whiteley (see Chap. 3).[16] Whiteley's seven assistants were all women and included Frances Micklethwait (see Chap. 3), who received an M.B.E. for her contributions to the war effort.[17]

The women chemists at St. Andrews worked on the production of synthetic drugs and bacteriological sugars, research on explosives and poison gases, and the improvement of industrial processes. The individuals who undertook this research sacrificed their own career advancement, as this report sent to Conway noted:

> It should be stated that the whole of this work was unpaid from Government sources, the workers receiving only their University salaries, in cases where they were members of staff, or the value of their Scholarships, if they held any such distinctions. Not only so, but the demand for chemists throughout the war was continuous, so that the

> workers who remained with me gave up many opportunities for professional advancement. I mention these facts as an index of public spirit with which these women gave their services, services which have not received any public recognition.[18]

All of these contributions by women chemists were "invisible." No mention was made of them in an article of 1917 by the Institute of Chemistry on the work accomplished by wartime (male) chemists.[19] More recently, in a lengthy review of the mobilisation of civilian chemists during the First World War,[20] the only mention of a woman chemist is: "And so, too, left without an offer from the [Reserved Occupations] Committee, the only woman of whom we have record, Margaret Turner, of the Chemical Laboratories at Aberystwyth."[21]

Yet Margaret K. Turner of the University College of Wales, Aberystwyth, did participate in the war effort. Turner had been hired at Aberystwyth at the beginning of the war with the rank of Demonstrator, and, in a history of the Department, it was reported that: "Miss Margaret Turner and, later, J. B. Whitworth took up posts as demonstrators and under the direction of the professor carried out large-scale preparations of intermediates for certain essential drugs."[22]

Turner wrote a stirring letter to the War Committee, volunteering for additional duties:

> I was one of the workers in the preparation of diethylamine some weeks ago and should be very glad to hear of any further help I could give. I can put all my time and energy at your service for the next 6 weeks, and am anxious to know whether the few helpers down here could not be allowed to contribute further to the needs of the country? I should be much obliged if you would inform me whether there is any other preparations we can make, as I, for one, am willing and eager to give up all ideas of holiday while there remains so much to be done.[23]

Phyllis McKie

Of all the women chemists, Phyllis Violet McKie[24] of the University College of Wales, Bangor, one of Orton's protégées, seems to have been the most productive during the war period. McKie was part of the team at Bangor producing paraldehyde.[25] In addition, she devised a new method for the preparation of the explosive tetranitromethane for the Ministry of Munitions, and she studied methods of preparation of saccharin and vanillin for war purposes.[26] Orton reported back to the War Committee:

> Miss McKie's investigations have been mainly concerned with a study of the methods of preparation of 'materials offensive and defensive' for the Department of Explosives Supply. Her work had been particularly successful as in the two cases investigated, new and far superior methods of preparation have been discovered and examined.[25]

McKie was on born 18 July 1893, the daughter of William McKie, Clerk at the Penryhyn Quarry Office. She was educated at the County School for Girls, Bangor, and entered the University College of Wales, Bangor, in 1912. McKie completed her B.Sc. in 1916, and was awarded an M.Sc. by research on the basis of her war work. A total of 12 publications resulted from the different directions of war research which she initiated at Bangor.

By 1920, McKie had moved to University College, London (UCL), before being appointed Demonstrator at Bedford College in 1921, the same year that she was awarded a Ph.D. (London) in chemistry. She then moved to Lady Margaret Hall, Oxford, on a Research Fellowship for the 1925/1926 academic year.

In 1926, McKie was appointed Head of the Chemistry Department at the Maria Grey Training College, where another Bangor alumna, Alice Smith, had been a Lecturer from 1914 until 1917 (see Chap. 7). Then, in 1929, McKie made yet another move, this time to become a Lecturer in Chemistry at Westfield College, London. The biographer of Westfield College, Janet

Sondheimer, noted: "In 1929 when chemistry teaching started the only place suitable was the attic... however in 1935 the lecturer Dr. Phyllis McKie, was able to transfer her department to the hut, left vacant by the removal of botany to the new laboratory."[27] When McKie left Westfield in 1943 to become Principal of St. Gabriel's Training College, chemistry lapsed at Westfield for several years. McKie remained Principal at St. Gabriel's until her retirement in 1956.

Ruth King

Picric acid was another explosive used in the war; and from 1917 to 1919, Ruth King[28] was assigned as a Wartime Research Worker at Chiswick Laboratory, Department of Explosive Supply, Ministry of Munitions, to research the optimum conditions for its synthesis.

King was born on 13 May 1894 and was educated at Uxbridge County School. She graduated with a B.Sc. (London) in chemistry from East London College in 1914 and was granted an M.Sc. in 1918, like McKie, on the basis of her wartime research. At the conclusion of the war, she was appointed Lecturer in Organic Chemistry at the University College of South West England (later the University of Exeter) as well as warden of Hope Hall, the women's residence.[29]

Like other women chemists given academic positions, King was assigned a high proportion of the teaching duties. In fact, from 1919 to 1945, she was the only organic chemist in the department at Exeter, where she remained until 1955 when she took early retirement to move to Canada to help care for her aged mother. Her mother, however, died while King was crossing the Atlantic; but, undaunted, King continued to Vancouver, where she obtained a post as Lecturer at the University of British Columbia, and remained there until 1961.

Women Chemists and the War Gases

We mentioned that Grace Leitch of Armstrong College (see Chap. 5) and Helen Gilchrist of the University of St. Andrews (see Chap. 7) had undertaken research on mustard gas. They were not the only woman chemists to work in this hazardous field. In addition to having pioneered the university production of pharmaceuticals, Martha Whiteley (see Chap. 3) of Imperial College (IC) also performed research on lachrymatory war gases. She recounted her wartime experiences in a speech at a luncheon at Royal Holloway College:

> Those were very exciting days, for our laboratories were requisitioned by the Ministry of Munitions who kept us busy analysing and reporting on small samples collected from the battlefields or from bombed areas at home. These included flares, explosives and poison gases. It was my privilege to examine the first sample of a new gas, used to such effect on the front that our troops had to evacuate from Armentières as it was reputed to cause blisters; it was called Mustard Gas. I naturally tested this property by applying a tiny smear to my arm and for nearly three months suffered great discomfort from the widespread open wound it caused in the bend of the elbow, and of which I still carry the scar. Incidentally, when shortly afterwards we were carrying on a research for a method of manufacturing the gas, my arm was always in requisition for the final test.[30]

Winifred Hurst (Mrs. Wright)

Winifred Grace Hurst[31] was one of Whiteley's team producing pharmaceuticals at Imperial College; however, her particular claim to fame is her subsequent work on devising protection against the German war gases. Born on 6 June 1891, Hurst was educated at King Edward VI School for Girls, Birmingham (KEVI). She obtained a B.Sc. (London) from Bedford College and then undertook research at IC between 1914 and 1918 on synthetic drugs for the navy, work for which she received the Diploma of Imperial College (DIC, equivalent to an M.Sc.). In 1918, Wright was attached to the Anti-Gas Department of the Ministry of Munitions, where she worked on gases for wartime defence. Together with the rest of the staff, she had no alternative other than to try the gases out on herself. Her superior, Colonel Harrison, died as a result of such experiments. Fortunately, she survived.

From 1919 until 1949, Hurst was affiliated to the Chemistry Department at Battersea Polytechnic. During that period, she married A. John Wright and had three daughters. In the Second World War, when bombing prevented her reaching her laboratory, she set up her chemical equipment on the kitchen table of her home in Esher, Surrey. She commented: "after that, I kept one table for cooking and one for chemistry."[31(b)]

In 1949, Hurst emigrated to South Africa to join her husband, who was a consultant to an electricity company on high tension lines. He suggested that she keep a record of the wild flowers destroyed by the erection of the pylons, this new direction leading to her book on *The Wild Flowers of Southern Africa*.[32] She started research on South African plants at the University of Natal, receiving her Ph.D. in 1954, which resulted in newspaper headlines as a "degreed grandmother"[31(b)] of 62. The studies of plants led her back into the field of chemistry, this time an analysis of alkaloids obtainable from various plants. Hurst died on 8 September 1978, aged 87 years.

Women Chemists in Industry

The vast majority of women in the wartime chemical industry were unskilled, simply working on specific synthesis tasks and following exact recipes to produce the enormous quantities of TNT, nitroglycerine, ammonium nitrate, and ammonium perchlorate that were required by the explosives industry.[33] In fact, the proportion of women in chemical factories was as high as 88%. The wages paid were usually two thirds of the male rate for the same task, as the argument was commonly made that three women were equivalent to two men.

The work was hard, often very dangerous, and it frequently led to debilitating effects from the toxic chemicals. TNT poisoning was among the worst health problems, the sufferers being called "canary girls" as a result of the yellow colour of their skin. The medical personnel had orders that only those most seriously affected by chemical poisoning were to be given time off from work; thus, many women suffered permanent health damage and some died as a result of their continued exposure to TNT.[34]

Many of these factories were enormous. The largest of all was the Gretna Explosives Factory in Scotland, a 10-mile-long complex for the synthesis of cordite, which was largely operated by women workers. The working conditions were extreme: some of the women developed lung damage from working with open vats of concentrated mineral acids; while others became comatose on a daily basis from ether vapour, having to be dragged outside until they recovered. Yet life in the factory was not totally negative: most of the women revelled in the camaraderie of the workplace and, for working-class women, the well-balanced, nutritious meals served in the works canteens were better food than many had eaten before the war.[35]

Women as Industrial Analytical Chemists

The majority of women chemists entered the analytical field. Women were probably more accepted for this work, as repetitious and exacting analyses were considered compatible with women's talents. Throughout the war, there was a demand for analytical chemists: some to determine purities of explosives and of their precursors; and others, the purities of pharmaceuticals.[36] Even graduating high school women chemistry students were taken on for the war effort, as was mentioned above.

A crucial task was to analyse samples of the iron and steel used in the production of, for example, ships and tanks. It is not surprising, then, that Sheffield, the centre of the British steel industry, became the focus for training of women in chemical analysis. In a response to Conway's questionnaire (see above), Fred K. Knowles of the Faculty of Metallurgy at the University of Sheffield noted that when the war started, men in the analytical and research laboratories of the industry were barred from joining the armed forces due to the essential nature of their occupation[37]; by the Autumn of 1916, however, the demand for "cannon fodder" became so great that even these individuals were drafted. With the situation drastically changed, women now became essential. Knowles continued:

> In these laboratories there is a large amount of routine repetition work which can bc carried out by semi-trained assistants, as distinct from chemists and physicists. To meet this emergency, special one month Intensive Courses for women were started in the Metallurgical Department of the Faculty of Applied Science, University of Sheffield: the aim being to give a training in accurate weighing, filtration, titration, general manipulation and calculations. At the end of the Course those students who passed an Examination in the rapid determination of the elements: — carbon, silicon, manganese, sulphur, phosphorus, readily found remunerative employment. The Classes commenced on the 6th November 1916, and continued practically for two full University years: during this time 96 women students entered for this work.[37]

Sheffield also provided specialised courses in other areas. For example, six women were trained as analysts for coke oven laboratories.[38]

Some of the steel companies welcomed the women analysts. The Chief Supervisor of the Women's Welfare Department of Thos. Frith & Sons Ltd. of Sheffield, J. H. A. Turner, wrote to Conway to inform her that 4 women had worked in the research laboratory and 16 in the general laboratory at the company, primarily on the analysis of iron and steels and in microphotography.

He added: "I understand that this Firm was one of the first (if not the first) in the Country to employ women at such work and the results have been quite satisfactory to the Heads of the two Laboratories."[39]

Not all companies were effusive in their praise of women chemists. William Rintoul of Nobel Explosives Company in Ayrshire reported to Conway: "Only routine work was entrusted to women. Our experience agrees with the generally accepted view that, in the main, women are unsuitable for the control and carrying out of research work unless under strict [male] supervision."[40]

Women chemists were employed as analysts at the National Physical Laboratory (NPL). Their work, too, was mainly in the analysis of iron and steel samples for the Admiralty. It is noticeable, though, that 10 of the 12 Junior Assistants at the NPL were female; while all of the Assistants, the Senior Assistants, and the Supervisor were male.[41] The reports listing women's contributions, such as that of the NPL, provide only names and assigned duties.

Two Glasgow women chemists were employed directly by the Admiralty as analytical chemists. Ada Hitchens (see Chap. 6) was assigned to work in the Admiralty Steel Analysis Laboratories, while Ruth Pirret (see Chap. 7) became a wartime researcher on marine engine boiler corrosion for the British Admiralty.[42]

Women in Biochemistry

The biochemist Dorothy Jordan Lloyd, researcher with F. Gowland Hopkins at Cambridge (see Chap. 8), was also given a specific task. On the outbreak of war, the Medical Research Committee assigned her to study culture media for meningococcus, one of the anærobic pathogens involved in trench diseases, and the causes and prevention of "ropiness" in bread.[43] Jordan Lloyd was one of several women with a background in biochemistry who were enlisted in the war effort.

Annie Homer

The expertise of Annie Homer[44] was considered so vital to the war effort that she was brought back from Canada. Born on 3 December 1882 in West Bromwich, Homer was educated at KEVI. She studied chemistry at Newnham from 1902 until 1905 and, like so many other women, obtained "mailboat degrees" (see Chap. 6) from Dublin. From 1907 to 1910, Homer was an Assistant Lecturer and a Demonstrator in Physical Science at Newnham, then a Demonstrator in Chemistry from 1910 to 1914, being awarded a D.Sc. in 1913. Homer was also a signatory of the letter to *Chemical News* in 1909 (see Chap. 2). Between 1907 and 1913, she authored 13 papers on organic synthesis, many involving the Friedel–Crafts reaction.

In 1914, Homer moved to Canada, becoming a Medical Research Fellow and Demonstrator in Biochemistry at the University of Toronto, and an Assistant Chemist at the Dominion Experimental Farm, Ottawa. At the beginning of the war, Homer was promoted to Assistant Director of the Antitoxin Laboratories, Toronto, and it was because of her expertise that she was brought back to Britain for the special work of reorganising the commercial production of antitoxins to meet war demands. In fact, her methods for the manufacture of high-grade commercial antitoxins were adopted in many parts of the world. Her appointment was as Assistant at the Lister Institute, though much of her research was accomplished at the Physiological Institute, University College, London. Despite completely changing her field from organic chemistry to biochemical toxicology, her publication rate never diminished, with 24 research papers appearing under her name between 1914 and 1920.

In the 1920s, Homer abandoned this part of her career, devoting her life instead to securing the development of oil and potash and other mineral resources in Palestine. Then, during the Second World War, she was involved in the search for oil in Palestine. Homer died on 1 January 1953.

Other War Work by Women Chemists

We have already discussed the work of May Leslie on the improvement of nitric acid synthesis (see Chap. 5), first at H. M. Factory, Litherland, Liverpool, then at H. M. Factory in Penrhyndeudraeth, North Wales.[45] Isabel Hadfield was consigned to research chemical problems relating to aeronautics (see Chap. 2). In addition, we know that towards the end of the war, Millicent Taylor was appointed to H. M. Factory, Oldbury, as a research chemist.[46] It seems likely that other explosives factories also employed women chemists, but they were not documented.

Other aspects of the war effort involved women scientists. For example, the women science students at King's College for Women undertook research on the manufacture of optical and laboratory glass.[47] This was carried out on behalf of the Glass Research Committee for the Ministry of Munitions.

Ruby Caroline Groves[48] was an Assistant in the Metallurgical Research Branch of the Royal Arsenal, Woolwich, during the war. She was born in 1891 and educated at KEVI. Entering the University of Birmingham, she graduated with a B.Sc. in chemistry in 1914 and an M.Sc. in 1915. At the end of the war, Groves was appointed as an Assistant Chemist in the Mineral Resources Department of the Imperial Institute, from which she published three papers on clays. She was promoted to Senior Assistant in 1929, dying on 14 December 1944, aged 53 years. Her sister, Ida Mary Groves, was also a chemist[49]; she, too, obtained her B.Sc. in chemistry from the University of Birmingham, and was subsequently employed as an Assistant Chemist at Birmingham General Hospital.

Lovelyn Eustice (Mrs. Bickerstaffe)

One of the women chemists with the widest government employment was Lovelyn Elaine Eustice.[50] Eustice was born on 6 August 1894 in Southampton, daughter of John Eustice,

Professor of Engineering and Vice-Principal of University College, Southampton, and Evelyn Margaret Gay. She was educated privately and then went to University College, Southampton, and Girton, subsequently being granted a B.Sc. (London) in 1917.

During the war years, Eustice was first a Chemist at the Government Rolling Mills, Southampton, in 1917, then moved to the Air Ministry in London in the same position in 1918. From 1919 to 1922, she was a Research Chemist for the Admiralty, Royal Naval College, Greenwich, before becoming a Science Mistress at Dudley High School in 1923. That same year, she married chemist Robert Bickerstaffe, and they spent the years 1927 to 1934 living in New York. Eustice became Director of the Research Bureau, Encyclopedia Britannica Co., in 1936, a position she held until 1941. She died on 10 August 1943.

Hilda Judd

Another former colleague of Martha Whiteley at Imperial College, Hilda Mary Judd,[51] daughter of John Wesley Judd, Professor of Geology at the Royal College of Science (later IC), undertook wartime research for the Silk Association. Judd obtained her B.Sc. (London) at IC in 1904. She stayed on at IC as a Research Chemist until 1906, co-authoring a series of publications on the synthesis of nitrogen-containing organic compounds. During this period, she was the college's representative on an expedition to Africa, as the College Magazine, *The Phoenix*, reported:

> It is with considerable relief that we welcome back to the College Miss Judd, our representative in the British Association Expedition to South Africa. Considering the dangers and difficulties through which she has passed, and the arduous work of attending all the meetings, she is looking remarkably well, and appears much better for the change.[52]

In 1906, Judd was appointed as Lecturer in Chemistry and Physics at Goldsmiths' College. She relinquished the position in 1916, returning to IC to undertake war work for the Silk Institute, silk being a highly important material in the war, particularly for the envelopes of observation balloons. With the war ended, she obtained a post as Researcher for the Food Investigation Board of the Department for Scientific and Industrial Research (DSIR). Two publications resulted from her studies, one in 1919 and the other in 1920.

Dorothea Hoffert (Mrs. Bedson)

Another researcher with the Food Investigation Board of DSIR was Dorothea Annie Hoffert.[53] Hoffert was born on 29 January 1893 in Ealing, the daughter of Hermann Henry Hoffert, H.M. Inspector of Secondary and Technical Schools, and Annie Ward. She attended Croydon High School (a GPDSC school), then Manchester High School for Girls, before entering Girton in 1910 to study chemistry. In 1913, she transferred to the University of Manchester, receiving a Dip. Ed. in 1914.

From 1914, Hoffert was a Junior Science Mistress at Bede School for Girls, Sunderland; then, in 1916, she was "requisitioned" for the war effort as a Research Worker under the Food Investigation Board of the DSIR at the City and Guilds of London Institute, Department of Technology. At the same time, Hoffert was engaged in research on dopes and varnishes for aeroplanes. She continued working in these roles after the war, receiving the silver medal in her final exam on Painters' Oils, Colours and Varnishes in 1920. That year, with a grant from DSIR, she returned to Cambridge to continue her chemistry studies.

From 1922 to 1926, Hoffert was a Research Assistant to Ida Smedley (see Chap. 2) at the Lister Institute; and during that 4-year period, she co-authored four publications with Smedley on carbohydrate and fat metabolism. In addition, she published two papers under her own name, one in *Chemistry and Industry* on an empirical rule for substitution in benzene derivatives. It was at the Lister Institute that she met Samuel Phillips Bedson,[54] and after their marriage in 1926 she discontinued research. She had three children.

Traditional Roles for Women

Some women chemists and biochemists were assigned to more traditional women's roles. The bacterial biochemist, Marjory Stephenson (Chap. 8), spent the war as a nurse.[55] She left her research position at University College, London, to join the British Red Cross in France and then Salonika. In Salonika, she was in charge of a nurses' convalescent home and also had responsibilities for invalid diets. She was mentioned in dispatches in 1917 and was awarded an M.B.E. for her war work.

Jesse Slater

Another women scientist initially given nursing duties was Jesse Mable Wilkins Slater.[56] Born on 24 February 1879, the daughter of John Slater, an architect, and Mary Emily Wilkins, she was educated at South Hampstead High School (a GPDSC school). Slater initially attended Bedford College, then transferred to Newnham as a Gilchrist Scholar in 1899. She excelled in both chemistry and physics, obtaining a B.Sc. (London).

Slater was invited to work with J. J. Thomson[57] at the Cavendish Laboratory in Cambridge, where she studied the decay products of thorium from 1903 to 1905. That year, she accepted a position as Science Teacher at KEVI, staying there until 1909, when she took up an offer of Science Teacher at Cheltenham Ladies' College. In 1913, she returned to Newnham, this time as Assistant Lecturer in Physics and Chemistry, being promoted to Lecturer in 1914.

Slater obtained leave from Newnham to undertake war-related duties, and, for the first 3 years, she was a part-time nurse. She was then called for full-time duty as a radiographer at British military hospitals in France, and later held the rank of Officier de l'Instruction Publique with the French army.[53] At the end of the war, she resumed her position at Newnham, where she stayed until 1926. Slater died on 25 December 1961.

Commentary

With few exceptions, the end of the war resulted in the termination of employment for women chemists. The government closed the explosives factories, while the male chemists returned from their war duties and reoccupied their former faculty and research positions. The respondent to Agnes Conway from the Sheffield Steel Company of Thos. Firth noted: "On the signing of the Armistice most of the women were replaced by returning soldiers, but two [of 16] in the General Laboratory have become so proficient that their services have been retained."[35]

The women chemists with specialised training stood the best chance of survival. For example, according to Knowles' letter to Conway, the graduates of the metallurgical analysis course at Sheffield seemed to survive:

> That women have been an undoubted success in this branch of industry, is proved by the fact that notwithstanding so many of the men (who are now demobilized) have resumed duty, a large proportion of the women who desired to stay on have retained their positions to the present time.[33]

As part of the final chapter, we will revisit the issue of the employment of women chemists in the post-First World War era.

Notes

1. Brooke, C. N. L. (1993). *A History of the University of Cambridge, Volume IV 1870–1990*. Cambridge University Press, Cambridge, p. 332.

2. Jordan, E. (1999). *The Women's Movement and Women's Employment in Nineteenth Century Britain*. Routledge, London.
3. The Editor (1915). The war and women: The influence of the world-conflict on women's status and work. *The Girl's Realm* **17**: 45–46.
4. Winter, J. M. (1989). *The Experience of World War I*. Oxford University Press, Oxford.
5. Anon. (1916). Pupils leaving. *Our Magazine: North London Collegiate School for Girls* 93.
6. "De Virginibus." (1916). *The Northerner: The Magazine of Armstrong College* **16**(2): 39.
7. Fisher, H. A. L. (1917). Duty of women students. *The Serpent* **6**: 52.
8. "De Virginibus." (1917). *The Northerner: The Magazine of Armstrong College* **17**(2): 29.
9. Anon. (1917). *Somerville Students' Association Thirtieth Annual Report and Oxford Letter* 2–33.
10. Draft Letter, Arthur Schuster, Royal Society, to all U.K. Universities, undated, November 1914, Royal Society Archives, Cmb 28.
11. Statement regarding position of drug work, 21 May 1917, Royal Society Archives.
12. Report presented to the Royal Society Council, Sectional Chemical Committee, December 1917, Royal Society Archives.
13. Conway, A. E. *Women's War Work Collection*. Imperial War Museum, London (WWWC-IWM).
14. Orton, K. J. P. (1918). *Annual Reports of the Heads of Department, 1917–1918*. University College of Wales, Bangor, pp. 8–9; T. Roberts, Archivist, University of Wales, Bangor, is thanked for supplying this information.
15. Chapman, A. W. (1957–1958). The early days of the chemistry department. *By Product: Journal of the University of Sheffield Chemistry Department* **11**: 2–5.
16. Letter, M. A. Whiteley to A. E. Conway, 3 October 1919, WWWC-IWM.
17. Burstall, F. H. (1952). Frances Mary Gore Micklethwaite (1868–1950). *Journal of the Chemical Society* 2946–2947.
18. Letter, (illegible signature) to War Committee, 23 August 1919, Royal Society Archives.

19. The Registrar (1917). Chemists in war. *Proceedings of the Institute of Chemistry* 29–34.
20. MacLeod, R. (1993). The chemists go to war: The mobilization of civilian chemists and the British war effort, 1914–1918. *Annals of Science* **50**: 455–481.
21. Note 20, MacLeod, p. 475.
22. James, T. C. and Davis, C. W. (1956). Schools of chemistry in Great Britain and Ireland–XXVII The University College of Wales, Aberystwyth. *Journal of the Royal Institute of Chemistry* **80**: 568–574.
23. Letter, M. K. Turner to War Committee, 31 August 1915, Royal Society Archives.
24. University of Wales, Bangor, Archives, Student Records. E. W. Thomas, Archivist, is thanked for supplying a copy of this document.
25. Letter, K. J. P. Orton to War Committee, 4 December 1915, Royal Society Archives.
26. Anon. (1920). Certificates of candidates for election at the ballot to be held at the Ordinary Scientific Meeting on Thursday, 2 December. *Journal of the Chemical Society* **117**: 92.
27. Sondheimer, J. (1983). *Castle Adamant in Hampstead: A History of Westfield College 1882–1982*. Westfield College, University of London, London, p. 102.
28. We thank the following for information on King: Nye A., Archivist, Queen Mary College, London; Personal communication, letter, K. Schofield, University of Exeter, 17 February 1998; Personal communication, letter, Erwin Wodarczak, Records Analyst/Archivist, University Archives, University of British Columbia, Vancouver, 3 March 1998; Anon (1922). Certificates of candidates for election at the ballot to be held at the ordinary scientific meeting on 15 December 1921. *Journal of the Chemical Society* **118**: 103.
29. Britton, H. T. S. (1956). Schools of chemistry in Great Britain and Ireland–XXVIII The University of Exeter. *Journal of the Royal Institute of Chemistry* **80**: 617–623.
30. Anon. (1953). Speech by Whiteley at the Summer Luncheon of the Royal Holloway College Association Summer Luncheon. *Royal Holloway College Association, College Letter* 48–49.
31. (a) Mercury reporter (1978). Top Natal scientist dies at 87. *Natal Mercury*, Durban, South Africa; (b) (1964). Woman Ph.D. at 62

now adds authorship to achievements. *The Daily News*, Durban, South Africa; (c) Anon. (1979). News review: Obituaries. *Chemistry in Britain* **15**: 5. Dr. Michael Laing is thanked for providing references 31(a) and 31(b).

32. Wright, W. G. (1963). *The Wild Flowers of Southern Africa: A Rambler's Pocket Guide*. Nelson, Natal.
33. Woollacott, A. (1994). *On Her Their Lives Depend: Munition Workers in the Great War*. University of California Press, Berkeley, California.
34. Thom, D. (1998). *Nice Girls and Rude Girls: Women Workers in World War I*. I. B. Tauris Publishers, London, pp. 122–143.
35. Rayner-Canham, M. F. and Rayner-Canham, G. W. (1996). The Gretna garrison. *Chemistry in Britain* **32**: 37–41.
36. *Home Office Report: Substitution of Women in Non-munitions Factories During the War*. His Majesty's Stationery Office, London, 1919.
37. Correspondence to Iron and Steel Works Analysts, F. K. Knowles, 21 August 1919, ED. 2.16/9 Women's War Work Collection, Department of Printed Books, Imperial War Museum, London.
38. Letter, L. T. O'Shea to W. M. Gibbons, 21 August 1919, WWWC-IWM.
39. Letter, J. H. A. Turner to A. E. Conway, 6 October 1919, WWWC-IWM.
40. Letter, W. Rintoul to A. E. Conway, undated, WWWC-IWM.
41. *The National Physical Laboratory: Report for the Year 1917–1918*. HMSO, London, 1918.
42. See, for example, Bengough, G. D., May, R. and Pirret, R. (1923). The cause of rapid corrosion of condenser tubes. *Engineering* 572–576.
43. Bate-Smith, E. C. (1947). Obituary notice: Dorothy Jordan Lloyd. *Biochemical Journal* **41**: 481–482.
44. (a) White, A. B. (ed.) (1979). *Newnham College Register, 1871–1971*, Vol. I: 1871–1923, 2nd ed. Newnham College, Cambridge, p. 33; (b) Anon. (1953). Obituary: Annie Homer. *Journal of the Royal Institute of Chemistry* **77**: 369; (c) Creese, M. R. S. (1998). *Ladies in the Laboratory: American and British Women in Science 1800–1900*. Scarecrow Press, Lanham, Maryland, p. 279; (d) Ogilvie, M. and Harvey, J. (eds.) (2000). *The Biographical Dictionary of Women in Science. Pioneering Lives from Ancient Times to the Mid 20th Century*. Routledge, New York, pp. 613–614.

45. Rayner-Canham, M. F. and Rayner-Canham, G. W. (1993). A chemist of some repute. *Chemistry in Britain* **29**: 206–208.
46. Baker, W. (1962). Millicent Taylor 1871–1960. *Proceedings of the Chemical Society* 94.
47. Marsh, N. (1986). *The History of Queen Elizabeth College: One Hundred Years of Education in Kensington.* King's College, London, p. 125.
48. Anon. (1945). Obituary: Ruby Caroline Groves. *Journal of the Royal Institute of Chemistry* 44–45; *Registers*, Royal Institute of Chemistry, 1926 to 1938; University of Birmingham, Student Records.
49. *Register*, Royal Institute of Chemistry, 1928 to 1932; University of Birmingham, Student Records.
50. Butler, K. T. and McMorran, H. I. (eds.) (1948). *Girton College Register, 1869–1946.* Girton College, Cambridge, p. 253.
51. (1920). Certificates of candidates for election at the ballot to be held at the Ordinary Scientific Meeting on Thursday, 2 December. *Journal of the Chemical Society* **117**: 90–91; Anon. (1951). *Old Students and Staff of the Royal College of Science*, 6th ed. Royal College of Science, London, p. 37.
52. Anon. (1905–1906). College news from a lady correspondent. *The Phoenix* **18**: 18. By permission of the Archives, Imperial College of Science, Technology and Medicine, London.
53. Butler, K. T. and McMorran, H. I. (eds.) (1948). *Girton College Register, 1869–1946.* Girton College, Cambridge, p. 214; (1922). Certificates of candidates for election at the ballot to be held at the Ordinary Scientific Meeting on Thursday, 15 June. *Journal of the Chemical Society* **119**: 46.
54. Downie, A. W. (1970). Samuel Phillips Bedson, 1886–1969. *Biographical Memoirs of Fellows of the Royal Society* **16**: 15–35.
55. Robertson, M. (1949). Marjory Stephenson 1885–1948. *Obituary Notices of Fellows of the Royal Society* **6**: 563–577.
56. We thank Ann Phillips, Archivist, Newnham College, for biographical information on Slater.
57. Rayleigh, Lord (1941). Joseph John Thomson, 1856–1940. *Obituary Notices of Fellows of the Royal Society* **3**: 586–609.

Chapter 13

The Interwar Period and Beyond

In earlier chapters, we identified particular niches in which a select few women chemists could find employment: academic appointments in women's colleges (see Chaps. 4 and 6), domestic chemistry (see Chap. 3), biochemistry (see Chap. 8), crystallography (see Chap. 9), and pharmacy (see Chap. 10). But what of the many hundreds of women chemists who graduated during the interwar period? Obviously, we cannot cover each individual; nevertheless, there were some specific career directions, and we will discuss them in this chapter together with biographies of women chemists who followed each of these paths.

The Interwar Period

Before reviewing their careers, it is important to see how life changed for women from the pre-First World War period of the pioneers to the interwar generation.

Life for Women Students

As the returning servicemen flooded back into the educational system, women at the co-educational universities went rapidly from a confident majority to a discrete minority. Ina Brooksbank, a student at St. Hugh's, Oxford, in 1917, vividly remembered the resentment displayed by male students who returned to the university after the First World War, and who felt that a "regiment of women" had taken over "their" university. This attitude was shared by some of the dons:

> We went down that term expecting only the lifting of a few restrictions, but on our return we found a different world. The city was full of men, bicycles and motor bicycles, often ridden in carpet slippers. We went to our usual lecture at Magdelen and found the hall full of men, seated, and women standing or sitting on the floor. Professor Raleigh entered, saw the situation and postponed the lecture at once. ... Another [don] announced that he didn't lecture to women, so out they had to go.[1]

Despite the hostility by some men, women's place at university was assured: the clock could not be turned back. The academic women of the interwar period lived in a completely different world to that of their mothers and grandmothers. Margaret Tuke, Principal of Bedford, articulated the difference, looking back from the vantage of 1928 at the bygone era:

> Not a few students of that time [the 1880s] worked ten to twelve hours a day and put into those hours an intensity of concentration difficult to realize in our own more easy-going times. ... upon her success, and that of her fellow-students, depended not their own reputation only but that of women as a whole.[2]

Tuke noted that the loss of "singleness of purpose" had to be balanced by the floodgates having opened, giving university access to the "average woman." She commented: "To-day the majority of students enter university as a necessity — sometimes an unwelcome necessity — at the wish of their parent, or because they realise that in doing so they will have a better start in life."[2]

The mid-to-late 1920s represented the zenith for the numbers of women students in academia. This phenomenon occurred not just in Britain, but also in other Western countries including the United States and the Netherlands.[3] In the context of the United States, Betty Vetter reported her findings: "Once before, women made a dent in the sturdy armor of the male science community. During the decade of the 1920s, [American] women earned 12 of every 100 Ph.D.s awarded in science and engineering, but this was a higher proportion than they ever would again until 1975."[4]

The decline in women's participation had been noticed as early as 1933 by Doreen Whiteley.[5] Whiteley found that the percentage of women students at English universities had declined from 31% in 1924/1925 to 27% in 1930/1931. She attributed the decline in part to the difficulty of girls obtaining scholarships, and in part to families putting a lower priority on their daughters' education rather than their sons' at a time of economic recession. Carol Dyhouse has shown that the decline continued to 23% in 1934/1935 and to 22% in 1937/1938.[6]

At the University of Manchester, Mabel Tylecote reported that the decline was particularly apparent in the sciences, adding: "An increased number of women entered the chemistry department in the early post-war years, but declined again afterwards."[7] She also noted increased hostility towards the women students: "Criticism of women students by no means abated and members of the Men's Union expressed themselves freely. ... Women were said to be seen at lectures taking down every word and to attend meetings which the men ignored. ... Antagonism was sharpened by the economic problems which had arisen."[7]

The Employment of Women

Though there may have been some validity to Whiteley's conclusion, in the postwar era there had been a dramatic societal change. The women who had entered employment during the war as heroines — particularly the munitions workers — were vilified in the postwar era, as Elizabeth Roberts commented: "They [women] were often regarded with open hostility by men who had realised for the first time that women were fully capable of carrying out jobs previously perceived as men's, thus presenting a real challenge."[8]

The feminist author Irene Clephane described in 1935 how public sentiment towards working women had changed dramatically and rapidly with the return of the job-hungry ex-servicemen:

> From being the saviours of the nation, women in employment were degraded in the public press to the position of ruthless self-seekers depriving men and their dependents of a livelihood. The woman who had no one to support her, the woman who herself had dependents, the woman who had no necessity, save that of the urge to personal independence and integrity, to earn: all of them became, in many people's minds, objects of opprobrium.[9]

Married women became a particular target. In 1918, the Civil Service had instituted a bar on the employment of married women; and during the 1920s, many authorities banned the hiring of married women teachers in schools and fired those then employed.[10]

It was not just a change in attitude to the employment of women; rather, it was as if the "New Woman" of the 1880s had never been, as Susan Kent commented: "The post-war backlash against feminism extended beyond the question of women's employment: a *Kinder, Küche, Kirche* ideology stressing traditional femininity and motherhood permeated British culture."[11] The feminist journalist Cicely Mary Hamilton had made the same observation back in 1935:

> Today, in a good many quarters of the field, the battle we had thought won is going badly against us — we are retreating where once we advanced; in the eyes of certain modern statesmen women are not personalities — they are reproductive

> faculty personified. Which means that they are back at secondary existence, counting only as "normal" as wives and mothers of sons."[12]

Adrian Bingham has pointed out that the view of young women in the interwar popular press was actually more complex and nuanced.[13] He contended that, though it was certainly true that there was "a fear and dislike of the "surplus" woman, who threatened the basis of political and social stability,"[13] at the same time the "modern young woman" was being encouraged to grasp new opportunities. This progressive image was intermingled with representations of women as mothers or beautiful companions of men.

Thus, the decline in women's entrance to universities in the late 1920s through to the 1950s seems to have been a multifaceted phenomenon, though the resurgence of traditionalist views of women's role in society must have been a factor. Again, this was an international experience. In the US context, Evelyn Fox Keller referred to the mid-20th century as "the nadir of the history of women in science," adding that, by the 1950s, "women scientists had effectively disappeared from American science."[14]

The Employability of Women Chemists

In 1927, the Institute of Chemistry issued a book, *The Profession of Chemistry*, which included a chapter on women in professional chemistry. The author, Richard Pilcher, reviewed the options open to women, noting that, though in theory any position available to men was equally available to women, women sometimes were left taking any vacancy that they could find:

> Some turn to secretarial work or to scientific journalism, but the majority take up teaching, for which they are often particularly suited. ... For the trained woman chemist who has no vocation for teaching, appointments are occasionally

> offered by the Research Associations ... In industry, and particularly in those industries where a large number of women are employed — such as food, margarine and jam factories — women are not infrequently engaged in analytical and research laboratories; but in other industries prospects for women are limited, because the higher positions call for experience in dealing with workmen, and moreover employers realize that, after a year or two, when the experience which a woman has gained in her work is becoming most valuable, her professional career may be terminated by marriage.[15]

The potential for marriage — "matrimonial mortality" — was one cloud that hung over every woman seeking employment. As we saw in Chap. 11, Elsie Phare, among many others, lost her position upon marriage. Though dismissal from a teaching or academic post was more cultural, industrial companies avoided hiring women on the grounds that the years of training would be wasted and lost when a women departed for marriage and children.

The Woman 'Super-Chemist' as Negative Role Model

Some women chemists stood out as exemplars, but, in a way, they served as a discouragement. Margaret Rossiter described the effect of Marie Curie's visit to the United States as raising the bar for women chemists to unattainable levels:

> Before long most professors and department chairmen were ... expecting that every female aspirant for a faculty position must be a budding Marie Curie. They routinely compared American women scientists of all ages to Curie, and finding them wanting, justified not hiring them on the unreasonable grounds that they were not as good as she, twice a Nobel Laureate![16]

The same phenomenon occurred in Britain. Ida Smedley (see Chap. 2) was used as a benchmark for women applicants for the Position of Reader in Chemistry at the King's College of Household and Social Science (see Chap. 3): "... it would be of great value to the Department to secure the services of a woman with the high scientific standing and personality of Dr. Ida [Smedley] Maclean."[17]

Smedley was not the only one to be chosen as the expectation for a woman chemist or biochemist. A 1929 article in the *Journal of Careers*[18] held up Martha Whiteley (Chap. 3) as a role model; while an article in the same journal in 1938[19] extolled Ida Smedley (Chap. 2), Marjory Stephenson (Chap. 8), Katherine Coward (see below), and particularly Dorothy Jordan Lloyd (Chap. 8) as the heights of careers to which women chemists and biochemists could aspire — but only those who were exceptional.

Jordan Lloyd stated in a 1933 careers article that, for women chemists, being among the best might still not be good enough: "When it comes to a permanent post, to obtain equal chances with a male rival, the woman must be obviously a little better. ... To be 'the best,' however, demands mental qualities of a high order and the best possible training."[20] Thus, young women chemists were given the message that only the truly exceptional and dedicated would find success in academia or industry, and that teaching was the only realistic option.

Women Chemists in Teaching Careers

During the interwar years, as we have mentioned above, teaching was the most common career for women science graduates.[21] Dyhouse commented: "Teaching, then, be it a vocation, the only realistic option or a last resort, remained the fate of the majority of women graduates in this period [pre-1939]."[22] By the 1930s, there were more university women graduates than potential employers, and even those who wanted to escape from teaching had few available options.

There was a second reason that women became teachers: from Dyhouse's statistics, about 30% of the women attending university in the interwar period were only able to afford to do so on grants for teacher training. This commitment was spelled out very clearly at some institutions, as Kathleen Uzzell recalled:

> When we were first at University we were called into a room where we were told we had to swear an oath to teach for five years, but it was pointed out it was a 'moral not a legal' oath. ... The promise to teach for five years meant a promise not to marry as there were no married female teachers except war widows.[23]

Hazel Reason

Of the many women chemists who entered into school teaching (see also Rose Stern in Chap. 2), we have chosen to include Hazel Alden Reason.[24] Reason graduated from Bedford College in 1924 with a B.Sc. (London) in chemistry, and then obtained a position as Senior Science Mistress at the County School for Girls, Guildford. During her free time, she worked towards an M.Sc. (London) on the history of science, which she completed in 1936.

Reason authored a book on the history of science, *The Road to Modern Science*,[25] which was first published in 1936. A second edition of the book appeared in 1940 (reprinted four times) and, indicating an amazing longevity, a third, revised edition in 1959. In the Foreword, she commented: "The primary object in writing this book was to present the story of scientific discovery in a form which would appeal to intelligent boys and girls." She added that she did not approve of the "great scientist" approach to the history of science; rather, her book covered "... the broad view of scientific discovery."[25]

College and University Chemistry Teaching

At the smaller universities and women's colleges, women chemists played a major role in the teaching duties — as a few of the many examples, Kathleen Balls at East London College (Chap. 3); Mary Lesslie and Violet Trew at Bedford College (Chap. 4); E. Eleanor Field at Royal Holloway College (Chap. 4); Grace Leitch at Armstrong College (Chap. 5); Emily Turner at University College, Sheffield (Chap. 5); Ida Freund at Newnham College and Beatrice Thomas at Girton College (Chap. 6); Ettie Steele at St. Andrews (Chap. 7); Ishbel Campbell at University College, Southampton (Chap. 7); and Isobel Agnes Smith at University College, Dundee (Chap. 7).

All of these women remained single throughout their careers. Some women academics did indeed see themselves following the equivalent to a religious vocation, as Elsie Phare recounted about her German Tutor at Newnham: "Miss Paues told me that she was married to scholarship (she wore a wedding ring), with the implication that that was the state to be wished."[26] Though scholarly celibacy might have suited a few, E. H. Neville — in an article in 1933 titled "This Misdemeanor of Marriage" — abhorred the fact that, for women to have an academic career, they were required to be: "enforced celibates, predestined spinsters, and women cunning enough to maintain complete secrecy in their sexual relations."[27]

We have shown that women became Demonstrators and Lecturers outside of the university system, such as at Battersea Polytechnic (Chap. 3) and the London School of Medicine for Women (Chap. 4). Women chemists taught at other institutions, and we have chosen the lives of three women as examples.

Hilda Hartle

Hilda Jane Hartle[28] was born on 11 September 1876 in Birmingham, the daughter of Edward Hartle and Anne Jane Warillow. Like so many others, she was educated at King Edward

VI High School for Girls (KEVI), Birmingham, and then obtained her university education at Newnham between 1897 and 1901.

Moving back to Birmingham, Hartle was a researcher with Percy Frankland at the University of Birmingham from 1901 to 1903. She then became Lecturer at Homerton College, Cambridge, a teachers' training college,[29] which had been founded as a congregational (nonconformist) institution, but later became nondenominational. While at Homerton, Hartle became a signatory of the 1904 petition for the admission of women to the Chemical Society and the 1909 letter to *Chemical News* (see Chap. 2).

In the early years, life for the women students at Homerton was very restricted, as Elizabeth Edwards described:

> Cambridge was a town dominated by the male values of its university, which barely tolerated its own women students, let alone those in an obscure teacher training institution. The university's misogyny had been underlined by a privilege which it had only been forced to relinquish as late as 1894. This was its right to imprison in its own private prison for a period for up to three weeks, women walking in the town — who could well have been Homerton students — whom it suspected of being prostitutes.[30]

Homerton was the only one of the early teachers' training colleges to teach chemistry,[31] and this, largely a result of Hartle. It was at Homerton that Hartle penned her attack on the domestic science movement (see Chap. 1). Awarded a Mary Ewart Travelling Scholarship for 1915/1916, she spent the year in the United States studying experimental methods at American schools and training colleges, observations she published in the *Times Educational Supplement* and elsewhere.

In 1920, Hartle was appointed Principal of the Brighton Municipal Training College for Teachers, a post she occupied until her retirement in 1941. The postretirement 30 years were spent actively working for numerous women's organisations.

Her obituarist, "M.E.G." (probably her longtime colleague at Homerton, Margaret Glennie), remarked: "She maintained a vivid interest in all these many activities almost to the end of her life and she also took an affectionate delight in her extensive correspondence with old pupils in all parts of the world."[28(a)] Hartle died on 20 May 1974, aged 98 years.

Peggy Lunam (Mrs. Edge)

Technical colleges also employed women chemists as Lecturers. Peggy Lunam[32] attended Armstrong College, where she was Student Treasurer of the Bedson (Chemistry) Club from 1929 to 1930. She earned a B.Sc. (Durham) in 1930 and an M.Sc. (Durham) in 1932. In 1935, she was a part-time Lecturer and Researcher in the Department of Chemistry at Constantine College, Middlesbrough.

Constantine Technical College had been founded in 1930 to support Middlesbrough's engineering, bridge, and shipbuilding industries.[33] At first, Constantine College concentrated on metallurgy, engineering, and chemistry, offering courses leading to University of London degrees, only later broadening its offerings to the Arts.

By 1938, Lunam had been promoted to Lecturer in Chemistry. In 1942, she married Herbert Allan Edge of the Research and Development Department, Imperial Chemical Industries (ICI) Agricultural Division, Billingham. She was noted as still being Lecturer at Constantine College in 1948, but later became Head of the Chemistry Department at Kirby Grammar School.

Frances Burdett

For women chemists, there seemed to be mobility between the different types of academic institutions, as is illustrated by Frances Burdett,[34] who taught at university, secondary school, and finally technical college. Burdett was born on 12 September

1882 and was educated at Wakefield Girls' High School. She entered the University College of North Wales, Bangor, in 1902; and after completing a B.Sc. in chemistry in 1905, she was appointed Assistant Lecturer in the Training Department of the College. While at the college, she undertook research with Kennedy Orton (see Chap. 7), part-authoring a publication on the influence of light on diazo reactions.

In 1907, she left Bangor to become Senior Science Mistress at the Park School, Glasgow, moving to a similar post at Carlton Street Secondary School, Bradford, in 1919, presumably to be closer to her family. Burdett stayed there for 1 year before she was appointed Assistant Lecturer in Chemistry at Bradford Technical College. She later became Lecturer in Chemistry at the same institution, a post she held until her retirement. She died on 24 May 1957.

Women in Chemical Industry

A reply from Dorothy Adams to a questionnaire on women's war work (see Chap. 12) pointed to bleak post-First World War opportunities for women chemists in industry:

> With regard to the prospects of scientifically trained women after the [First World] war my experience has led me to the conclusion that there will be practically no scope for them in industry. There is, and will continue to be for some time, a far larger supply of male Chemists than will be needed. Under such circumstances women with the same qualifications will stand the poorest chances of employment. As teachers and lecturers there is still some demand for such women, but in industry there is next to none. I have been led to this conclusion by my experience in endeavouring to obtain a fresh post myself. I do not stand alone in my opinion, Mr. Pilcher, the Registrar of the Institute of Chemistry whom I consulted on the subject told me exactly the same things as I learnt later from my own experience.[35]

The Employability of Women Chemists

In 1938, the *Journal of Careers* published a series of articles on the prospects for employment for women science graduates, and the author was blunt about potential marriage being a significant deterrent to hiring women:

> This question of marriage is undeniably a deterring factor in the employment of women scientists in industry. Firm after firm, among the large number which the Journal of Careers has consulted, raises it as an objection. Even a woman who did brilliant work for some years, of a quality which is still remembered by men colleagues in terms of highest praise, apparently closed the door to other women in that particular firm, for it is recorded "but she left to get married and we haven't employed a woman since."[36]

A second reason for not hiring women, or restricting them to routine work, was the claim that women did not have the mental aptitude for research, or that imaginative women were far rarer than imaginative men. The same author also commented on this perspective:

> That women lack the research type of mind and are therefore not suited to industrial research is a generalisation often advanced by firms which do not employ them. It is also advanced by firms which do employ them and which therefore restrict them to analytical work and to routine work as technicians.[36]

The advantage of men was that they would "take their job home with them"[36] and that it was during evenings and nights at home that inspiration would strike; whereas women, when leaving work, would have their minds stray onto other matters.

Women had one employment advantage: they were inexpensive to hire. It was common for the salary paid to a women to be much less than that paid to a man (see Culhane's experience below). In his opening remark, the author recounted one "typical" experience:

> Some months ago a leading firm telephoned the Secretary of one of the Colleges of London University and asked her to recommend an experienced woman chemist. "We have an immediate vacancy," said the voice at the other end of the telephone. "We should like to appoint a man, but we are only offering £150 a year, and we'll never get a man for that."[19]

In an interview in the *Journal of Careers* on the prospects for women in science in 1929, Jordan Lloyd made clear her beliefs that only the very best should consider an industrial career. She also described the perceived shortcomings of her own gender:

> Again, there are a number of industrial chemistry posts for which second or third rate ability is sufficient, and generally speaking a man of second or third-grade is to be preferred to a woman of that mental calibre, because he can be used for a wider range of duties and he is not usually, though he may be, quite as inert mentally. Women are sometimes taken on for routine posts because they can be offered a lower salary than men and because they sometimes show a placid contentment in routine posts and do not crave for responsibility or any duties beyond those for which they are specifically appointed.[18]

Nevertheless, despite the gloomy forecasts, Horrocks has shown that during the interwar period many women chemists did find employment in industry, particularly the food, pharmaceutical, cosmetics, textiles, and photographic industries.[37] We will conclude this section with one case study, the life of Kathleen Culhane. For so many of the forgotten women chemists, scanty information remains on their life and work; but for Culhane, we have a rich narrative that epitomises the struggle of women seeking an industrial chemistry career during the interwar period.

Kathleen Culhane (Mrs. Lathbury)

Kathleen Culhane[38] was born on 14 January 1900, the fourth of six children, her mother dying when she was quite young. Fortunately, her father, J. W. S. Culhane, a medical doctor, insisted on equal opportunity for his daughters and his sons, and he became Culhane's idol. She attended Hastings and St. Leonards College, Sussex, where the only science at the time was botany, then Hastings School of Science, before entering Royal Holloway College (RHC) in 1918. It was at RHC that she discovered that chemistry was her real interest, and she graduated in 1922 with an Honours degree in chemistry.

Wanting to enter the chemical industry, Culhane was extremely frustrated that employers would not take an attractive young woman seriously for chemist positions. In fact, she was only considered for interviews when she signed her applications as "K. Culhane" rather than "Kathleen Culhane." However, once her gender became apparent at the interviews, she failed to obtain any of the positions (the ruse was more successful for her chemist daughter in the 1950s). Marriage was scarcely a survival option, for, as we mentioned in Chap. 12, a significant proportion of middle-class single males of her generation had been killed in the war. Finally, she obtained work as a school teacher and, later, a private tutor.

Joining the Institute of Chemistry proved to be the turning point in her fortunes. Through the Institute, she met John R. Marrack of the Hale Clinical Laboratory of the London Hospital. Marrack allowed her to gain experience in medicinal chemistry by permitting her to do emergency blood sugar determinations in her free time, but without pay. After 2 years of combining teaching with unpaid analytical work, Culhane obtained an industrial chemistry position with Neocellon, Wandsworth, a manufacturer of lacquers and enamels. However, her delight was diminished after being told by the company that the only reason for hiring her was that they could not afford the salary of a male chemist.

The completion of a study of enamel coatings for light bulbs coincided with an offer of a job back at the Clinical Laboratory, this time as a paid chemical advisor and insulin tester. Not long after, she accepted a position in the physiology department of the large chemical and pharmaceutical company, British Drug Houses (BDH). However, as time progressed, her initial enthusiasm waned:

> I gradually discovered that it was not the intention to employ me as a chemist but as a woman chemist ... I was expected to do all the boring, routine jobs ... while anything interesting was handed to one of the men ... The routine work increased enormously in quantity and I took pride in perfecting my technique ... thinking it must surely win promotion that way. This did not materialize so, by superhuman efforts and late work, I got some research done which was successful and I was allowed to publish it ... The problems I worked on were of my own finding ... I managed to avoid being disliked and was merely regarded as eccentric.[39]

In addition, the senior staff lunch room was male-only, and Culhane recalled how she had to eat her lunch with the women cleaners and clerks.[39]

At the time, there were no chemical tests for insulin, and Culhane significantly improved the physiological testing procedures. Four researchers were chosen by the League of Nations Health Organization Committee, Culhane being one (though against the wishes of some Committee members), to compare independently the physiological activity of amorphous and crystalline insulin. Her results had significant differences from those of the other three researchers, and the Committee concluded that hers must have been in error. She was asked to withdraw them, but this she refused to do, convinced of their correctness. Only later was it shown that her results were indeed more accurate.[38(a)]

Culhane commenced a study of vitamins in 1933, though she continued with her work on insulin as well. As part of her diverse research studies, she gave a presentation on the need to standardise products containing added vitamins, arguing particularly for enhanced levels of vitamins in margarine. A British newspaper reported on the meeting, describing her as "Miss Kitty Culhane, the Girl Pied Piper of Science" as she had mentioned the use of mice in vitamin research, and the newspaper added how an "abstruse lecture on vitamins" had been delivered by "a pretty girl with blue eyes and bobbed hair."[40]

That same year, she married Major G. P. Lathbury. Having mentioned her intent to marry to her supervisor, Culhane was amazed that the directors had to give special approval for the employment of a woman after marriage. The approval was granted in her case due to the importance of her work. She resigned her position as senior chemist in 1935 due to pregnancy. Her inexperienced male successor was given a higher starting salary than what she had been earning at the time of her resignation.[38(a)]

With the arrival of war in 1939, Culhane offered her services to the war effort, having sent her small daughter to the country. As she later wrote: "Although it was often publicly stated that industry was short of scientists the Appointments Board were unable to tell me of a single opening for which a woman would be considered."[40] Persistence finally resulted in a position as an assistant wages clerk, where the senior clerk offered to teach her percentage calculations.

After more badgering of officials, Culhane was appointed Manager of a statistical quality control department at a Royal Ordnance Factory; and on the basis of her work, she was made a Fellow of the Royal Statistical Society in 1943. Because salaries determined travel status, earning less than half the salary paid to males in the same position had a secondary effect — on train journeys to London, her male colleagues travelled First Class, while she had to sit alone in a Third Class compartment. Amazingly, she was not embittered by her experiences, instead regarding them as a source of amusement.

After the war, she retired from science and took up a second career as an artist, becoming a member of the Haslemere and Farnham Art Society, and producing paintings that were included in professional exhibitions. It was of great joy to her that her daughter and one of her grandsons became graduate chemists. She died on 9 May 1993, aged 93 years.

Women Chemists in the Food Industry

A significant number of women chemists found employment in the food industry, as Horrocks reported:

> The food industry offered the largest proportion of posts, and in Lyons provided the firm employing the greatest number of women chemists. Many of them were graduates of King's College of Household and Social Science (KCHSS). Indeed, eighty-five KCHSS graduates were employed by eighteen different food firms and four analytical chemists' practices during the period 1910 to 1949. Over half of these — forty-four — worked for Lyons. Other food manufacturers known to have employed women chemists included Glaxo, United Dairies, Chivers, CWS, Lever Brothers, Fullers, Peek Frean, Robertson's, Schweppes and Vitamins Ltd.[37]

We have chosen four individuals in this category: Ethel Beeching (Mars Confections Ltd.), Mamie Olliver (Chivers and Co.), Winifred Adams (Horlicks Ltd.), and Enid Bradford (Marmite Food Extract Company, Ltd. and J. Lyons & Co. Ltd.).

Ethel Beeching

Born on 29 June 1900 in Islington, London, Ethel Irene Beeching[41] was the daughter of C. L. T. Beeching, Secretary of the Institute of Certified Grocers. She was educated at James Allens' School for

Girls, Dulwich, and then studied at Bedford, obtaining a B.Sc. (London) in chemistry in 1923. After holding a teaching position for a year, she joined James Spencer's research group at Bedford (see Chap. 4), while teaching part-time at St. Philomena's College, Carshalton, a convent boarding school. She completed an M.Sc. (London) in 1925 on the topic of the magnetic susceptibility of some metals and their oxides.

In 1926, Beeching accepted a position in the laboratories of the British Association for Research in the Cocoa, Chocolate, Confectionary and Jam Trades; however, in January 1927, she became a Works Chemist to Alfred Hughes & Sons Ltd. of Birmingham, at which time she became a Member of the Analytical Society. Beeching later went to Mars Confections Ltd., Slough, and worked her way up from Analytical and Control Chemist to Chief Chemist. She stayed at Mars until retirement. Following in her father's footsteps, she became a Director of the Institute of Grocers. Beeching also joined the Women's Chemists' Dining Club (see Chap. 2), being a Committee Member in 1953. She died on 28 July 1967, aged 67 years.

Mamie Olliver

In the later years of our coverage, Mamie Olliver,[42] a researcher in the food industry, had the highest profile among women chemists. She was also one of the first women to gain success and recognition in the field of industrial chemistry. Olliver was born on 10 April 1905 in Coventry, where she attended Barr's Hill Secondary School. She entered King's College of Household and Social Science in 1923, completing a B.Sc. (Household and Social Science) in 1926.[43] Then, she was hired by Excel Co. Ltd. of London as a chemical analyst; simultaneously, she worked towards a B.Sc. in chemistry at Birkbeck College that she completed in 1928, followed by an M.Sc. in biochemistry in 1930.

In 1930, Olliver joined Chivers & Co. at their research facilities in Histon, near Cambridge, being promoted to their Chief Chemist in the same year. During her time there, she designed and organised the central research and quality control laboratories. She herself was an active researcher, authoring 13 publications, many related to the vitamin C content of foods. Her most important research was the discovery of blackcurrant juice as a valuable source of vitamin C. Following the merger of Chivers and Schweppes in 1959, she was appointed Research Investigator and Consultant for the Schweppes Group of Companies. She was forced into retirement in 1965 at the age of 60.

Olliver was very active with professional societies. She was a member of the Biochemical Society, Nutrition Society, Society of Chemical Industry, Society for Public Analysts, and also Fellow of the Chemical Society. Elected Member of Council of the Society for Analytical Chemistry, she had the third-highest vote among the eight candidates for the six positions.[44] During the Second World War, Olliver was an active member of several committees of the Ministries of Food and Health. The first woman member of the Council of the Royal Institute of Chemistry (RIC) in 1948, and serving on many of the RIC's committees, she was elected Vice-President in 1951. Olliver was particularly active in promoting women in chemistry.[45]

Her longtime friend, Doris Kett, recalled:

> I first met Mamie on the education committee of RSC when I was the association of science education representative, several years before I became a member myself. We were the only two women on the committee. She welcomed me warmly and was most helpful and we soon became friends and I visited her at her bungalow where she lived until she died. … She was not married and had no brothers or sisters or close relatives but was very fond of a God-daughter who was her heiress as far as I know. She was a regular worshipper at her local church in Histon where she was well known. Histon is a small picturesque village just outside Cambridge. When the RSC moved to Cambridge she frequently had lunch there as they thought a lot of her.[46]

Olliver retained a keen interest in all aspects of food technology until she died on 17 January 1995, just a few days before her 90th birthday.

Winifred Adams

The first woman chemist hired by Horlicks Ltd. was Winifred Elizabeth Adams.[47] Adams, born on 10 November 1909 in Wandsworth, London, was the daughter of Herbert Edward Adams, a teacher at Dulwich College, and Winifred Rackham, a preparatory school Mistress. She was educated at Clapham High School (a GPDSC school), then studied at the Chelsea Polytechnic before entering Newnham to study biochemistry in 1929.

Graduating in 1933, Adams was hired by Horlicks Ltd. of Slough as an Assistant Chemist. She was the first woman to be appointed to their Research Staff, and "an object of some suspicion to her male colleagues in the early years."[47(a)] Promoted to Senior Chemist in 1940 and then to Chief Chemist in 1948, she had to retire in 1964, the mandatory retirement age for women staff then being 55.

Outside of her professional life, Adams was a good musician, and she formed a quartet with her parents and her only sister (a professional cellist). She was an avid traveller, particularly to Canada and the Shetland Islands. After retirement, she moved to Brighton and became active with conservation groups, dying on 15 July 1994, aged 84 years.

Enid Bradford (Mrs. Bentley)

The chemist and crystallographer, Winifred Booth Wright (see Chap. 9), was one of the many women scientists who worked at J. C. Lyons & Co., another being Enid Agnes Margaret Bradford.[48] Bradford obtained a B.Sc. in chemistry from King's College, London, in 1939. She commenced her career at Metal Box Co. Ltd., from where she authored a publication on the determination of fats in flour and pastries. By 1941, she held a position as a Research Chemist to the Marmite Food Extract Company, Ltd.

Later in the Second World War, Bradford worked at the Charterhouse Rheumatism Clinic, London, from where she co-authored papers on winter sources of vitamin C and on the determination of riboflavin in blood. After the war, she was employed by J. C. Lyons & Co. Ltd., from where she authored four papers for *The Analyst*: two on riboflavin in tea, one on the microbiological assay of vitamins, and one on the use of a single tap source to simultaneously run three different applications. Later in life, Bradford married Mr. Bentley and became a Consultant. She died on 20 August 1981, aged 79 years.

Women Chemists in Biomedical Laboratories

In her article titled "Biochemistry as a Career for Women" in the *Journal of Careers*, Dorothy Jordan Lloyd made it clear that the biochemical sciences were among the most demanding in preparation:

> Biochemistry at most of the universities in this country is not included in the syllabus of any degree course in science, but must be studied as a post-graduate subject for a further one or two years. The preliminary training for a career in biochemistry, therefore, usually calls for four or five years of hard work. Even after this, two years spent working on a research problem in a first-class laboratory and the attainment of a Ph.D. are a very desirable supplementary training.[20]

In Chap. 8, we discussed the biochemists at Cambridge with Hopkins, but there were also women who entered biochemistry through other paths, and we have provided a selection below.

Katherine Coward

One of the earliest biochemists was Katherine Hope Coward,[49] who was born on 2 July 1885, daughter of a teacher. Unlike other biochemists who started with chemistry, Coward studied botany at the University of Manchester, obtaining a B.Sc. in 1906 and an M.Sc. in 1908.

In 1920, Coward entered University College, London (UCL), to study biochemistry. Awarded a Beit Fellowship, she undertook research on vitamin A with Jack Drummond,[50] a total of 22 publications being authored or co-authored from her work. She was elected Fellow of the Chemical Society in 1923, one of her nominators being Katherine Burke (see Chap. 3). Receiving her D.Sc. in biochemistry in 1924, she travelled to the United States on a Rockefeller Travelling Scholarship to continue her studies on vitamin A at the Department of Agricultural Chemistry of the University of Wisconsin at Madison.

Returning to Britain in 1926, Coward took charge of the newly formed vitamin-testing department of the Pharmaceutical Society's Pharmacological Laboratories, and she was also appointed Reader in Biochemistry at the University of London in 1933. Her research broadened to include the study of each of the vitamins, resulting in a total of 62 additional papers, five of the later publications being co-authored with Elsie Woodward (Mrs. Kassner; see Chap. 10). One of her obituarists, "J.H.B.," described the crucial role of her contributions:

> The work she did when in charge of the vitamin-testing Department of the Pharmaceutical Society's Pharmacological Laboratories was of outstanding importance. There were already qualitative methods for various vitamins, but there was a grave lack of methods which were quantitative. Dr Coward was a mathematician with a knowledge of, and a liking for, statistical methods, and was therefore ideally suited for the work of defining such quantitative methods as were

> needed. Between 1926 and 1938, she and her colleagues completed a series of papers, which were of great value for the growing number throughout the country who were concerned with these problems.[49(e)]

Coward was professionally active, being a member of the Vitamin Committee of the British Pharmacopoeia Commission from 1933 to 1953 and a member of the Committee of the Biochemical Society from 1932 until 1936; and in 1937, she was elected as an honorary member of the Pharmaceutical Society. Her advice was sought by committees of the League of Nations and the World Health Organisation. She retired in 1950, but continued to be active, as her obituarist, G. S. Cox, recalled:

> After she retired in 1950 she was only too happy to be involved in assisting with problems in statistics, and she was a frequent visitor to the pharmacology laboratories at No 17, [The Pavement; see Chap. 10], giving freely of her time to help research workers. She had a great gift with figures and could perform the most complicated calculations in her head. As a retirement gift she was given a hand-operated calculating machine and, some months after her retirement, when I asked her how she liked it her comment was "It's all right, but it is far too slow; I can do the calculations quicker in my head". And I am sure she could![49(f)]

She died on 8 July 1978, aged 93 years.

Winifred (Freda) Wright

In her short life, Winifred Mary Wright[51] became a highly published biochemist. Wright was born on 1 January 1900 in Brighton, the daughter of Arthur Wright, a consulting electrical engineer, and Edith M. Wassel. She was educated at West Hill House, Eastbourne, then attended Bedford for 1 year before entering Girton in 1920. Completing the requirements for a chemistry degree in 1924, she was awarded a Yarrow Studentship, allowing her to undertake research at Cambridge. Initially, her research was with Eric Rideal,[52] then on her own, on the topics of low-temperature oxidation at charcoal surfaces and the decomposition of hydrogen peroxide under different conditions, resulting in a total of eight publications and providing the basis of her Ph.D. in 1928.

Wright worked for a year with C. G. L. Wolf at Addenbrooke's Hospital, Cambridge, her research culminating in a lengthy paper on the serological diagnosis of cancer. In 1930, she was appointed assistant in the Pharmacology Department at UCL; in addition to lecturing, she collaborated with Harry Ing[53] in a study of the curariform action of quaternary ammonium salts.

One of the obituarists of Wright, "F.M.H." (probably her contemporary at Cambridge, Frances Mary Hamer; see below), commented: "Her specialisation in the field of physical chemistry did not prevent her from throwing herself whole-heartedly into those different problems involving animal tissues, and it is perhaps typical that, only shortly before her death, she embarked upon a medical training."[51(b)] Her death came on 21 June 1932, at the early age of 32.

Mollie Barr

The Wellcome Physiological Research Laboratories, Beckenham, Kent, hired a significant number of women chemists and biochemists; as an example, we have chosen Mollie Barr.[54] Barr was born on 11 June 1906, the daughter of Horace C. Barr, a burial practitioner, and Edith Barr of Belvedere, Kent. She was educated at several schools, the last being Blackheath High School, London (a GPDSC school).

Barr completed a B.Sc. in chemistry at Bedford in 1927, and was then awarded a 3-year grant from the Department of Scientific and Industrial Research (DSIR) to work towards her Ph.D. on molten salt electrolysis under James Spencer. However, Barr terminated her research in 1928 with an M.Sc.

and went to Wellcome as a biochemist working with Alexander Glenny.[55] Most of her research was on immunology, particularly diphtheria; and between 1931 and 1955, Barr authored or co-authored a total of 33 publications.

Elsie Widdowson

Elsie May Widdowson[56] became a pioneer in the scientific analysis of food, particularly on the importance of diet in infant development. She was born on 21 October 1906 in Dulwich, London, and attended Sydenham High School (a GPDSC school). It was the tradition at Sydenham High School for the matriculants to enter either Bedford or Royal Holloway, but Widdowson had heard good reports of IC from three girls who had entered a year ahead of her, so she decided to follow them.

Widdowson completed a B.Sc. (Honours) in chemistry, during which time she had commenced research with Samuel Schryver[57] in the biochemistry laboratory, separating amino acids from plant and animal sources. While at IC, she was offered a 3-year research position working with Helen Archbold (see Chap. 3) on the reducing sugars in apples. Her interest was more in human biochemistry; thus, when the grant expired, she enrolled in a 1-year postgraduate diploma course in dietetics at King's College of Household and Social Science.

As a preliminary to the course, Widdowson was sent to the main kitchen at King's College Hospital to learn about large-scale food preparation. It was there that she met Robert McCance,[58] who was studying the loss of nutrients from food during cooking. From her experience with apples, Widdowson challenged McCance's figures for carbohydrates, contending that some had been lost by hydrolysis. Impressed by the correctness of her arguments, McCance obtained a grant for her.

Widdowson realised that there was a need for tables of the chemical content of foods. She persuaded McCance to allow her to determine the food composition for those items not previously analysed; then, the two of them compiled the tables, noting values before and after cooking, and published them as *The Chemical Composition of Foods*.[59] This monograph has been in print to the present day, the current, sixth edition being published by the Royal Society of Chemistry as *McCance and Widdowson's The Composition of Foods*.[60]

During Widdowson's life, she authored or co-authored over 600 publications, mainly with McCance; in fact, an account of their 60-year collaboration has been published.[61] Here, we can only give a superficial coverage of the many avenues of food chemistry research in which she worked. For example, their studies together between 1934 and 1938 were on salt deficiency in humans, the absorption and excretion of iron, diet variations between individuals, and how kidney function differed between babies and adults.

In 1938, McCance was invited to become Reader of Medicine at the University of Cambridge. At his request, the Medical Research Council agreed to Widdowson moving with him. During their first year at Cambridge, they studied strontium absorption by means of injecting each other with larger and larger doses and determining the fraction excreted through the kidneys versus the bowels. Unfortunately, the last batch of strontium lactate was contaminated by bacteria and they both fell seriously ill in the laboratory. As the previous doses had had no significant physiological effect, they did not have their usual observer with them, but by good fortune someone came by and called for help. Even during their sickness, they continued to collect their urine and fæces samples for later analysis.

When the Second World War started, Widdowson and McCance realised that dietary aspects of food rationing would become of high importance. Again, they experimented upon themselves (and some of their colleagues), going on a diet that, at the time, was thought to be far too little to maintain health. They also identified the importance of the addition of a calcium supplement to the diet, particularly through addition to bread.

After the war, Widdowson worked for the next 20 years on animal nutrition. In 1968, she finally received the formal recognition she deserved, being appointed Head of the Infant Nutrition Research Centre at the Medical Research Council's Dunn Nutrition Laboratory. Retiring in 1973, Widdowson then accepted a research opportunity at the Department of Investigative Medicine at Addenbrooke's Hospital, where she continued working until she was 82. She believed in "active" retirement, including Presidency of the British Nutrition Foundation from 1986 until 1996. Among the many honours she received was election as Fellow of the Royal Society in 1976 and appointment as C.B.E. in 1979. She died on 14 June 2000, aged 93 years.

Women Chemists in the Photographic Industry

As Horrocks noted,[37] the two leading photographic companies, Ilford and Kodak, had positive attitudes to the employment of women chemists, and as a result gained two outstanding women researchers: Frances Hamer and Nellie Fisher.

Frances Hamer

Frances Mary Hamer,[62] born on 14 October 1894 in Kentish Town, London, was the daughter of William Heaton Hamer, Medical Officer of Health for London, and Agnes Conan. Her mother, together with all of her aunts on both sides of the family, had attended North London Collegiate School (NLCS); and she was named after Frances Mary Buss, the founder of NLCS (see Chap. 1), who later became Hamer's godmother.[63] Hamer herself was enrolled at NLCS, matriculating in 1916, then entering Girton later that year.

While still an undergraduate student at Girton, Hamer joined the research group of William Pope.[64] As a matter of wartime urgency, Pope had been asked to investigate the structure of, and reliable synthetic method for, photographic sensitizers.[65] In 1905, a German company had synthesized a dyestuff, pinacyanol, which, when incorporated into photographic plates, improved sensitivity towards the red end of the visible spectrum; this discovery became of vital military importance, as the air reconnaissance in the latter part of the First World War was undertaken at dawn, when the light had a strong reddish bias. Thus, plates of the battlefronts taken by British planes were far inferior to those taken by the German planes of the allied battle movements. In Pope's group, Hamer worked with William Mills (see Chap. 11) to determine the structure of pinacyanol and find a reliable method of synthesis. She was successful in both ventures, and nearly all of the pinacyanol used in the panchromatic film came from the Cambridge laboratories. Following the end of the war, Mills and Hamer were permitted to publish their synthetic procedure.

An essay on her pinacyanol work gained her the Gamble Prize in 1921, and that same year she was awarded the Yarrow Scientific Research Fellowship to continue her work in Cambridge, where she remained until 1924. She worked for a few months at the Davy Faraday Research Laboratory of the Royal Institution, and then was hired by the photographic company of Ilford Ltd.[66] The 6 years she spent at Ilford were marred by difficult relations with the Research Director, Frank Forster Renwick. Matters came to a head in 1930, when Hamer complained that her new laboratory was so hot in summer it was impossible to work; Renwick refused to take any action and Hamer resigned.

By good fortune, Kodak Ltd. was looking for an organic chemist at the time, as was recounted in the Harrow Research Laboratory (Kodak Ltd.) Album:

> I [the unknown author] wanted a first-class organic chemist, and heard that Dr. Frances Mary Hamer, well-known for her work on sensitizing dyes at Cambridge with Pope and Mills, and who had been at the Ilford Research laboratories, in 1930 had fallen out with her directors over the matter of ventilation in her laboratory. I wrote and told [C. E. Kenneth] Mees. He wrote back and said: "Take her out to tea and sound her out".

> So I did, and she came, and Dr. Renwick accused me of pinching one of his people. Well, first of all Dr. Hamer was unpinchable, and in any case the whole thing was perfectly ethical.[67]

Thus, Hamer became Kodak's Head of the Organic Chemistry Research Department.

During her years at Ilford and Kodak, Hamer authored over 70 research publications and was responsible for a large number of patents, having made major contributions to the synthesis of cyanine dyes and discovering new classes of sensitisers, including some that were extremely effective in the infrared region. Hamer had a valuable assistant, Nellie Fisher (see below); while between 1938 and 1942, they were joined at Kodak by Hamer's old friend, Edith Pope (see below), who worked as Journal Abstractor and Indexer.

Hamer was professionally active, having served on the Councils of the Chemical Society, the Royal Institute of Chemistry, and the Royal Photographic Society (RPS). It was the RPS which recognised her achievements: first, in 1948, the Henderson Award for her investigations over many years into the constitution and synthesis of photographic sensitising dyes; then, in 1963, the RPS Progress Medal and election to an Honorary Fellowship of the Society. Despite her major contributions to the chemistry of photography, and to her own disappointment, she was never elected Fellow of the Royal Society. With hindsight, this was not surprising for, in addition to the handicap of being a woman, Hamer's applied research was outside of mainstream academic chemical circles, and thus she would have had few champions among the Fellows of the time.

In 1945, Hamer returned to academic research and became a Honorary Lecturer at IC while still being a Research Chemist and Consultant at Kodak, finally retiring in 1959. Upon retirement, she undertook the full-time work of writing the definitive monograph on the cyanine dyes, *The Cyanine Dyes and Related Compounds*.[68]

Hamer was an enthusiastic walker, but a serious accident in 1964 and another one in 1971 meant that her mobility was reduced to local walks with the aid of a walking stick. However, she was able to maintain her other interest, gardening, until shortly before her death on 29 April 1980 at the age of 85.

Nellie Fisher

Nellie Ivy Fisher,[69] initially an assistant to Hamer, became a well-known researcher in her own right following her move to Australia. Fisher was born on 15 October 1907, one of a family of six brothers and sisters, and entered Imperial College in 1926. Completing a B.Sc. in chemistry in 1929, she spent 1 year as a researcher at IC before joining Frances Hamer as a Research Assistant at Ilford Research Laboratories. She subsequently received a Ph.D. for her research on the synthesis of isocyanine dyes.

Fisher followed Hamer to Kodak in 1934, and they worked together co-authoring a total of seven publications and several patents. In 1939, Fisher was asked to transfer to Kodak's laboratories in Australia.[70] Kodak (Australia) needed a specialist organic research chemist to provide expertise in preparing emergency quantities of vital spectral sensitisers, in case supplies were restricted during the Second World War.

Arriving in Melbourne in late 1939, Fisher became actively involved in many manufacturing problems using her knowledge of dyes as filters. In 1945, she was placed in charge of the Manufacturing Analytical and Services Laboratory, where she stayed until her retirement in 1962. Her former colleague, Nigel Beale, recalled:

> She inspired the laboratory staff in all areas and provided encouragement and enthusiasm to young trainees who served part of their program time in the control area. Successful careers were launched as a result of a positive atmosphere and some graduates aspired to management status. Dr Fisher was

a great walker and enjoyed activities with the Melbourne Walking Club.[70]

She died on 10 August 1995 in Melbourne.

Women Chemists in Other Research Laboratories

We have shown above that certain companies were receptive to the hiring of women, but the majority were not. The *Journal of Careers* article on careers for women scientists in 1938 reported that many of the large companies employing chemists did not employ women and were very open about it, one firm replying that: "all applications from women scientists are 'automatically ruled through'."[36] An Institute of Chemistry survey showed that of 963 available positions during 1935 and 1936, only 80 were open to women.[71] In this section, we will provide a selection of other research career pathways followed by women chemists.

Employment by a Government laboratory was an option, including DSIR. One woman chemist to follow this path was Louisa Mary Hargreaves, who was born on 21 October 1906, the daughter of a civil servant, living in Eltham.[72] She was educated at the City of London School for Girls, then entered Bedford in 1926. After graduating with a B.Sc. in chemistry in 1929, she accepted a position at DSIR at Teddington, researching water pollution. She married Mr. Williams in 1938.

Rona Robinson

The earliest industrial chemist we could identify was Rona Robinson.[73] Robinson was born on 26 June 1884, and entered the University of Manchester in 1902. She completed a B.Sc. in chemistry in 1905, followed by an M.Sc. in 1907. During the next 8 years, she undertook research on dyes in her own private laboratory at Moseley Villa, Metford Road, Withington, Manchester, together with some research at the Royal Institution Laboratory, Manchester.

In 1915, Robinson joined J.B. & W.R. Sharpe Ltd. as an analytical and research chemist. In addition, she was the works chemist responsible for transferring chemical reactions that she had devised to large-scale production. The following year, she was promoted to Chief Chemist. She left in 1920 to become Chief Chemist to Clayton Aniline Co. Ltd., where three patents were issued with her as inventor, two of which were on aldehyde–amino condensation products. She held the post at Clayton Aniline until her retirement. She died on 7 April 1962.

Edith Pawsey (Mrs. Murch)

The life history of Edith Hilda Pawsey[74] is particularly interesting in that she worked for industry, then government, before entering the "women's role" of scientific librarian (see below). Pawsey was born in 1897 in Manchester, and later studied at Portsmouth Municipal College. She obtained a B.Sc. in chemistry by private study in 1916, despite being assigned a tutor who refused to teach women. In 1917, she obtained an Honours B.Sc. in chemistry.

From 1918 to 1933, Pawsey worked as an Assistant Research Chemist at the South Metropolitan Gas Company, her research on coal gas earning her an M.Sc. (London) in 1919. Pawsey was at the Gas Company the same time as Eunice Bucknell (see Chap. 2), and Bucknell was the first-named nominator of Pawsey for her admission to the Chemical Society.

Upon her marriage to chemist W. O. Murch, Pawsey was required to resign her position. Over the subsequent years, she had two daughters; but the exigencies of war overruled marital status and, during the Second World War, she was employed in the Armament Research Establishment, Fort Halstead.

In 1946, Pawsey joined the organic chemistry section of the Chemical Research Laboratory at the DSIR. When, in 1949, she was asked to take over the chemistry library temporarily, she did the job so well that she was appointed the permanent Librarian, resulting in her becoming an Associate of the Library Association. Pawsey first retired in 1962, but was soon rehired by the National Physical Laboratory to help in the library, retiring for a second and final time in 1968. She died on 14 March 1995, aged 98 years.

Gwenyth Gell and Mavis Gell (Mrs. Tiller)

An interesting phenomenon was sisters who both became chemists, such as the Badger sisters, May and Louise (see Chap. 11); and the Groves sisters, Ruby Caroline and Ida Mary (see Chap. 12). Another pair was Gwenyth Gell[75] and Mavis Gell.[76] The Gells, daughters of John Gell, an electrical engineer, both pursued careers in industrial chemistry. The younger sister, Gwenyth, was born on 30 October 1905; and was educated at St. Aiden's, Stroud Green, London, and at the Northern Polytechnic Institute, London. She entered Bedford in 1923, completing a B.Sc. in general science in 1926, followed by a B.Sc. in chemistry in 1927. In 1928, she was a Research Chemist with C. Blackwell of Bermondsey; then later, she was a teacher at McCaven Derby High School.

The elder sister, Mavis, was born in 1901; and she also attended Bedford, where she obtained a B.Sc. in chemistry in 1923. In 1927, Gell was an Assayer at Broken Hill, Australia. She subsequently returned to Britain for, in the 1932 and 1936 *Registers* of the Royal Institute of Chemistry (RIC), Mavis Gell was listed as a Temporary Worker in the Metallurgy Department of the National Physical Laboratory. At some point in her career, she was awarded a C.B.E., though it is not clear for what accomplishment. She died on 28 June 1989 in Wellington, New Zealand, aged 88 years.

Women Chemists as Scientific Librarians and Indexers

Rossiter has described how, in the United States, one avenue of employment for women scientists was science librarian or archivist work.[77] For interwar Britain, Helen Plant has shown that technical librarianship and information work became a significant career option for women science graduates.[78] We have already described how Agnes Lothian became librarian and historian in pharmacy (see Chap. 10).

Among those who proceeded directly from university to chemical librarianship was Mary Eynon Miller. Miller, daughter of a draper in Matlock, Derbyshire, attended Whalley Range High School and then Manchester High School for Girls.[79] After completing a B.Sc. in chemistry at the University of Manchester in 1926, she commenced research with Robert Robinson towards an M.Sc. Miller accepted a position as Chemical Librarian with Messrs. Castner and Kellner at Western Point, near Runcorn, which subsequently became Imperial Chemical Industries (General Chemicals Ltd.).

Enid Pope (Mrs. Hulsken)

Like Miller (above), Enid Marian Pope[80] proceeded directly from university to library work. Pope was born on 19 November 1911 in Birmingham, the daughter of Thomas Henry Pope, a research chemist, and Florence George. She was educated at Wallasey High School in Cheshire and then Croydon High School (a GPDSC school) before entering Newnham in 1930. She completed the requirements for a degree in chemistry in 1933, and received a B.Sc. (London) in 1934.

That year, Pope accepted a post as an Abstractor and Editor of Abstract Journals for the Research Association of British Rubber Manufacturers, leaving them in 1938 to work with Kodak as Journal Abstractor and Indexer at their Research Laboratories in Harrow, joining her longtime friend, Hamer (see above). She took a position as Research Librarian at the Distillers Research Department, Epsom, in 1944 — the same year she married Johannus Martinus Hulsken, a Chief Officer in the Dutch Merchant Navy. Pope continued working after marriage; but in 1948, she resigned her position prior to the birth of her son. In later years, she was a self-employed Abstractor and Indexer. She died on 25 October 1987.

Margaret Dougal

The first publications of Margaret Douie Dougal[81] were a series of four articles on the teaching of chemistry, which appeared in *Chemical News* in 1893, but she also carried out research in inorganic chemistry under Thomas Edward Thorpe at the Royal College of Science. T. E. Thorpe had contended that there was a need for an index to the publications of the Chemical Society and, for 15 years, until 1909, Dougal worked as Compiler and Indexer, preparing *A Collective Index of the Transactions, Proceedings and Abstracts of the Chemical Society*. The first two volumes, covering the 20 years from 1873 to 1892, took Dougal over 5 years to produce.

After Dougal had completed this work, James Dewar made a point of congratulating her in his Presidential Address to the Chemical Society.[82] Emphasising the tremendous usefulness of the indexes to members, he went on to note that, although the task had been of unexpected length and complexity because of the lack of a system in the previous annual indexing, Margaret Dougal's compilation was "an example of thoroughness and accuracy to her successors."[82]

Margaret Le Pla

Dougal's replacement in 1910 was Margaret Le Pla.[83] Le Pla, born on 17 April 1885, the daughter of Henry Le Pla, a Nonconformist Minister at Southall, Middlesex, was educated at South Hampstead High School (a GPDSC school). After graduation from Bedford in 1906, she was appointed Demonstrator in Chemistry and also Research Assistant to James Spencer.

In 1910, Le Pla accepted a position with the Chemical Society as an Indexer, particularly of the *Journal of the Chemical*

Society, her work being carried out in a room adjoining the Reading Room of the Library of the Chemical Society at Burlington House. In the early part of the 20th century, the *Journal of the Chemical Society* carried an Abstracts supplement; and when the publication of the supplement was transferred to the Bureau of Chemical Abstracts, Le Pla indexed these also, herself moving to the Bureau offices.

When she retired from indexing the Abstracts, she continued to index the *Journal of the Chemical Society* and *Annual Reports of the Chemical Society*. In addition, she indexed Thorpe's *Dictionary of Applied Chemistry*. Amazingly, these tasks left her with idle time and, her obituarist reported, Le Pla endeavoured to fill it:

> Her energies, however, sought further outlets, and she took over the re-organization of the Research Library of Messrs. C.C. Wakefield & Co., and the care of this library was placed in her hands. Although in her later years she did not give herself much time for recreation, Margaret Le Pla loved music, and for many years she sang in the Old Vic Opera Company. Latterly, she took great interest in the functions of the Women Chemists Dining Club [see Chap. 2], of which she was a member.[83(a)]

She died on 26 January 1953, still working hard.

Margaret Whetham (Mrs. Anderson)

Both Dougal and Le Pla had been indexers of chemical publications, but Margaret Dampier Whetham[84] left her biochemical background to index publications across many disciplines. Born on 21 April 1900 in Cambridge, Whetham was the daughter of William Cecil Dampier Whetham (later, Sir William Dampier), Lecturer in Physics at the University of Cambridge, and Catherine Durning Holt, former science student at Newnham and author of several books.

Whetham was home-schooled and then attended University College, Exeter, before entering Newnham in 1918, where she read science for 3 years. In 1920, while still a student, she commenced research work with Marjory Stephenson (see Chap. 8), undertaking pioneering work on the washed cell suspension technique for analysing cells. During her 7 years working with Stephenson, Whetham co-authored four research papers. She was also the co-founder of the Cambridge in-house biochemistry magazine, *Brighter Biochemistry*, which lasted from 1923 until 1930 and served to give insights into life under "Hoppy" (see Chap. 8). Whetham, herself, penned the following verse:

A monograph by MS
Would do much to relieve the distress
Caused by all these inferior
Books on bacteria.[85]

In 1927, Whetham married Alan Bruce Anderson, a clinical pathologist, and gave up her research work. Over the ensuing years, she had five children. She returned to the workforce in 1948 as Abstractor for *British Chemical Abstracts*; then in 1950, she switched to abstracting for Food Science Abstracts for 7 years. From being an abstractor, she became an indexer, and over her remaining career she compiled indexes for a total of 567 books. She became active with the Society of Indexers, finally becoming Vice-President. Whetham wrote numerous journal articles, authored *Book Indexing*,[86] and co-authored with her father a compilation: *Cambridge Readings in the Literature of Science*.[87] She died in 1997.

Women Chemists as Factory Inspectors

We mentioned in Chap. 4 that Ruth Drummond, a Demonstrator at Bedford, became a Factory Inspector. Becoming a Lady Factory Inspector was one path of employment for women graduates — particularly those with a scientific background.

The first Lady Factory Inspectors had been appointed by the Home Secretary in 1893.[88] It was a tough task for a middle-class woman, requiring her to venture alone into unsavoury neighbourhoods, cope with aggressive plant managers, and attempt to offer some specific recommendations to alleviate the plight of the working poor. As Adelaide Anderson,[89] Head of the Lady Inspectors, wrote in 1905 to the Chief Inspector:

> A woman, as a Factory Inspector, in an Industrial district away from her own family and social surroundings, as well as her women colleagues, can find no normal associates in, or through, her work. Her work compels her to lead a life that is quite different from that of other women, and the slightest deviation from extreme caution and prudence may subject her to injurious criticism.[90]

The male Inspectors were often less than supportive, disliking the zeal of these well-qualified and determined women.

Dorothy Fox (Mrs. Richards)

One of the later Lady Factory Inspectors was Dorothy Lilian Fox.[91] Born on 24 March 1905, Fox was the daughter of George Frederick Fox, a manufacturer of motor accessories in Birmingham. She was educated at KEVI and then entered Bedford in 1924, having applied late because her father would not permit her to leave home. In addition to academic success, she was a gifted hockey player and sculler for the University of London women's teams. After obtaining a B.Sc. in physiology in 1927 and a B.Sc. in chemistry in 1928, she looked for employment, but was unsuccessful.

Fox returned to Bedford, where she undertook research with Eustace Turner, completing a Ph.D. in 1930 on diaryl ethers. Having heard from her father about Lady Factory Inspectors, she obtained a position as Factory Inspector in Reading. She continued as a Factory Inspector until 1939, being the first woman to inspect large chemical works such as the Imperial Chemical Industries (ICI).

It was in 1939 that Fox married H. G. H. Richards, a consultant pathologist, and she changed her career direction again. She assisted Richards with his medical research work; then, with the start of the Second World War, she took a course in building construction and spent the war working on the design and construction of factory air raid shelters and canteens. Following the war, she became active in local politics, being first elected as a Lincolnshire County Councillor in 1967. She remained on the Council until 1982, dying 2 years later on 14 January 1984, aged 78 years.

Women Chemists in the 1940s

Though some women chemists were drafted for war work, most male scientists retained their positions during the Second World War, unlike the First. David Edgerton has noted: "Crucially, there was no systematic replacement of men by women, least of all in science, technology and medicine"[92] — as had been the case in the First World War.

Life for Women Students

Even during the 1940s, the presence of women students in universities was still being questioned. At the University of Edinburgh, a claim was made in *The Student: Edinburgh University Magazine* that 90% of women at university were husband-hunters. In response, one of the women students, "L.M.S.," retorted that 90% were serious, though she felt the need to add, but not too serious: "But far am I from being a 'careerist' or 'blue stocking,' for if all work and no play makes Jack a dull boy it must equally make Jill a dull girl."[93]

Later in that decade, a male contributor, "S.G.W.," made clear his view that women students should see their future as a supporting role for men:

> We like the intelligence and intellectual awareness of the modern woman, but put forward a plea that her gifts and knowledge

> should be directed towards stimulating and fostering thought and action (rather in the same way as did the ladies of the eighteenth century salon), instead of using them for the hard grind of research and study, which is more properly the sphere of men.[94]

An article in a 1944 issue of the Imperial College magazine *The Phoenix* concluded that intelligent women should aim for marriage and domestic life rather than chemistry, contending: "You should think twice (intelligently), before choosing it [a career in chemistry]. Remember that your hands will be stained and your nails spoilt beyond repair or manicure ..."[95] What is of note is the lack of rebuttal from women students, unlike the vehement responses of earlier generations of women.

Employment of Women Chemists in the Second World War

The Second World War forced companies that, until then, had refused to employ women chemists into hiring them. Among these companies were Courtaulds, the polymer company, and Imperial Chemical Industries (ICI), which the *Journal of Careers* singled out for special criticism:

> Imperial Chemical Industries, Ltd., which has extensive laboratories in its various works and the largest organic chemistry research unit (21 laboratories and 200 research workers and their assistants) in the country as well as an important agricultural station, does not generally favour the employment of women graduates to research or analytical work in its laboratories, although it has a small number engaged in library and abstracting work, patent work, and secretarial duties in research departments, and two or three women graduates have been employed and are still employed in agricultural fertilizer research.[36]

However, it should be noted that the ICI did appoint Beryl Hamilton as a Chemical Analyst in 1938 (see Chap. 6).

In 1941, an article in *Chemistry and Industry* detailed the types of work in the chemical industry that women could undertake. The author made the following opening remarks:

> Early in the [Second World] war the view was that the employment of women in chemical works would necessarily be severely limited. However, certain firms which have put the matter to practical test have found that women can be substituted for men to a far greater extent than was thought possible, provided attention is given to certain points.[96]

Among the long lists of tasks found suitable for women were taking of samples for analysis, and routine chemical testing in the laboratory and plant. What seems surprising is that the experiences of the First World War and the contributions of women then (see Chap. 12) had been totally forgotten.

Jessie Mole

One of the first beneficiaries of ICI's reversal of its opposition to women researchers was Jesse Dorehill Crampton Mole.[97] Mole, daughter of Ernest Mole, a retired police officer, was educated at Notre Dame High School, Clapham, and the City of London School for Girls. She entered Bedford in 1934, graduating with a B.Sc. in chemistry in 1936.

Mole then undertook research towards a Ph.D. with Eustice Turner at Bedford, which she was awarded in 1938. She continued research with Turner for another year, investigating molecular dissymmetry due to restricted rotation in diphenylamines and triphenylamines, before obtaining an appointment with ICI, first at Widnes, then in the Technical Services Department in Liverpool. One of her nominators in 1941 for admission as Fellow of the Chemical Society was ICI Librarian, Mary Eynon Miller (see above).

Mole returned to London in 1950 as Scientific Officer with the Ministry of Supply, Atomic Energy Division. Then, in 1955, she obtained a transfer to the Atomic Research Establishment, Harwell, where she was Assistant to the Director until she retired in 1975.

From her research days at Bedford, Mole had kept her chemicals. Margaret Jamieson (see Chap. 4) recalled Mole returning to Bedford in 1979:

> Just before I [Jamieson] retired, she [Mole] brought in an old suitcase, hastily packed with her research chemicals when War broke out — it had not been opened. We put it in a good draught and opened it. Sure enough, the old bottles contained arsenicals whose preparation she had not had time to publish.[98]

Vera Furness

The change in policy by Courtaulds was fortunate for Vera Isabella Furness.[99] Born on 2 June 1921, Furness attended a teacher training college and became a secondary school teacher. While teaching full-time, she studied for a B.Sc. (London) in chemistry through the Birmingham Central Technical College (later University of Aston), which she completed in 1946. Hired as a development chemist at BX Plastics, she undertook research towards an M.Sc., which she finished in 1948, followed by a Ph.D. in 1952.

While working on her doctorate, Furness was a part-time Lecturer at the Birmingham Central Technical College; and though she was offered a full-time teaching post when her Ph.D. was completed, she decided her interest lay in industrial chemistry. In 1953, she was hired by Courtaulds, one of her tasks being to find a method for commercially dying the acrylonitrile polymer, Courtelle. This she accomplished by developing a copolymer that could be dyed more successfully. Furness was involved in both chemical and mechanical challenges of the commercial production of this internationally successful polymer, and was responsible for at least six patents.

Furness's later research included the development of a process to make carbon fibres on an industrial scale for the Royal Aircraft Establishment. She progressed to become Section Leader, responsible for 100 graduate researchers. Her final appointment was Director and General Manager of the Courtaulds factory at Campsie, Northern Ireland, in 1978, from which she retired in 1981. She died on 8 June 2002, aged 81 years.

Elizabeth Frith (Mrs. Tuckett)

Another polymer researcher was Elizabeth Mary Frith.[100] Born on 25 September 1920 in Streatham, London, Frith was the daughter of Canon Herbert Charles Frith, of Chichester, and Nora Gabain. Educated at Brighton and Hove High School (a GPDSC school), St. Mary's Hall, Brighton, and Benenden School, she entered Girton in 1938. She completed Part II of the Natural Science Tripos in 1941 and was immediately awarded a 1-year research scholarship, which led her into a study of polymers in the Department of Colloid Science at Cambridge. From 1942 until 1946, she undertook research for the Ministry of Aircraft Production. This work overlapped with the award of a Hertha Ayrton Fellowship from 1944 until 1948.

In 1944, Frith married polymer chemist Ronald Francis Tuckett, and they had one son. The following year, five papers on her polymer studies were published (as Frith), three under her name alone and the other two jointly with Tuckett. In 1951, Frith and Tuckett co-authored a book, *Linear Polymers*.[101]

Antoinette (Tony) Patey (Mrs. Pirie)

It was joining a team to study the effect of war gases on the eye that led to the subsequent fame of Antoinette Patey.[102] Patey was born on 4 October 1905 in London, the daughter of William James Patey, a pharmacist, and Florence Keen. She was educated at Wycombe Abbey School, where she had an inspiring chemistry teacher who even let her camp in the chemistry laboratory so that she could complete her experiments.

Matriculating in 1924, she entered Newnham, obtaining first-class honours in biochemistry. Joining the research group of Hopkins (see Chap. 8), in 1930 she co-edited the in-house magazine *Brighter Biochemistry* with another of Hopkins's research students, Norman Wingate Pirie.[103] They added to the reputation of "Hoppy's group" as a "marriage bureau" by marrying, having one son and one daughter. Both Patey and Pirie received their Ph.D. degrees at the same time in 1933. Patey's research had been on bacteriophage, particularly the effect of lysosyme on mucopolysaccharides. Karl Meyer in New York heard of her work and invited her to undertake research with him in the Department of Opthalmology, the commencement of her career in the biochemistry of the eye.[104]

When the Second World War began, Patey joined the research team of Ida Mann[105] at the Imperial Cancer Research Fund's Mill Hill laboratories. The team studied the effect of poison gases on the cornea and how to protect the eye from injury during a gas attack. When Mann moved to Oxford in 1942, Patey accompanied Mann as her assistant and, in 1946, they co-authored the book *The Science of Seeing*.[106]

Then, in 1947, Patey was named Mann's successor as Reader in Opthalmology and Head of the Nuffield Laboratory of Opthalmology, the first nonclinician appointed to an Oxford Medical School. During these years, she assembled a team of researchers who studied the biochemical processes of the eye as a means of understanding its many diseases. The research topics included lens metabolism, lens proteins, and particularly the nature of cataracts.

In 1956, Patey co-authored the classic work, *Biochemistry of the Eye*.[107] She also established the International Committee for Eye Research, the forerunner of the International Society for Eye Research; and she was the first woman to be awarded the prestigious Proctor Medal of the

Association for Research in Opthalmology, an honour she received in 1968.[104]

During the 1950s, Patey became concerned about the health hazards of the testing of nuclear weapons in the atmosphere. In addition to speaking at meetings, in 1957 Patey edited the book, *Fall Out: Radiation Hazards from Nuclear Explosions*,[108] a compilation of contributions from nine scientists across a range of disciplines, which became an authoritative source on the subject. Patey died on 11 October 1991 at Oxford.

Patricia Green (Mrs. Clarke)

It was her war work which initiated the choice of career for Patricia Hannah Green.[109] Green was born on 29 July 1919 in Pontypridd, Wales, and was educated at Howell's School, Llandaff (see Chap. 6). Interested in biological chemistry, she entered Girton in 1937 to study biochemistry, graduating in 1940, the same year she married Michael Clarke. She left Cambridge for war work at the Armament Research Department of the Ministry of Supply, initially at Woolwich Arsenal, and then at Swansea, researching the chemistry of explosives.

In 1944, Green's talents as a biochemist were finally utilized when she accepted a position with B. C. J. G. Knight's group at Wellcome Research Laboratories in Beckenham, Kent. Serious infections of war wounds were being caused by the pathogen *Clostridium oedomatiens*, and her research on growth and toxin production of the organism was designed to find an improved method of immunising against it.

Green discontinued research upon the birth of her two children in 1947 and 1949. She started work again, part-time, in 1951 with S. T. Cowan at the National Collection of Type Cultures of Bacteria at the Central Public Health Laboratory, Colindale, London, where they developed a series of methods for the identification of bacteria based on enzyme reactions. In 1953, she was appointed Assistant Lecturer in the Department of Biochemistry at UCL, where she commenced research on bacterial adaptation, a subject to which Marjory Stephenson (see Chap. 8) had introduced her in 1940 at Cambridge.

It was her research on microbial biochemistry and mutations that was to garner Green promotion at UCL through to Professor of Microbial Biochemistry in 1974 and admission to Fellowship of the Royal Society in 1976, becoming Vice-President of the Royal Society for the 1981/1982 year. Coming full circle, in 1981, she was invited to give the Marjory Stephenson Memorial Lecture to the Society for General Microbiology. Green formally retired in 1982; but like so many other women, her work was her life, and in 1986 she accepted an appointment as Royal Society Kan Tong-Po Professor at the Chinese University of Hong Kong to assist with the organisation of teaching and research in biotechnology.

The Late- and After-Career of Some Women Chemists

In the late 19th and early 20th centuries, university authorities considered that women students were in particular moral danger and needed an authority figure to oversee them during their free hours.[110] A unique role for older woman academics was therefore created — that of lady superintendent or warden of women's halls of residence. Of the women chemists discussed in this book, 11 held appointments of this type.

The role was also found in the United States, as Margaret Rossiter had noted, the incumbent often a senior woman scientist, one difference being that the position was called Dean of Women.[111] The duties of the warden of University Hall, University of St. Andrews, made it clear that the incumbent had to be an academic: "For the proper discharge of all these advisory duties, as well as her administrative duties as head of a woman's College Hall, the warden must not only have academic experience, but must also keep herself constantly in touch with educational movements and ideas."[112]

At Bedford, Mary Crewdson (see Chap. 4) was appointed warden of Northcutt House and then Lindsell Hall, retiring in 1954 and being succeeded by Mary Lesslie (see Chap. 4), who was given the title of Dean of Lindsell Hall, herself retiring in 1968. Ruth Drummond (see Chap. 4) was a warden when needed at Bedford, then moved to the University of Birmingham as warden at Ashbourne Hall. Kathleen Balls (see Chap. 3) at Queen Mary College was appointed lady superintendent in 1924, while Dorothea Grove (see Chap. 3) became warden of the Women's Hall of Residence in 1954. At the University of Sheffield, Dorothy Bennett (see Chap. 5) was tutor for women from 1926 and warden of University Hall for Women (later Fairfax House) from 1934, resigning from both in 1947.

Ruth King (see Chap. 12) was appointed warden of Hope Hall at University College of South West England (later the University of Exeter) in 1918; while Ettie Steele was warden of Chatten House (later McIntosh Hall), University of St. Andrews, from 1930 until 1959. Two of the women chemists became sub-wardens: Grace Leitch (see Chap. 5) at Easton Hall, Armstrong College, from 1921 to 1941; and May Leslie (see Chap. 5) at Weetwood Hall, University of Leeds, from 1935 to 1937. Finally, Millicent Taylor (see Chap. 5) was acting warden during Spring and Summer Terms 1934 and 1935 at Clifton Hill House, University of Bristol.

Though the role of warden could be interpreted in a negative light, at the time, these single women, who had little opportunity for academic advancement, at least had a position and an income on which they could rely into their later years.

Commentary

The 20th century had started so promisingly for women: educational opportunities were taken for granted. The First World War had given access to a wide range of job opportunities for academic women, particularly those with a chemical background. Women's university enrolments — including chemistry — continued to rise into the 1920s. Then came the reversal. Marjorie Nicholson, a 1914 graduate of the University of Michigan, eloquently captured the sentiment of feminist academic women on both sides of the Atlantic:

> We of the pre-[First World] war generation used to pride ourselves sentimentally on being the "lost generation," used to think that, because war cut across the stable path on which our feet were set, we were an unfortunate generation. But as I look back upon the records, I find myself wondering whether our generation was not the only generation of women which ever found itself. We came late enough to escape the self-consciousness and belligerence of the pioneers, to take education for granted. We came early enough to take equally for granted professional positions in which we could make full use of our training. This was our double glory. Positions were everywhere open to us; it never occurred to us at the time that we were taken only because men were not available. ... The millennium had come; it did not occur to us that life could be different. Within a decade shades of the prison house began to close, not upon the growing boy, but upon the emancipated girls.[113]

For women chemists, there were gains and losses. As we showed in Chap. 2, women chemists were finally admitted to the Chemical Society. On the other hand, women's positions as university and college teaching staff (see above) ceased to exist as each of the original incumbents retired. The 1920s and 1930s did offer opportunities in industrial chemistry for women graduates, but only with certain companies or organisations, as we have shown earlier; and whatever the organisation, there was always the assumption that "matrimonial mortality" made training women a waste of time.

In the latter half of the 20th century, it seems as though women chemists had to start their crusade for acceptance all over again. The progress women chemists had made between

1880 and 1925 had been forgotten. Even as late as 1958, a woman student, J. Lemon, at Imperial College lamented: "From the staff, the attitude [towards women students] is gentle contempt or amusement. From the [male] students, we are an object of amazement, except when they are in difficulty (such as ironing shirts)."[114]

Notes

1. Brooksbank, I. (1991). Bingles and bicycles. *Oxford Today* **3**(2): 35.
2. Tuke, M. J. (1928). Women students in the universities: Fifty years ago and to-day. *Contemporary Review* **133**: 71–77.
3. Rayner-Canham, M. F. and Rayner-Canham, G. W. (1996). Women in chemistry: Participation during the early twentieth century. *Journal of Chemical Education* **73**: 203–205.
4. Vetter, B. (1984). Changing patterns of recruitment and employment. In Hass, V. B. and Perucci, C. C. (eds.), *Women in Scientific and Engineering Professions*, University of Michigan Press, Ann Arbor, Michigan, p. 59.
5. Whiteley, L. D. (1925). *The Poor Student and the University: A Report on the Scholarship System with Particular Reference to Awards Made by Local Educational Authorities*. London, pp. 23–25.
6. Dyhouse, C. (2002). Going to university in England between the wars: Access and funding. *History of Education* **31**: 1–14.
7. Tylecote, M. (1941). *The Education of Women at Manchester University 1883 to 1933*. Manchester University Press, Manchester, pp. 120–121.
8. Roberts, E. (1988). *Women's Work 1840–1940*. Macmillan Education, London, p. 67.
9. Clephane, I. (1935). *Towards Sex Freedom*, cited in: Braybon, G. (1981). *Women Workers in the First World War: The British Experience*. Croom Helm, London, pp. 185–186.
10. Beddoe, D. (1989). *Women Between the Wars, 1918–1939: Back to Home and Duty*. Pandora, London, p. 82.
11. Kent, S. K. (1988). The politics of sexual difference: World War I and the demise of British feminism. *Journal of British Studies* **27**: 232–253.

12. Hamilton, C. (1935). *Life Errant*. J. M. Dent, London, p. 251.
13. Bingham, A. (2004). *Gender, Modernity, and the Popular Press in Inter-War Britain*. Clarendon Press, Oxford, p. 7.
14. Fox Keller, E. (1987). The gender/science system: Or, is sex to gender as nature is to science? *Hypatia* **2**: 40.
15. Pilcher, R. (1927). *The Profession of Chemistry*. Institute of Chemistry of Great Britain and Ireland, London, p. 92.
16. Rossiter, M. (1982). *Women Scientists in America: Struggles and Strategies to 1940*. Johns Hopkins Press, Baltimore, Maryland.
17. Marsh, N. (1986). *The History of Queen Elizabeth College: One Hundred Years of Education in Kensington*. King's College, London, p. 124.
18. Anon. (1929). Prospects for women in science. *Journal of Careers* **9**(89): 18–19.
19. Anon. (1938). Prospects of employment for women science graduates: Part I of a survey of opportunities in government, industrial and other research laboratories. *Journal of Careers* **17**(182): 88–93.
20. Jordan Lloyd, D. (1933). Biochemistry as a career for women. *Journal of Careers* **12**(135): 20–22.
21. Anon. (1936). Demand for science graduates maintained: Women graduates still mainly teachers. *Journal of Careers* **9**: 37–38.
22. Dyhouse, C. (1997). Signing the pledge? Women's investment in university education and teacher training before 1939. *History of Education* **26**: 207–223.
23. Uzzell, K., cited in: Note 22, Dyhouse, p. 217.
24. Anon. (1936). Forms of recommendation for fellowship. The ballot will be held at the ordinary scientific meeting on Thursday, February 20th. *Proceedings of the Chemical Society* 7.
25. Reason, H. A. (1936). *The Road to Modern Science*, 1st ed. G. Bell & Sons Ltd., London; (1940). 2nd ed.; (1959). 3rd ed.
26. Phare, E. (1982). From Devon to Cambridge. *Cambridge Review* 149.
27. Neville, E. H. (1933). This misdemeanor of marriage. *Universities Review* **6**: 5–8.
28. (a) "M.E.G." [Glennie, M.] (1975). Hilda Jane Hartle, 1876–1974. *Newnham College Roll Letter* 43; (b) White, A. B. (ed.) (1979). *Newnham College Register, 1871–1971*, Vol. I: 1871–1923, 2nd ed. Newnham College, Cambridge, p. 141.

29. Simms, T. H. (1979). *Homerton College 1695–1978: From Dissenting Academy to Approved Society in the University of Cambridge*. Trustees of Homerton College, Cambridge, p. 50.
30. Edwards, E. (2000). Women principals, 1900–1960: Gender and power. *History of Education* **29**(5): 407.
31. Edwards, E. (2001). *Women in Teacher Training Colleges, 1900–1960: A Culture of Femininity*. Routledge, London.
32. University of Newcastle-upon-Tyne, student records; Armstrong College Old Student Association, *Year Book*, 1933, 1935; King's Old Student Association, *Year Book*, 1938; *Register*, Royal Institute of Chemistry, 1948; Anon. (1983). Personal news: Deaths. *Chemistry in Britain* **19**: 990.
33. Leonard, J. W. (1981). *Constantine College*. Teeside Polytechnic, Middlesbrough.
34. Anon. (1957). Obituary: Frances Burdett. *Journal of the Royal Institute of Chemistry* **81**: 710; *Register*, Royal Institute of Chemistry, 1948.
35. Letter, D. Adams to A. E. Conway, 24 December 1918, *Women's War Work Collection*, Imperial War Museum, London.
36. Anon. (1938). Prospects of employment for women science graduates: Part III — Industrial research laboratories. *Journal of Careers* **17**(185): 289–296.
37. Horrocks, S. M. (2000). A promising pioneer profession? Women in industrial chemistry in inter-war Britain. *British Journal for the History of Science* **33**: 351–367.
38. (a) Bramley, R. (1991). Kathleen Culhane Lathbury. *Chemistry in Britain* **27**: 428–431; (b) From Bramley, R., unpublished manuscript, Mrs. Kathleen Lathbury B.Sc., C.Chem., F.R.S.C.; (c) Royal Holloway College, student records; (d) Haines, M. C. (2001). Lathbury, Kathleen née Culhane. *International Women in Science: A Biographical Dictionary to 1950*. ABC-CLIO, Santa Barbara, California, p. 165.
39. Cited in: Note 38(a), Bramley, p. 428.
40. Cited in: Note 38(a), Bramley, p. 430.
41. Bedford College, Student Records; Anon. (1945). List of applications for fellowship. *Proceedings of the Chemical Society* 81; Anon. (1967). Personal news: Deaths. *Chemistry in Britain* **3**: 502.
42. McCombie, T. (1995). Mamie Olliver (1905–1995). *Chemistry in Britain* 31.

43. Note 17, Marsh, p. 268.
44. Anon. (1955). Proceedings of the Society for Analytical Chemistry, Annual General Meeting. *The Analyst* **80**: 325.
45. Olliver, M. (1955). Women in chemistry. *Journal of the Royal Institute of Chemistry* **79**: 413–420.
46. Personal communication, Letters, Doris Kett, 6 April 2000 and 14 June 2000.
47. (a) Richards, A. (1995). Elizabeth Adams 1909–1994, N.C. 1929. *Newnham College Roll Letter* 60; (b) White, A. B. (ed.) (1981). *Newnham College Register, 1871–1971*, Vol. II: 1924–1950, 2nd ed. Newnham College, Cambridge, p. 60.
48. Anon. (1941). List of applications for fellowship. *Proceedings of the Chemical Society* 62; Anon. (1981). Personal news, deaths. *Chemistry in Britain* **17**: 556.
49. (a) University of Manchester, student records; (b) University College, University of London, *Annual Report. 1925–1926*; (c) Anon. (1937). A new honorary member. *Pharmaceutical Journal* **138**: 484; (d) Anon. (1978). Deaths: Coward. *Pharmaceutical Journal* **221**: 134; (e) "J.H.B." (1978). Deaths: Coward. *Pharmaceutical Journal* **221**: 177; and (f) Cox, G. S. (1978). Deaths: Coward. *Pharmaceutical Journal* **221**: 195.
50. Young, F. G. (1954). Jack Cecil Drummond, 1891–1952. *Obituary Notices of Fellows of the Royal Society* **9**: 99–129.
51. (a) Rideal, E. K. (1932). Winifred Mary Wright. *Journal of the Chemical Society* 2998–2999; (b) "F.M.H." (1933). Winifred Mary Wright. *The Girton Review* (Lent Term): 5–6; (c) Butler, K. T. and McMorran, H. I. (eds.) (1948). *Girton College Register, 1869–1946*. Girton College, Cambridge, p. 321.
52. Eley, D. D. (1976). Eric Keightley Rideal, 11 April 1890–25 September 1974. *Biographical Memoirs of Fellows of the Royal Society* **22**: 381–413.
53. Schild, H. O. and Rose, F. L. (1976). Harry Raymond Ing, 31 July 1899–23 September 1974. *Biographical Memoirs of Fellows of the Royal Society* **22**: 239–255.
54. Bedford College, student records.
55. Oakley, C. L. (1966). Alexander Thomas Glenny, 1882–1965. *Biographical Memoirs of Fellows of the Royal Society* **12**: 163–180.

56. Ashwell, M. (2002). Elsie May Widdowson, 21 October 1906–14 June 2000. *Biographical Memoirs of Fellows of the Royal Society* **48**: 483–506.
57. "V.H.B." (1932). Obituary notices. Samuel Barnett Schryver — 1869–1929. *Proceedings of the Royal Society, Series B* **110**: xxii–xxiv.
58. Widdowson, E. M. (1995). Robert Alexander McCance, 9 December 1898–5 March 1993. *Biographical Memoirs of Fellows of the Royal Society* **41**: 263–280.
59. McCance, R. A. and Widdowson, E. M. (1940). *The Chemical Composition of Foods*. Medical Research Council Special Report Series no. 235. HMSO, London.
60. Food Standards Agency (2002). *McCance and Widdowson's The Composition of Foods*, 6th summary ed. Royal Society of Chemistry, Cambridge.
61. Ashwell, M. A. (ed.) (1993). *McCance and Widdowson — A Scientific Partnership of 60 Years*. British Nutrition Foundation, London.
62. (a) Delius, P. (1963). Dr. Frances Mary Hamer. *British Journal of Photography* **110**: 260; (b) Jeffreys, R. Q. and Gauntlett, M. D. (1981). Frances Mary Hamer 1894–1980. *Chemistry in Britain* **17**: 31; (c) Butler, K. T. and McMorran, H. I. (eds.) (1948). *Girton College Register, 1869–1946*. Girton College, Cambridge, p. 678.
63. Anon. (n.d.). Frances Mary Hamer, unpublished manuscript. North London Collegiate School Archives.
64. Gibson, C. S. (1941). Sir William Jackson Pope, 1870–1939. *Obituary Notices of Fellows of the Royal Society* **3**: 291–324.
65. Mann, F. G. (Michaelmas Term 1964). Book reviews: The chemistry of photographic sensitisers. *Girton Review* 9–11. This article was supposedly a review of Hamer's book (see below); but instead, Mann used the space for a biography of Hamer.
66. Harrison, G. B. (1954). The laboratories of Ilford Limited. *Proceedings of the Royal Society of London, Series B, Biological Sciences* **142**: 9–20.
67. *Harrow Research Laboratory Album, 1928–1976*, unpublished manuscript. North London Collegiate School Archives, pp. 14–15.
68. Hamer, F. M. (1964). *The Cyanine Dyes and Related Compounds*. Interscience Publishers, New York.

69. Anon. (1930). Forms of recommendation for fellowship. The ballot will be held at the ordinary scientific meeting on Thursday, February 20th. *Proceedings of the Chemical Society* **8**; Anon. (1951). *Old Students and Staff of the Royal College of Science*, 6th ed. Royal College of Science, London, http://www.asap.unimelb.edu.au/bsparcs/biogs/P003044b.htm, accessed 26 Jan 2008.
70. Personal communication, Letter, Nigel Beale, Melbourne, Australia, 2000.
71. Anon. (1937). Salaries and employment in chemistry. *Journal of Careers* **16**: 169.
72. Bedford College, Student Records; *Register,* Royal Institute of Chemistry, 1932, 1936, 1938, 1948.
73. University of Manchester, Student Record; Anon. (1963). Obituary: Rona Robinson. *Journal of the Royal Institute of Chemistry* **87**: 66.
74. Truter, M. (1995). Edith Hilda Murch 1897–1995. *Chemistry in Britain* **31**: 902; Anon. (1996). Corrigendum. *Chemistry in Britain* **32**: 55.
75. Bedford College, Student Records; *Register*, Royal Institute of Chemistry, 1928, 1932.
76. Bedford College, Student Records; *Register*, Royal Institute of Chemistry, 1928–1948; Anon. (1989). Personal news, deaths. *Chemistry in Britain* **25**: 1249.
77. (a) Rossiter, M. W. (1986). Women and the history of scientific communication. *Journal of Library History* **21**: 39–59; (b) Rossiter, M. W. (1996). Chemical librarianship: A kind of "Women's Work" in America. *Ambix* **43**: 46–58.
78. Plant, H. (2005). Women scientists in British industry: Technical library and information workers, c.1918–1960. *Women's History Review* **14**: 301–321.
79. University of Manchester, Student Records.
80. White, A. B. (ed.) (1981). *Newnham College Register, 1871–1971*, Vol. II: 1924–1950, 2nd ed. Newnham College, Cambridge, p. 81.
81. Creese, M. R. S. (1998). *Ladies in the Laboratory: American and British Women in Science 1800–1900*. Scarecrow Press, Lanham, Maryland, p. 266; Royal College of Science, Student Records.
82. Dewar, J., Sir. (1899). Presidential address. *Journal of the Chemical Society* **75**: 1168.

83. (a) Cummings, A. E. (1953). Margaret Le Pla: 1885–1953. *Journal of the Chemical Society* 3335–3336; (b) Cummings, A. E. (1953). Margaret Le Pla. *Chemistry and Industry* 130–131.
84. (a) White, A. B. (ed.) (1979). *Newnham College Register, 1871–1971*, Vol. I: 1871–1923, 2nd ed. Newnham College, Cambridge, p. 38; (b) Haines, C. M. C. (2001). Anderson, Margaret Dampier née Whetham. *International Women in Science: A Biographical Dictionary to 1950*, ABC-CLIO, Santa Barbara, California, p. 7.
85. Anderson, M. D. (1929). *Brighter Biochemistry* **6**: 47.
86. Anderson, M. D. (1971). *Book Indexing*. Cambridge University Press, Cambridge.
87. Whetham, W. C. D. and Whetham, M. D. (1924). *Cambridge Readings in the Literature of Science*. Cambridge University Press, Cambridge.
88. (a) Jones, H. (1988). Women health workers: The case of the first women factory inspectors in Britain. *Social History of Medicine* **1**: 165–181; (b) McFeeley, M. D. (1988). *Lady Inspectors: The Campaign for a Better Workplace, 1893–1921*. Blackwell, London.
89. Zimmeck, M. (2004). Anderson, Adelaide (1863–1936). *Oxford Dictionary of National Biography*, Oxford University Press, http://www.oxforddnb.com/view/article/37113, accessed 23 Dec 2007.
90. Cited in: Note 88(a), Jones, p. 175.
91. Bedford College, Student Records; *Hampshire Chronicle*, 9 February 1963; *Lincolnshire Echo*, 17 January 1984; Anon. (1984). Personal news: Deaths. *Chemistry in Britain* **30**: 304.
92. Edgerton, D. (2006). *Warfare State: Britain, 1920–1970*. Cambridge University Press, Cambridge, p. 175.
93. "L.M.S." (1940–1941). University women: A protest. *The Student: Edinburgh University Magazine* **37** (new ser.): 238.
94. "S.G.W." (1948–1949). Varsity women. *The Student: Edinburgh University Magazine* **45** (new ser.): 58.
95. "Alex." (1944). The intelligent woman's guide to chemistry, dancing and housekeeping. *The Phoenix* (new ser.) **52**: 26.
96. Anon. (1941). War-time employment of women in the chemical industry. *Chemistry and Industry* 873–875.

97. Bedford College, Student Records.
98. Harris, M. M. (2001). Chemistry. In Crook, J. M. (ed.), *Bedford College: Memories of 150 Years,* Royal Holloway and Bedford College, p. 87.
99. Haines, C. M. C. (2001). Furness, Vera I. *International Women in Science: A Biographical Dictionary to 1950*, ABC-CLIO, Santa Barbara, California, p. 107.
100. Butler, K. T. and McMorran, H. I. (eds.) (1948). *Girton College Register, 1869–1946*. Girton College, Cambridge, p. 689.
101. Frith, E. M. and Tuckett, R. F. (1951). *Linear Polymers*. Longmans, Green & Co., London.
102. van Heyningen, R. E. (2004). Pirie [née Patey], Antoinette (1905–1991). *Oxford Dictionary of National Biography*, Oxford University Press, http://www.oxforddnb.com/view/article/40595, accessed 4 Nov 2004; Ogilvie, M. and Harvey, J. (eds.) (2000). *The Biographical Dictionary of Women in Science. Pioneering Lives from Ancient Times to the Mid 20th Century*. Routledge, New York, p. 1027; White, A. B. (ed.) (1979). *Newnham College Register, 1871–1971*, Vol. I: 1871–1923, 2nd ed. Newnham College, Cambridge, pp. 41–42.
103. Pierpoint, W. S. (1999). Norman Wingate Pirie, 1 July 1907–29 March 1997. *Biographical Memoirs of Fellows of the Royal Society* **45**: 398–415.
104. Kinoshita, J. H. (1968). On the presentation of the Proctor Medal of the Association for Research in Opthalmology. *Investigative Opthalmology* **7**(6): 626–627.
105. Tiffany, J. M. (2004). Mann, Dame Ida Caroline (1893–1983). *Oxford Dictionary of National Biography*, Oxford University Press, http://www.oxforddnb.com/view/article/405951, accessed 13 Jan 2008.
106. Mann, I. and Pirie, A. (1946). *The Science of Seeing*. Penguin Books, Harmondsworth.
107. Pirie, A. and van Heyningen, R. (1956). *Biochemistry of the Eye*. Charles. C. Thomas, Springfield, Illinois.
108. Pirie, A. (ed.) (1957). *Fall Out: Radiation Hazards from Nuclear Explosions*. McGibbon & Kee, London.
109. Haines, M. C. (2001). Clarke, Patricia Hannah née Green. *International Women in Science: A Biographical Dictionary to*

1950, ABC-CLIO, Santa Barbara, California, pp. 62–64; Guide to the manuscript papers of British scientists: C, http://www.bath.ac.uk/ncuacs/guidec.htm#ClarkePA, accessed 26 Dec 2007.

110. Dyhouse, C. (1995). The British Federation of University Women and the status of women in universities, 1907–1939. *Women's History Review* **4**: 465–485.
111. Rossiter, M. W. (1980). "Women's Work" in science, 1880–1910. *Isis* **71**: 381–398.
112. Cited in: Dyhouse, C. (1998). *No Distinction of Sex? Women in British Universities, 1870–1939*. UCL Press, London, p. 254.
113. Nicholson, M. (1938). The rights and privileges pertaining thereto. *Journal of the American Association of University Women* **31**(3): 136.
114. Lemon, J. (1958). A woman's point of view. *The Phoenix* (new ser.) **73**: 10–11.

Index